FreeCAD 0.21 Black Book
(Colored)

By
Gaurav Verma
Matt Weber
(CADCAMCAE Works)

Edited by
Kristen

ISBN # 978-1-77459-117-8

NOTICE TO THE READER

DEDICATION

To teachers, who make it possible to disseminate knowledge
to enlighten the young and curious minds
of our future generations

To students, who are the future of the world

THANKS

To my friends and colleagues

To my family for their love and support

Training and Consultant Services

At CADCAMCAE WORKS, we provide effective and affordable one to one online training on various software packages in Computer Aided Design(CAD), Computer Aided Manufacturing(CAM), Computer Aided Engineering (CAE), Computer programming languages(C/C++, Java, .NET, Android, Javascript, HTML and so on). The training is delivered through remote access to your system and voice chat via Internet at any time, any place, and at any pace to individuals, groups, students of colleges/universities, and CAD/CAM/CAE training centers. The main features of this program are:

Training as per your need

Highly experienced Engineers and Technician conduct the classes on the software applications used in the industries. The methodology adopted to teach the software is totally practical based, so that the learner can adapt to the design and development industries in almost no time. The efforts are to make the training process cost effective and time saving while you have the comfort of your time and place, thereby relieving you from the hassles of traveling to training centers or rearranging your time table.

Software Packages on which we provide
basic and advanced training are:

CAD/CAM/CAE: CATIA, Creo Parametric, Creo Direct, SolidWorks, Autodesk Inventor, Solid Edge, UG NX, AutoCAD, AutoCAD LT, EdgeCAM, MasterCAM, SolidCAM, DelCAM, BOBCAM, UG NX Manufacturing, UG Mold Wizard, UG Progressive Die, UG Die Design, SolidWorks Mold, Creo Manufacturing, Creo Expert Machinist, NX Nastran, Hypermesh, SolidWorks Simulation, Autodesk Simulation Mechanical, Creo Simulate, Gambit, ANSYS and many others.

Computer Programming Languages: C++, VB.NET, HTML, Android, Javascript and so on.

Game Designing: Unity.

Civil Engineering: AutoCAD MEP, Revit Structure, Revit Architecture, AutoCAD Map 3D and so on.

We also provide consultant services for Design and development on the above mentioned software packages

For more information you can mail us at:
cadcamcaeworks@gmail.com

Table of Contents

Chapter 2 : Sketching

Chapter 3 : Solid Modeling with Part Design

Chapter 4 : Solid Modeling with Part Workbench

Chapter 8 : Path Workbench

Chapter 9 : FEM Workbench

Chapter 10 : Tech Drawing

Chapter 12 : Miscellaneous Tool

Chapter 13 : Assembly Workbench

Chapter 14 : Introduction to Computational Fluid Dynamics

Preface

FreeCAD is a parametric (history based) open source CAD/CAM/CAE software which finds its uses in various fields of designing. The software has a modular structure which allows to plugin extensions (modules) to add functionality to the core application. An extension can be as complex as a whole new application programmed in C++ or as simple as a Python script or self-recorded macro. You have complete access to almost any part of FreeCAD from the built-in Python interpreter, macros or external scripts, be it geometry creation and transformation, the 2D or 3D representation of that geometry (scenegraph) or even the FreeCAD interface. The software allows import/export to standard formats such as STEP, IGES, OBJ, STL, DXF, SVG, STL, DAE, IFC or OFF, NASTRAN, VRML in addition to FreeCAD's native FCStd file format. FreeCAD's installer allows flexible installations on Windows systems. Packages for Ubuntu systems are also available and updated regularly. There is a huge library of external workbenches available for FreeCAD on web. You can download free copy of software from https://www.freecadweb.org/ web link.

The **FreeCAD 0.21 Black Book** is the 4th edition of our series on FreeCAD. This book is written to help beginners in creating some of the most complex solid models. The book follows a step by step methodology. In this book, we have tried to give real-world examples with real challenges in designing. We have tried to cover most of the topics utilized in industries for designing. The book covers almost all the information required by a learner to master the FreeCAD. The book starts with sketching and ends at advanced topics like Path (CAM), FEM (Simulation), Sheet Metal, CFDOf, Assembly4, and so on. Some of the salient features of this book are :

In-Depth explanation of concepts

Every new topic of this book starts with the explanation of the basic concepts. In this way, the user becomes capable of relating the things with real world.

Topics Covered

Every chapter starts with a list of topics being covered in that chapter. In this way, the user can easily find the topics of his/her interest easily.

Instruction through illustration

The instructions to perform any action are provided by maximum number of illustrations so that the user can perform the actions discussed in the book easily and effectively. There are about 2000 illustrations that make the learning process effective.

Tutorial point of view

At the end of concept's explanation, the tutorial make the understanding of users firm and long lasting. Almost each chapter of the book has tutorials that are real world projects. Moreover most of the tools in this book are discussed in the form of tutorials.

Project

Projects and exercises are provided to students for practicing.

For Faculty

If you are a faculty member, then you can ask for video tutorials on any of the topic, exercise, tutorial, or concept. As faculty, you can register on our website to get electronic desk copies of our latest books. Faculty resources are available in the **Faculty Member** page of our website (**www.cadcamcaeworks.com**) once you login. Note that faculty registration approval is manual and it may take two days for approval before you can access the faculty website.

Formatting Conventions Used in the Text

All the key terms like name of button, tool, drop-down etc. are kept bold.

Free Resources

Link to the resources used in this book are provided to the users via email. To get the resources, mail us at ***cadcamcaeworks@gmail.com*** with your contact information. With your contact record with us, you will be provided latest updates and informations regarding various technologies. The format to write us mail for resources is as follows:

Subject of E-mail as ***Application for resources of .... book***.
Also, given your information like
Name:
Course pursuing/Profession:
Contact Address:
E-mail ID:

Note: We respect your privacy and value it. If you do not want to give your personal informations then you can ask for resources without giving your information.

About Authors

The author of this book, Matt Weber, has written more than 16 books on CAD/CAM/ CAE available in market. He has coauthored SolidWorks Simulation, SolidWorks Electrical, SolidWorks Flow Simulation, and SolidWorks CAM Black Books. The author has hands on experience on almost all the CAD/CAM/CAE packages. If you have any query/doubt in any CAD/CAM/CAE package, then you can contact the author by writing at cadcamcaeworks@gmail.com

The author of this book, Gaurav Verma, has written and assisted in more than 17 titles in CAD/CAM/CAE which are already available in market. He has authored Autodesk Fusion 360 Black Book, AutoCAD Electrical Black Book, Autodesk Revit Black Books, and so on. He has provided consultant services to many industries in US, Greece, Canada, and UK. He has assisted in preparing many Government aided skill development programs. He has been speaker for Autodesk University, Russia 2014. He has assisted in preparing AutoCAD Electrical course for Autodesk Design Academy. He has worked on Sheetmetal, Forging, Machining, and Casting designs in Design and Development departments of various manufacturing firms.

For Any query or suggestion

If you have any query or suggestion, please let us know by mailing us on *cadcamcaeworks@gmail.com*. Your valuable constructive suggestions will be incorporated in our books and your name will be addressed in special thanks area of our books on your confirmation.

Page Left Blank Intentionally

Chapter 1

Starting with FreeCAD

Topics Covered

The major topics covered in this chapter are:

- *Overview of FreeCAD*
- *Installing FreeCAD*
- *Starting FreeCAD*
- *File Menu*
- *View Menu*
- *Standard Views*
- *Freeze Display*
- *Draw Style*
- *Tools Menu*
- *Navigating in the 3D view*
- *FreeCAD Interface*

OVERVIEW OF FREECAD

FreeCAD is an open-source parametric 3D modeling application, made primarily to design real-life objects. Parametric modeling describes a certain type of modeling, where the shape of the 3D objects you design are controlled by parameters. For example, the shape of a brick might be controlled by three parameters: height, width, and length. In FreeCAD, as in other parametric modelers, these parameters are part of the object, and stay modifiable at any time, after the object has been created. It connects your entire product development process in a single program that works on all the three major platforms viz. Windows, Mac OS, and Linux; refer to Figure-1.

Figure-1. Application interface

FreeCAD is not designed for a particular kind of work or to make a certain kind of objects. Instead, it allows a wide range of uses and permits users to produce models of all sizes and purposes from small electronic components to 3D-printable pieces and all the way up to buildings. Each of these tasks have different dedicated sets of tools and workflow available. Being open-source, FreeCAD benefits from the contributions and efforts of a large community of programmers, enthusiasts, and users worldwide. FreeCAD is essentially an application built by the people who use it, instead of being made by a company trying to sell you a product. And of course, it also means that FreeCAD is free, not only to use but also to distribute, copy, modify, or even sell.

FreeCAD also benefits from the huge, accumulated experience of the open-source world. In its bowels, it includes several other open source components, as FreeCAD itself can be used as a component in other applications. It also possesses all kinds of features that have become a standard in the open-source world, such as supporting a wide range of file formats being hugely scriptable, customizing, and modifiable. All this is made possible through a dynamic and enthusiast community of users.

INSTALLING FREECAD

- Connect your PC with the internet connection and then go to **https://www. freecadweb.org/** as shown in Figure-2.

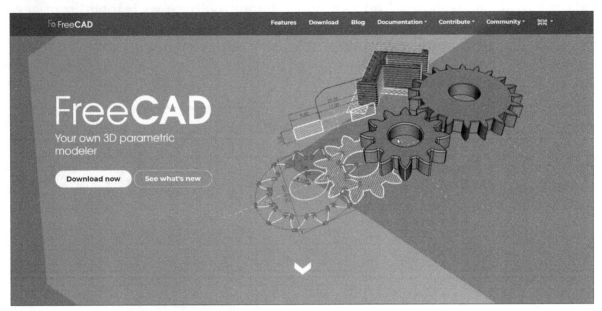

Figure-2. FreeCAD website

- Click on the **Download now** button, the **Downloads** tab of this website will be displayed; refer to Figure-3.

Figure-3. Download tab of FreeCAD website

- Select desired Operating System and the type of processor installed on your PC. The software will begin to download.
- Open the downloaded setup file and follow the instructions as per the setup window. The software will be installed in a couple of minutes.

FREECAD INTERFACE

The FreeCAD interface is based on the modern Qt, a well known graphical user interface **(GUI)** toolkit and has a state of the art organization. Some aspects of the interface can be customized. You can add custom toolbars with tools from several workbenches or tools defined in macros and you can create your own keyboard shortcuts. But the menus and default toolbars that come with FreeCAD and its workbenches cannot be changed. Different parts of the FreeCAD interface are discussed next; refer to Figure-4.

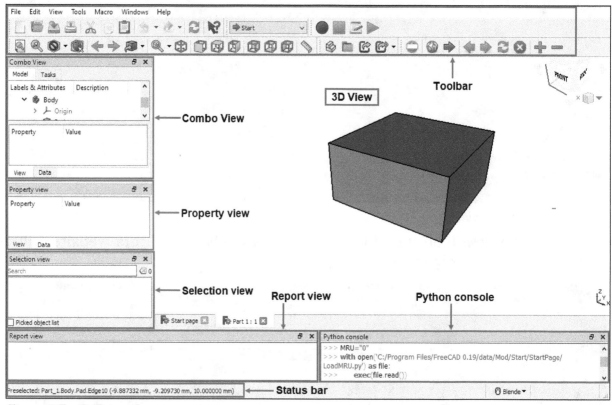

Figure-4. FreeCAD interface

Toolbars : The standard toolbars that appear in the interface are:

- The tools in the **File** toolbar are used to work with files, open documents, copy, paste, undo, and redo actions.
- The **Workbench** toolbar contains a single widget to select the active workbench.
- The tools in **Macro** toolbar are used to record, edit, and execute macros.
- In the **View** toolbar, the tools are used to control how objects appear in the 3D view.
- The tools in the **Structure** toolbar are used to organize objects in the document and create links to additional documents.

3D View : The **3D view** is the main component of the interface. It can be undocked out of the main window if you can have several views of the same document (or same objects) or several documents opened at the same time. You can select objects or parts of objects by clicking them and you can pan, zoom, and rotate the view with the mouse buttons.

Combo View : The **Combo view** is one of the main panels in the FreeCAD interface. It is located on the left side of screen by default. It is composed of two tabs. The **Model** tab shows you the tree view which is a representation of the document's content including 2D and 3D geometry with their parametric history but also supporting objects that contain data saved in the document. The **Tasks** tab is where FreeCAD will prompt you for values specific to the tool you're currently using at the time—for example, entering a 'length' value when the **Line** tool is being used. It will close automatically after pressing the **OK** (or **Cancel**) button. Also, by double-clicking the related object in the combo view, most tools will allow you to reopen the task panel in order to modify the settings.

Property editor : The **Property editor** appears when the **Model** tab of the **Combo View** is active in the interface. It allows managing the publicly exposed properties of the objects in the document. Generally, the property editor is intended to deal with just one object at one time. The values shown in the property editor belong to the selected object of the active document. Despite this, some properties like colors can be set for multiple selected objects. If there are no elements selected, the property editor will be empty.

Selection view : The **Selection view** is a panel in the interface by default located below the combo view. Just like the property editor, it shows more information about the currently selected objects. A selection can be made by picking an object in the 3D view or in the tree view. Multiple bodies can be selected by holding **Ctrl** key.

Report view : The **Report view** is a panel that shows text messages from FreeCAD processes and tools. The report view is normally hidden but it is a good idea to leave open as it will list any information, warnings, or errors to help you decipher (or debug) what you may have done wrong.

Python console : The **Python console** is a panel that runs an instance of the Python interpreter. It can be used to control FreeCAD processes. It creates and modifies objects and their properties. The Python console in FreeCAD has basic syntax highlighting able to differentiate with various styles and colors, comments, strings, numeric values, built in functions, printed text output, and delimiters like parentheses and commas.

Status bar : The **Status bar** is a simple information bar that appears at the bottom of the FreeCAD interface. When the mouse pointer is over a button or menu, the usage information of that command is displayed both in a textual pop-up and in the status bar. It shows the zoom level in the right corner next to the mouse navigation style. The status bar also shows the last pre-selected object or element of an object and the coordinates of the mouse pointer when the last pre-selection occurred; refer to Figure-5.

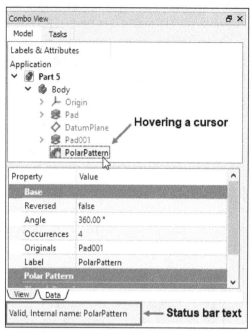

Figure-5. Valid Internal name Polar Pattern status bar text

STARTING FREECAD

- To start **FreeCAD** from **Start** menu, in Windows 7, click on the **Start** button in the **Taskbar** at the bottom left corner, click on **All Programs** folder and then on **FreeCAD** folder. If you are using Windows 8 or later then click on **FreeCAD** folder in the **Start** menu and select the **FreeCAD** icon; refer to Figure-6. Click on the program name to start the application.

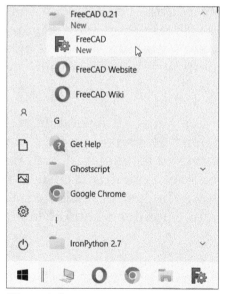

Figure-6. Start menu

- On starting the application, the initial screen of the application will be displayed as shown in Figure-7.

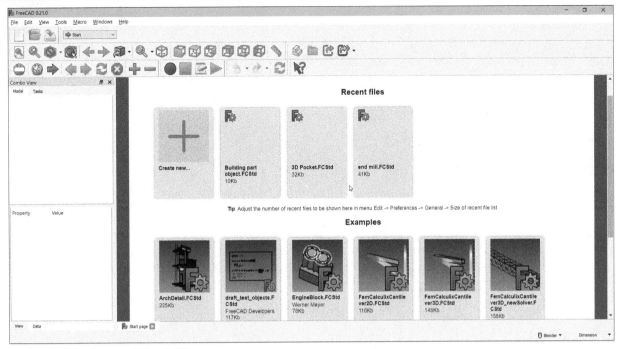

Figure-7. Initial screen of FreeCAD

FILE MENU

The options in the **File** menu are used to manage files and related parameters. Various tools of **File** menu are discussed next.

New

The **New** tool is used for initiating a new document file. The procedure to use this tool is discussed next.

- Click on the **New** tool from the **File** menu or **Ribbon** or press **Ctrl+N** key from keyboard; refer to Figure-6. The new document file will open; refer to Figure-9.

Figure-8. New tool

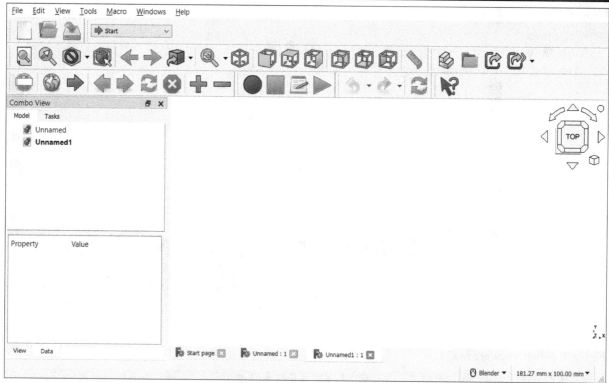

Figure-9. New document file

Open

The **Open** tool is used to open FreeCAD files and import files of other CAD applications. The procedure to use this tool is discussed next.

- Click on the **Open** tool from the **File** menu or **Ribbon** or press **CTRL+O** key from keyboard; refer to Figure-10. The **Open document** dialog box will be displayed; refer to Figure-11.

Figure-10. Open tool

Figure-11. Open document dialog box

- Select desired format of file from the **File Type** drop-down to define format of file to be opened.
- Select desired file which you want to open and click on **Open** button from the dialog box. The file will open.

Open Recent

The **Open Recent** cascading menu displays a list of recent files which can be opened.

- Hover the cursor on the **Open Recent** cascading menu of the **File** menu to display the recent files list; refer to Figure-12. Select desired file name to open respective file.

Figure-12. Open Recent cascading menu

Close

The **Close** tool is used to close the current file without closing the application. The procedure to use this tool is discussed next.

- Click on the **Close** tool from **File** menu or press **Ctrl+F4** key from keyboard; refer to Figure-13. The current file will be closed.

Figure-13. Close tool

Close All

The **Close All** tool is used to close all the files currently open without closing the application. The procedure to use this tool is discussed next.

- Click on the **Close All** tool from the **File** menu; refer to Figure-14. All the files which are currently open will be closed.

Figure-14. Close All tool

Save

The **Save** tool is used to save the current file opened in the application. The procedure to use this tool is discussed next.

- Click on the **Save** tool from **File** menu or **Ribbon** or press **Ctrl+S** key from keyboard; refer to Figure-15. The **Save FreeCAD Document** dialog box will be displayed; refer to Figure-16.

Figure-15. Save tool

Figure-16. Save FreeCAD Document dialog box

- Specify desired location and name for the file, and then click on the **Save** button from the dialog box to save the file.

Save As

Using the **Save As** tool, you can save the file with different name. The procedure to use this tool is discussed next.

- Click on **Save As** tool from **File** menu; refer to Figure-17. The **Save FreeCAD Document** dialog box will be displayed as shown in Figure-16.

Figure-17. Save As tool

- Rest of the procedure is same as discussed for the **Save** tool.

Save a Copy

The **Save a Copy** tool is used to create a copy of the file with new name. The procedure to use this tool is discussed next.

- Click on the **Save a Copy** tool from the **File** menu; refer to Figure-18. The **Save FreeCAD Document** dialog box will be displayed as shown in Figure-16.

Figure-18. Save a Copy tool

• Rest of the procedure is same as we have discussed earlier.

Save All

The **Save All** tool is used to save all the documents currently open in the application. The procedure to use this tool is discussed next.

• Click on the **Save All** tool from the **File** menu; refer to Figure-19. The **Save FreeCAD Document** dialog box will be displayed as shown in Figure-16.

Figure-19. Save All tool

• Rest of the procedure is same as discussed earlier.

Revert

The **Revert** tool is used to close the active document and reopen the last saved version of this document. Using this tool, you will lose all the changes you have made since the last file save. The procedure to use this tool is discussed next.

• Click on the **Revert** tool from the **File** menu; refer to Figure-20. The **Revert document** dialog box will be displayed asking you for the confirmation to discard all the changes; refer to Figure-21.

Figure-20. Revert tool

Figure-21. Revert document dialog box

- Click on the **Yes** button from the dialog box to discard all the changes.

Import

Many times, importing reduces lots of extra work of rebuilding the base sketch for models. In FreeCAD, we can directly use the CAD files of other software for creating or manipulating the model. Note that if you are a user of Image workbench then in FreeCAD 0.21, the **Image** workbench has been removed and its functionality is added in the **Import** tool so you can import image files as well by using this tool. The procedure to use this tool is discussed next.

- Click on the **Import** tool from the **File** menu; refer to Figure-22. The **Import file** dialog box will be displayed; refer to Figure-23.

Figure-22. Import tool

Figure-23. Import file dialog box

- Select desired format of non-native file which you want to open from the **File Type** drop-down; refer to Figure-24.

Figure-24. File type drop down

- Select desired file which you want to import and click on the **Open** button from the dialog box. The model will be updated in this graphic area.

Export

The **Export** tool is used to export existing file in various different formats. The procedure to use this tool is discussed next. IGES and STEP are the most common formats for import-export in CAD software.

- Select the solid part or sketch which you want to export from the **Model** tab of the **Combo View** as shown in Figure-25.

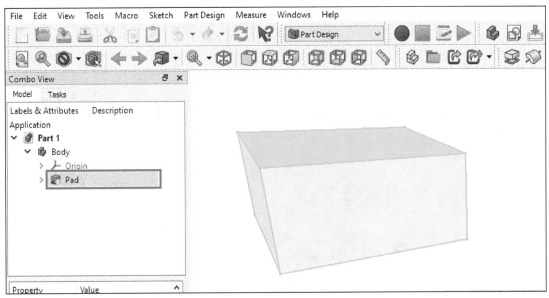

Figure-25. Solid model selected to export

- Click on the **Export** tool from **File** menu; refer to Figure-26. The **Export file** dialog box will be displayed; refer to Figure-27.

Figure-26. Export tool

Figure-27. Export file dialog box

- Select desired format of file from the **Save as type** drop-down.
- Specify desired name for the file in the **File name** edit box and click on **Save** button from the dialog box to export the file.

Merge project

The **Merge project** tool is used to add the contents of a FreeCAD file into the active document. The procedure to use this tool is discussed next.

• Click on the **Merge project** tool from the **File** menu; refer to Figure-28. The **Merge project** dialog box will be displayed; refer to Figure-29.

Figure-28. Merge project tool

Figure-29. Merge project dialog box

• Select desired file which you want to add in the active document and click on **Open** button from the dialog box. The model of selected file will be added in the active document.

Project information

The **Project information** tool is used to display general information related to current project in **Projection information** dialog box. Some of this information can be edited in the dialog box. The procedure to use this tool is discussed next.

• Click on the **Project information** tool from the **File** menu; refer to Figure-30. The **Project information** dialog box will be displayed; refer to Figure-31.

Figure-30. Project informtion tool

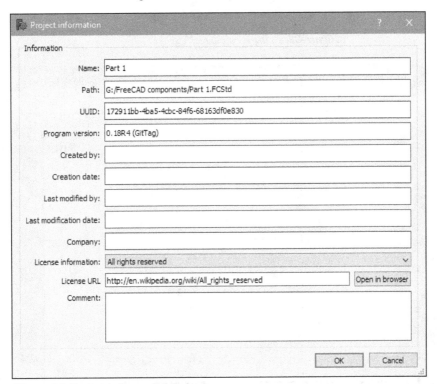

Figure-31. Project information dialog box

- Specify and edit the information about the active document as desired and click on **OK** button from the dialog box.

Print

As the name suggests, the **Print** tool is used to get print out of the model/drawing on paper using printer connected to the system.

Print Preview

The **Print Preview** tool is used to check the print before sending command to printer. The procedure to use this tool is discussed next.

- Click on the **Print preview** tool from the **File** menu; refer to Figure-32. The **Print Preview** dialog box will be displayed along with the preview of the model; refer to Figure-33.

Figure-32. Print preview tool

Figure-33. Print Preview dialog box

- If the preview is satisfactory then click on the **Print** button otherwise, click on the **Close** button from the toolbar displayed at the top of the modeling area and modify the model; refer to Figure-33.

Export PDF

The **Export PDF** tool is used to export or save the active document in PDF file format. The procedure to use this tool is discussed next.

- Click on the **Export PDF** tool from **File** menu; refer to Figure-34. The **Export PDF** dialog box will be displayed; refer to Figure-35.

Figure-34. Export PDF tool

Figure-35. Export PDF dialog box

• Specify desired file name for saving the file as PDF and click on **Save** button from the dialog box.

View Menu

The **View** menu provides tools to change the 3D view, the view properties of objects in the model, and tools related to the display of interface components. Various tools of **View** menu are discussed next.

Create new view

The **Create new view** tool is used to create a new 3D view for the active document. It can be useful if you want to inspect the model from multiple directions or at different zoom levels. The procedure to use this tool is discussed next.

• Click on the **Create new view** tool from the **View** menu; refer to Figure-36. A new window will open with the 3D view of an active model; refer to Figure-37.

Figure-36. Create new view tool

Figure-37. New 3D view of a model

Orthographic view

The **Orthographic view** tool switches the camera in the active 3D view to orthographic view mode. In this mode, objects that are far from the camera do not appear smaller than those that are closer. The procedure to use this tool is discussed next.

- Click on the **Orthographic view** tool from the **View** menu; refer to Figure-38. An orthographic view of the model will be displayed; refer to Figure-39.

Figure-38. Orthographic view tool

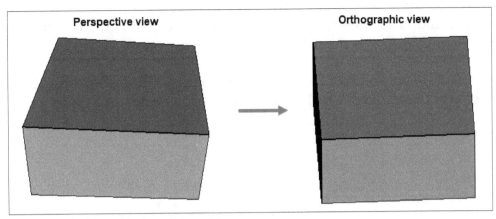

Figure-39. Orthographic view of the model

Perspective view

The **Perspective view** tool switches the camera in the active 3D view to perspective view mode. In this mode, objects that are far from the camera appear smaller than those that are closer. The procedure to use this tool is discussed next.

- Click on the **Perspective view** tool from the **View** menu; refer to Figure-40. A perspective view of the model will be displayed; refer to Figure-41.

Figure-40. Perspective view tool

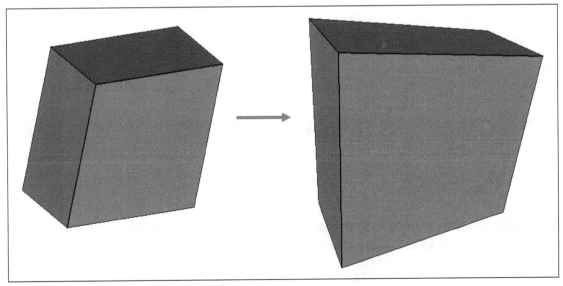

Figure-41. Perspective view of the model

Fullscreen

The **Fullscreen** tool switches the FreeCAD's main window in fullscreen mode. The procedure to use this tool is discussed next.

- Click on the **Fullscreen** tool from the **View** menu; refer to Figure-42. The view of the model will be displayed in Fullscreen window; refer to Figure-43.

Figure-42. Fullscreen tool

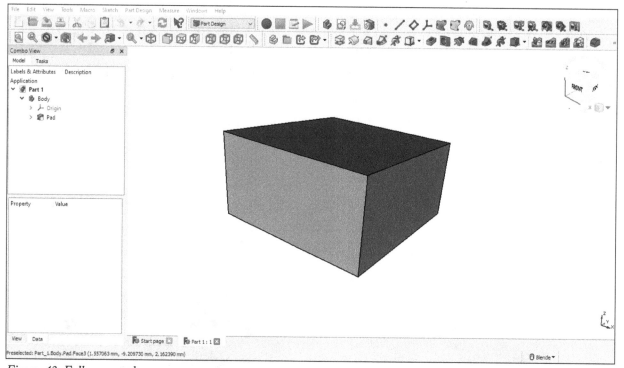

Figure-43. Fullscreen mode

STANDARD VIEWS

Standard views provide tools to orient the part, assembly, or sketch in one of the preset standard views. The standard views display the model or drawing through one, two, or four viewports.

Various tools available in the **Standard views** cascading menu (refer to Figure-44) in the **View** menu are discussed next.

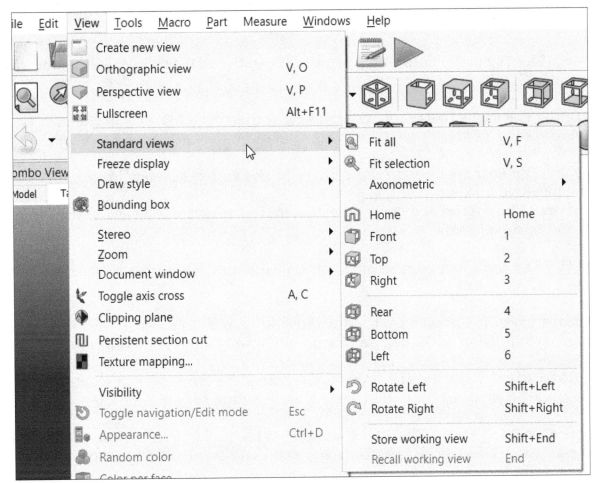

Figure-44. Standard views cascading menu

- **Fit all** : The **Fit all** tool zooms and pans the camera so that all visible objects fit inside the active 3D view.

- **Fit selection** : The **Fit selection** tool zooms and pans the camera so that all selected objects fit inside the active 3D view.

- **Axonometric** : The **Axonometric** cascading menu has three view tools:

    Isometric View : The **Isometric** tool realigns the camera in the active 3D view to obtain an isometric view. For a truly isometric view, the 3D view must be in orthographic mode but the tool also works, if the view is in perspective mode.

    Dimetric View : The **Dimetric** tool realigns the camera in the active 3D view to obtain a dimetric view.

    Trimetric View : The **Trimetric** tool realigns the camera in the active 3D view to obtain a trimetric view.

Note that in Isometric view all dimensions are the same scale and all projectors have equal angles, usually 30°. In Diametric view, two axes use the same scale, while the third (usually the vertical axis) is determined separately. In Trimetric view, all three projectors have different angles.

- **Front** : The **Front** tool points the camera in the active 3D view in the direction of the positive Y axis.

- **Top** : The **Top** tool points the camera in the active 3D view in the direction of the negative Z axis.

- **Right** : The **Right** tool points the camera in the active 3D view in the direction of the negative X axis.

- **Rear** : The **Rear** tool points the camera in the active 3D view in the direction of the negative Y axis.
- **Bottom** : The **Bottom** tool points the camera in the active 3D view in the direction of the positive Z axis.

- **Left** : The **Left** tool points the camera in the active 3D view in the direction of the positive X axis.

- **Rotate Left** : The **Rotate Left** tool rotates the camera in the active 3D view around the view direction in 90-degree increments towards the left (counterclockwise).

- **Rotate Right** : The **Rotate Right** tool rotates the camera in the active 3D view around the view direction in 90-degree increments towards the right (clockwise).

- **Store working view** : The **Store working view** tool stores the camera setting of the active 3D view as temporary view. You can also use **SHIFT+END** shortcut key to perform the same task.

- **Recall working view** : The **Recall working view** tool restores the earlier stored working view. You can also use **END** key to perform the same task.

FREEZE DISPLAY

FreeCAD can store camera settings for up to 50 'frozen views'. Frozen view is the view with current specified orientation which you can restore after the view get disturbed. Frozen views are not stored in the document and if not saved with the **Save views** menu option, will be lost when the FreeCAD application closes. The tools that deal with frozen views are available in the **Freeze display** cascading menu of the **View** menu; refer to Figure-45.

Figure-45. Freeze display cascading menu

Save Views

The **Save views** tool saves all existing frozen views in a file with the ***.cam** extension. To use this tool, one or more frozen views must exist. A frozen view can be created with the **Freeze view** menu option which will be discussed later. The procedure to use this tool is discussed next.

• Click on the **Save views** tool from the **Freeze display** cascading menu. The **Save frozen views** dialog box will be displayed; refer to Figure-46.

Figure-46. Save frozen views dialog box

• Specify desired name for the file in the **File name** edit box and click on **Save** button from the dialog box to save the frozen view.

Load views

The **Load views** tool loads the frozen views from a file with the ***.cam** extension. Note that when you load a frozen view then all existing frozen views will be deleted. The procedure to use this tool is discussed next.

• Click on the **Load views** tool from the **Freeze display** cascading menu.
• If the frozen views are saved earlier then the **Restore views** dialog box will be displayed; refer to Figure-47, asking you to confirm that you want to lose all existing frozen views.

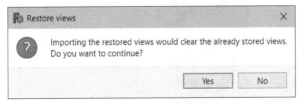

Figure-47. Restore frozen views dialog box

- Click on the **Yes** button from the dialog box, the **Restore frozen views** dialog box will be displayed; refer to Figure-48.

Figure-48. Restore frozen views dialog box

- If the frozen views are not saved earlier then also the **Restore frozen views** dialog box will be displayed.
- Select the file which you want to restore and click on **Open** button from the dialog box.

Freeze view

The **Freeze view** tool stores the current camera settings in a new frozen view. You can create maximum 50 frozen views. The procedure to use this tool is discussed next.

- Click on the **Freeze view** tool from **Freeze display** cascading menu. The frozen view of the model will be created; refer to Figure-49.

Figure-49. Frozen view of the model

Clear views

The **Clear views** tool deletes all existing frozen views. Click on the **Clear views** tool from **Freeze display** cascading menu, all the existing frozen views will be deleted.

Restore view

For each frozen view, a **Restore view** option is added with which the frozen view can be restored. The procedure to use this tool is discussed next.

- Click on the **Restore view** tool from **Freeze display** cascading menu, the frozen view of the model will be restored; refer to Figure-50

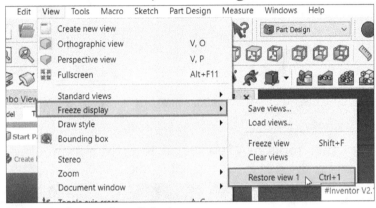

Figure-50. Restore views dialog box

DRAW STYLE

The tools in **Draw style** cascading menu are used to display objects in graphics area of different view styles like wireframe, solid, hidden lines, and so on; refer to Figure-51 and Figure-52

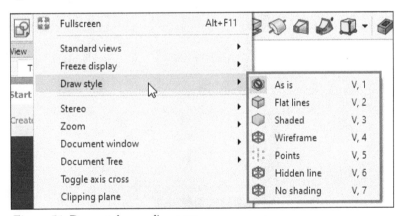

Figure-51. Draw style cascading menu

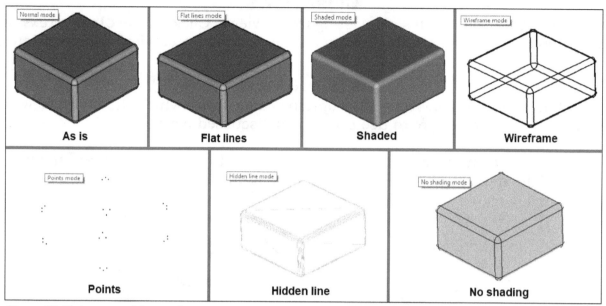

Figure-52. Various type of draw styles

You will learn about other tools of **View** menu later in this book.

TOOLS MENU

The **Tools** menu provides tools for debugging models, customizing FreeCAD's behavior as well as adding auxiliary tools; refer to Figure-53. Various tools available in the **Tools** menu are discussed next.

Figure-53. Tools menu

Edit parameters

The **Edit parameters** tool gives access to the parameters that control the program. By clicking on this tool, the **Parameter Editor** dialog box will be displayed as shown in Figure-54. The **Parameter Editor** controls the behavior of FreeCAD and its workbenches. The parameters are stored in a file called **user.cfg**.

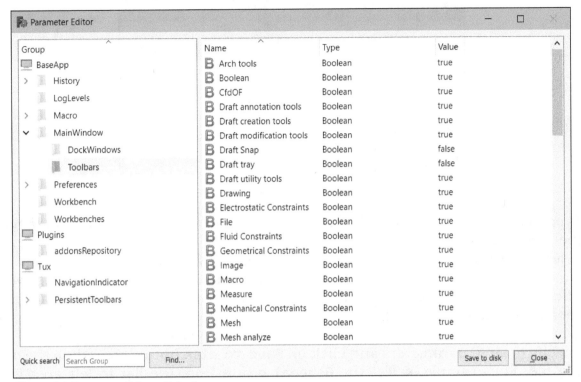

Figure-54. Parameter Editor dialog box

- Type desired name of the group in the **Quick search** edit box which you want to modify. The specified group name will be highlighted with red color in the left panel of the dialog box; refer to Figure-55.

Figure-55. Searched group highlighted

- You can search object in the dialog box using advanced filters by using the **Find** dialog box. Click on the **Find** button next to the **Quick search** edit box in the dialog box. The **Find** dialog box will be displayed; refer to Figure-56.

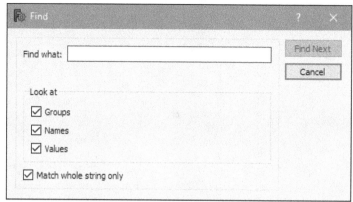

Figure-56. Find dialog box

- Type desired string name which you want to find in the **Find what** input box of the **Find** dialog box.
- Select the **Groups**, **Names**, and **Values** check boxes from **Look at** area of the **Find** dialog box to find the string values of groups, names, and values, respectively.
- Click on the **Find Next** button. The searched strings will be displayed in the **Parameter Editor** dialog box.
- Click on the **Cancel** button to close the **Find** dialog box.
- Specify desired parameters and click on **Save to disk** button to update the **user. cfg** file. For example, select the **General** group from the list in left area of the dialog box and modify the **FileOpenSavePath** parameter to change default directory where software will look to save/open the files.
- Click on the **Close** button to close the dialog box. You will learn more about the parameters in this dialog box later based on chapter requirement.

Save Image

The **Save image** tool is used to create an image file or a screenshot from the active 3D view. The procedure to use this tool is discussed next.

- Click on the **Save image** tool from the **Tools** menu. The **Save image** dialog box will be displayed; refer to Figure-57.

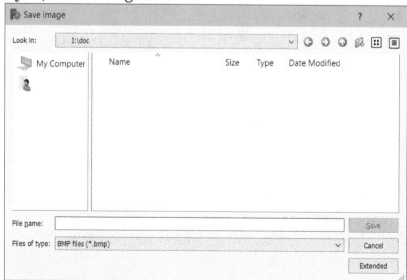

Figure-57. Save image dialog box

- Click on the **Extended** button from the dialog box. An additional panel will be added as shown in Figure-58.

Figure-58. Extended options in Save image dialog box

- Select standard size for the image from **Standard sizes** drop-down list or specify width and height of the image in the **Width** and **Height** edit boxes, respectively in the **Image dimensions** area of the dialog box.
- Select desired **Aspect ratio** button to set the width to height ratio of the image.
- Select desired option from **Background** drop down list in the **Image properties** area of the dialog box to set background for the image.
- Select desired option from **Creation method** drop-down list in the **Image properties** area to set the creation method for the image.
- Select **Insert MIBA** radio button from **Image comment** area to add MIBA information to the file. The options in the **Image comment** area of the dialog box will be activated only on selecting **JPG**, **JPEG**, and **PNG** image file types.
- Select the **Insert comment** radio button to type a comment in the text field to embed a comment in the file.
- Select the **Add watermark** check box to add a watermark of the FreeCAD logo in the image file.
- Specify desired file name and click on **Save** button from the dialog box.

Load Image

The **Load image** tool is used to load earlier saved image files in the graphics area. The procedure to use this tool is given next.

- Click on the **Load image** tool from **Tools** menu. The **Load image** dialog box will be displayed as shown in Figure-59.

Figure-59. Choose an image file to open dialog box

- Select desired image file from the dialog box and click on the **Open** button. The image file will load in a new document tab.
- Right-click in the drawing area and select **Fit to window**, **zoom in**, and **zoom out** options to perform respective operations.
- To pan an image that is the larger than current view area, you can either use the scrollbars, or hold down the middle mouse button and move the cursor.

Scene Inspector

The **Scene inspector** tool displays an overview of all nodes in the scene graph of the active 3D view. It is more a utility for programmers than for average users. It can be used to find out why the rendering is slow or why something is not rendered properly. The procedure to use this tool is discussed next.

- Click on the **Scene inspector** tool from **Tools** menu. The **Scene Inspector** dialog box will be displayed as shown in Figure-60. This dialog box is modeless, it means it can stay open while you continue working in FreeCAD.

Figure-60. Scene Inspector dialog box

- Click on the **Refresh** button to update the overview or click on **Close** button to close the dialog box.

Dependency Graph

The **Dependency Graph** tool displays dependencies between objects of the active document in a dependency graph. In this graph, objects are listed in reverse chronological order, with the first created object at the bottom.

The dependency graph is purely a visualization tool therefore it cannot be edited. It automatically updates, if changes are made to the model.

Note that to use this tool, a third party software named **Graphviz** needs to be installed.

If the software is not pre-installed, install it after downloading from the **Graphviz** website (https://graphviz.org). If the software is installed in an unconventional location, FreeCAD will ask you for software installation path.

The procedure to use this tool is discussed next.

- Click on the **Dependency graph** tool from **Tools** menu. The **Graphviz not found** dialog box will be displayed, if **Graphviz** is not installed in standard location; refer to Figure-61.

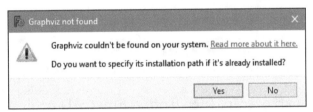

Figure-61. Graphviz not found dialog box

- Click on **Yes** button from the dialog box. The **Graphviz installation path** dialog box will be displayed asking for software installation folder; refer to Figure-62.

Figure-62. Graphviz installation path dialog box

- Browse to the folder where you have installed the software in your system (generally the path is: C:\Program Files\Graphviz\bin) and click on **Select Folder** button from the dialog box. A new window named **Dependency graph** will open displaying dependencies between objects; refer to Figure-63.

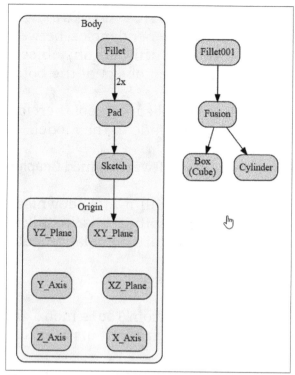

Figure-63. Dependency graph window

Project Utility

The **Project utility** tool assists in repairing damaged project files.

By using this tool, you can extract files from a FreeCAD project file **(*.FCStd)**, which is in fact a **ZIP** file and after manual edits, create a new project file from them.

The procedure to use this tool is discussed next.

*   Click on the **Project utility** tool from the **Tools** menu. The **Project utility** dialog box will be displayed; refer to Figure-64.

Figure-64. Project utility dialog box

*   Click on the **Browse** [...] button for **Source** edit box to define location of source file for extracting/creating the project in the **Extract project/Create project** area, respectively.
*   Similarly, click on the **Browse** [...] button for **Destination** edit box to specify the destination location for extracting/creating the project in the **Extract project/Create project** area, respectively.

- Click on the **Extract** button to create zip file and click on the **Create** button to save project file.
- Select the **Load project file after creation** check box to open the created file.
- Click on the **Close** button to close the dialog box.

Measure distance

The **Measure distance** tool is used to create a distance object that measures and displays the distance between two points. The procedure to use this tool is discussed next.

- Click on the **Measure distance** tool from **Tools** menu. The measurement scale symbol will get attached to the cursor.
- Select the first and second dimension points of object to measure the distance between the points. The measured distance dimension will be displayed; refer to Figure-65.

Figure-65. Dimension created on the object

Add text document

The **Add text document** tool creates an object capable of holding arbitrary text. This element can be used to write general information or documentation about the model.

- Click on the **Add text document** tool from the **Tools** menu. A new document file named **Text document** will be opened; refer to Figure-66.

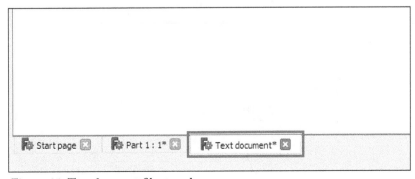

Figure-66. Text document file opened

- Specify desired text in the document file and close the tab after saving the file.

View turntable

The **View turntable** tool continuously rotates the camera in a desired angle and speed for desired time. The procedure to use this tool is discussed next.

- Click on the **View turntable** tool from **Tools** menu. The **View Turntable** dialog box will be displayed; refer to Figure-67.

Figure-67. View Turntable dialog box

- Move the **Angle** slider to specify the angle by which you want to rotate the model.
- Move the **Speed** slider to adjust rotation speed of the model.
- Select the **Fullscreen** check box to switch 3D view to fullscreen mode.
- Select the **Enable timer** check box to specify the duration of rotation.
- Click on **Play/Stop** button to start or stop the rotation of model.
- Click on **Close** button to close the dialog box.

Units calculator

The **Units calculator** tool is used to convert values from one unit system to another. The procedure to use this tool is discussed next.

- Click on **Units calculator** tool from **Tools** menu. The **Units calculator** dialog box will be displayed; refer to Figure-68. This dialog box is modeless, it means it can stay open while you continue working in FreeCAD.

Figure-68. Units calculator dialog box

- Specify desired values in the first two edit boxes to perform conversion.
- In first edit box, specify the value to be converted and in the **as:** edit box, specify the unit value to be converted.
- Result will be displayed in **=>** edit box. For example, if you want to convert **55** mm to feet then specify **55** in first edit box and **ft** in the **as:** edit box; refer to Figure-67. You can use in for inch, ft for feet, cm for centimeter, m for meter, and so on to define units of length. Similarly, you can use units of weight, density, speed, current, and so on.

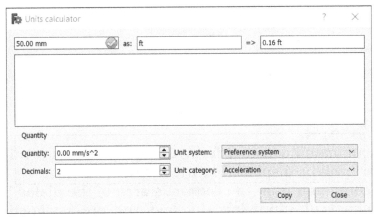

Figure-69. Result of conversion

Customize

Customize tool gives access to several customization options. The customization depends on the workbenches that have been loaded in the current FreeCAD session. The procedure to use this tool is discussed next.

- Click on the **Customize** tool from **Tools** menu. The **Customize** dialog box will be displayed; refer to Figure-70.

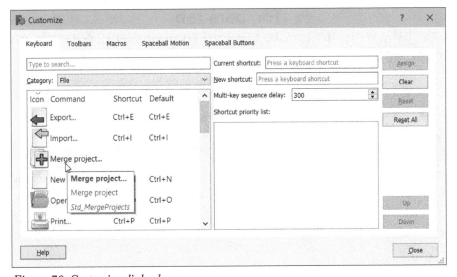

Figure-70. Customize dialog box

The options in various tabs of the **Customize** dialog box are discussed next.

Keyboard tab

Using the options in **Keyboard** tab, you can define custom keyboard shortcuts for using the tool. The **Keyboard** tab of **Customize** dialog box is shown in Figure-71.

Figure-71. Keyboard tab

- Select desired command category from **Category** drop-down list. Commands of selected category will be displayed in the **Commands** area.
- Select desired command from the **Commands** area of the dialog box.
- The **Current shortcut** input box displays the pre-assigned shortcut for the selected tool.
- To specify a new shortcut for the tool, click in the **New shortcut** input box and press new shortcut keys from the keyboard.
- Click on **Clear** button to remove specified shortcut for the selected command.
- Click on **Reset** button to remove a custom shortcut for the selected command.
- Click on **Reset All** button to remove all custom shortcuts specified for the commands.
- Click on the **Close** button to exit the dialog box.

Toolbars tab

In the **Toolbars** tab, you can create custom toolbars and perform modification in toolbars; refer to Figure-72.

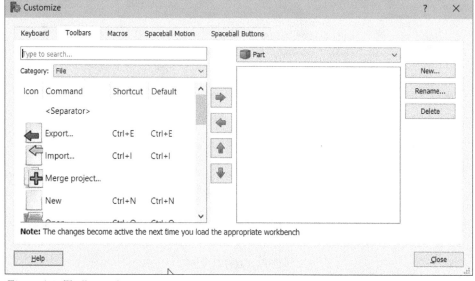

Figure-72. Toolbars tab

- In the drop-down list at right side in the dialog box, select a workbench whose custom toolbar is to be modified.
- Click on the **New** button from right side of the dialog box. The **New toolbar** dialog box will be displayed asking you to specify the toolbar name; refer to Figure-73.

Figure-73. New toolbar dialog box

- Specify desired name in the **Toolbar name** edit box and click on the **OK** button. The new toolbar will be created in the right area of the dialog box.
- Select the newly created toolbar from the right area to change its name and click on **Rename** button, the **Rename toolbar** dialog box will be displayed; refer to Figure-74.

Figure-74. Rename toolbar dialog box

- Specify desired new name of the toolbar and click on **OK** button.
- If you want to delete the toolbar then select the toolbar from right area and click on **Delete** button.
- To disable a toolbar, deselect the check box in front of toolbar name in the right panel. The toolbar will be invisible in the FreeCAD interface.
- To add a command in the toolbar, select the created toolbar in the right area. If no toolbar is selected, the command will be added to the first toolbar available in the drop-down list in right side of the dialog box.
- Select desired command category from the drop-down list available in the left side of the dialog box. The list of related commands will be displayed.
- Select desired command from the left panel which you want to add to the toolbar and click on ➡ button.
- To remove a command, select it from the right panel and click on ⬅ button.
- To change a command position, select it and click on ⬆ button or ⬇ button.
- Click on the **Close** button to exit the dialog box.

Macros tab

In the **Macros** tab, you can manage macros created earlier. FreeCAD uses a dedicated folder for user macros and only macros in that folder can be set up.

- Select desired macro from the **Macro** drop-down list. The parameters related to macro will be displayed; refer to Figure-75.

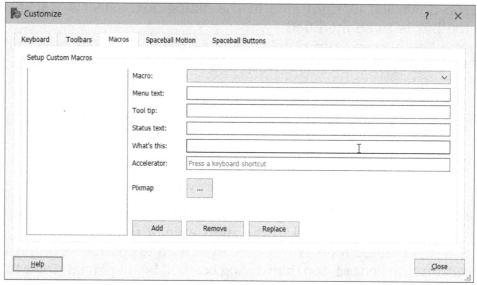

Figure-75. Macros tab

- Specify desired name to identify the macro command in the **Menu text** edit box.
- Specify a desired tooltip text in the **Tool tip** edit box which will display near the cursor when you hover the cursor on the macro icon.
- Specify desired text in the **Status text** edit box which will display in the status bar when you hover the cursor on the macro icon.
- Specify the wiki page name for the macro if available on the internet in the **What's this** edit box.
- Specify the keyboard shortcut for a macro in the **Accelerator** edit box.
- To add an icon for the macro command, click on **Pixmap** button. The **Choose Icon** dialog box will be displayed; refer to Figure-76.

Figure-76. Choose Icon dialog box

- Select desired icon from the panel, the **Choose Icon** dialog box will be closed automatically. Click on **Icon folders** button from the **Choose Icon** dialog box if you want to choose an icon from the local drive.
- Click on the **Add** button from the **Customize** dialog box to add modified macro in the macro list available for document.
- To remove a macro, select it from the left area of the dialog box and click on **Remove** button.

- To change the macro command, double-click on the command from the left panel. Make desired changes and click on **Replace** button.
- Click on the **Close** button to exit the dialog box. You will learn more about macro creation and management later in this book.

Similarly, you can use the **Spaceball Motion** and **Spaceball buttons** tabs to customize functions of mouse with spaceball feature.

Addon Manager

With the **Addon manager** tool, you can install and manage external workbenches and macros provided by the FreeCAD community. Note that you need an internet connection to download the addon resources.

- Click on the **Addon manager** tool from **Tools** menu. The **Addon Manager** dialog box will be displayed; refer to Figure-77.

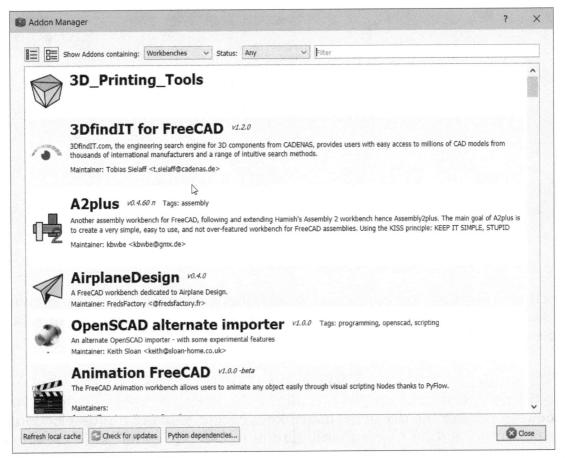

Figure-77. Addon Manager dialog box

- Select desired addon which you want to install from the list available in **Workbenches** tab.
- Click on **Install/update selected** button to install the addon.
- To uninstall the addon, select the addon from the list which you want to uninstall and click on **Uninstall selected** button from the dialog box.
- If you have installed or updated a workbench then by clicking on **Close** button, a new **Addon manager** dialog box will open informing you to restart FreeCAD for the changes to take effect.
- Click on **OK** button from the dialog box and restart FreeCAD.

Navigating in the 3d view

Navigating in the FreeCAD 3D view can be done with a mouse, a Space Navigator device, the keyboard, a touchpad, or a combination of these. FreeCAD can use several navigation modes which determine how the three basic view manipulation operations (pan, rotate, and zoom) are done as well as how to select objects on the screen. Navigation modes are accessed from the **Preferences screen** or directly by right-clicking anywhere in the graphic area; refer to Figure-78.

Figure-78. Navigation modes in 3D view

Each of these modes attribute different mouse buttons, mouse+keyboard combinations, or mouse gestures to these four operations. The procedure to use all these modes is discussed next.

CAD Navigation

This is the default navigation style. It allows the user a simple control of the view and does not require the use of keyboard keys except to make multi-selections. The mouse gestures used for object manipulation in **CAD** navigation style are shown in Figure-79.

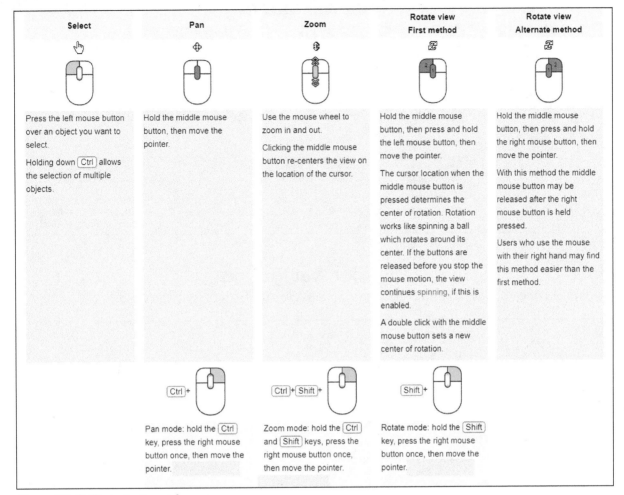

Figure-79. CAD navigation style

OpenInventor Navigation

In **OpenInventor** navigation, you must hold down the **Ctrl** key in order to select objects. The procedure to use this navigation style is shown in Figure-80.

Figure-80. OpenInventor navigation style

Revit Navigation

The procedure to use **Revit** navigation style is shown in Figure-81.

Figure-81. Revit navigation style

Blender Navigation

The procedure to use **Blender** navigation style is shown in Figure-82.

Figure-82. Blender navigation style

MayaGesture Navigation

In **Maya-Gesture** navigation, panning, zooming, and rotating the view require the **Alt** key together with a mouse button; therefore, a three-button mouse is required. The procedure to use this navigation is shown in Figure-83.

Figure-83. MayaGesture navigation style

Touchpad Navigation

In **Touchpad** navigation, panning, zooming, and rotating the view require a modifier key together with the touchpad. The procedure to use this navigation is shown in Figure-84.

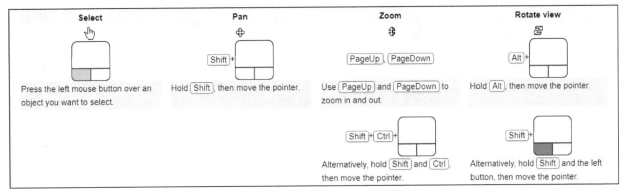

Figure-84. Touchpad navigation style

Gesture Navigation

The procedure to use **Gesture** navigation style is shown in Figure-85.

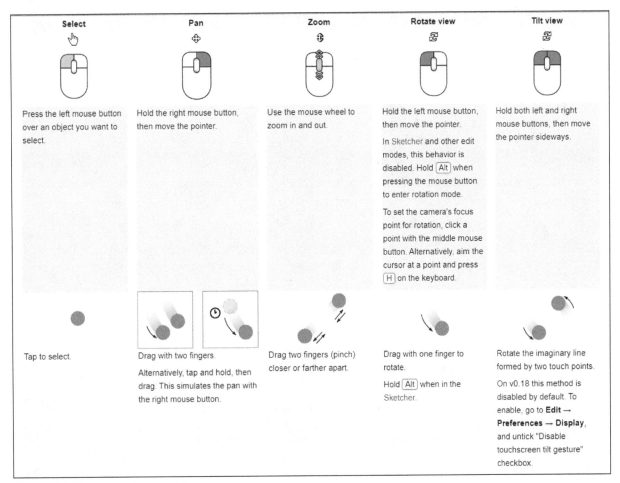

Figure-85. Gesture navigation style

OpenCascade Navigation

The procedure to use **OpenCascade** navigation is shown in Figure-86.

Figure-86. OpenCascade navigation style

Chapter 2

Sketching

Topics Covered

The major topics covered in this chapter are:

- *Introduction*
- *Starting Sketch*
- *Sketch Creation Tools*
- *Sketch Editing Tools*
- *Sketcher Constraints*
- *Sketcher Tools*
- *Sketcher B-Spline Tools*
- *Sketching Practical and Practice*

INTRODUCTION

In Engineering, sketches are based on real dimensions of real world objects. These sketches work as building blocks for various 3D operations. In this chapter, we will be working with sketch entities like; Line, Circle, Arc, Polygon, Ellipse, and so on to form base feature for various 3D operations. Note that the sketching environment is the base of 3D models so you should be proficient in sketching.

In this chapter, we will be working in **Sketcher Workbench**. We will learn about various tools of toolbar used in 2D sketching.

STARTING SKETCH

- To start a new sketch file, click on the **New** button from **File** menu and select **Sketcher** workbench from **Switch between workbenches** drop-down in the **Toolbar**; refer to Figure-1. The tools to create 2D sketching will be displayed in toolbar in inactive mode; refer to Figure-2.

Figure-1. Sketcher workbench

Figure-2. Sketching tools

- Click on **Create sketch** button from the toolbar. The **Choose orientation** dialog box will be displayed asking you to select the plane; refer to Figure-3.

Figure-3. Choose orientation dialog box

- Select desired radio button from **Sketch orientation** area to define which plane will be used as sketching plane.
- Select the **Reverse direction** check box to reverse the plane direction.

- Specify the offset distance for plane in **Offset** edit box. The plane for sketching will be at specified offset distance normal to selected base plane.
- Click on the **OK** button to close the dialog box. The selected plane will become parallel to the screen and act as current sketching plane. Now, we are ready to draw sketch on the selected plane.

First, we will start with the sketch creation tools and later we will discuss the other tools.

SKETCH CREATION TOOLS

After selecting the plane, all the sketch creation tools will become active. These tools are discussed next.

Point

The **Point** tool creates a point in the current sketch which can be used for constructing geometry elements. The procedure to use this tool is discussed next.

- Click on **Create point** tool from **Toolbar** in the **Sketcher** workbench; refer to Figure-4. You will be asked to specify location to the point.

Figure-4. Point tool

- Click at desired locations in the 3D view area. The points will be created; refer to Figure-5.

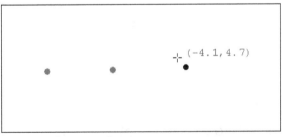

Figure-5. Creation of point

- Press **ESC** key from keyboard or press **RMB** to exit the tool.

Line

The **Line** tool is used to create a line by picking two points in the 3D view. The procedure to use this tool is discussed next.

- Click on **Create line** tool from **Toolbar** in **Sketcher** workbench; refer to Figure-6. You can also press **GL** key from keyboard to select the **Line** tool. You will be asked to specify starting point of the line.

Figure-6. Line tool

- Click at desired location to specify start point and move the cursor away in desired direction in which you want to create the line; refer to Figure-7. You will be asked to specify location of end point of the line.

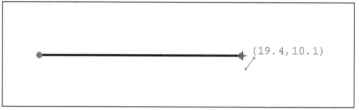
Figure-7. Creation of line

- Click at desired location to specify the end point and right-click to exit the tool. If you want to create a line for construction purpose then select the line you have created and click **RMB** on an empty area of the 3D view, a shortcut menu will be displayed.
- Click on the **Toggle construction geometry** option from the shortcut menu, the normal line will become construction line and vice-versa; refer to Figure-8. Note that construction geometry is displayed in blue color.

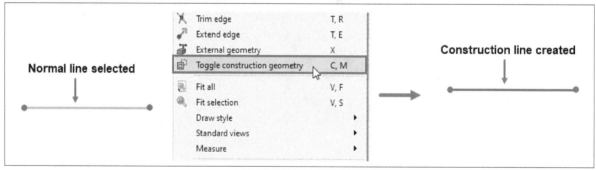
Figure-8. Creation of construction line

- Press **ESC** key or press **RMB** to exit the tool.

Arc

The tools available in **Create arc** drop-down of **Toolbar** in the **Sketcher** workbench are used to create an arc using three points. There are two tools in **Create arc** drop-down; **Center and end points** and **End points and rim point**; refer to Figure-9. The procedures to use these tools are discussed next.

Figure-9. Create an arc drop-down

Creating center and end points arc

The **Center and end points arc** tool is used to create an arc using three points (the center point, the start angle along the radius, and the end angle). The procedure to use this tool is discussed next.

- Click on **Center and end points arc** tool from **Create arc** drop-down. You will be asked to specify center point of the arc.
- Click at desired location in the 3D view area to specify center point of the arc; refer to Figure-10.

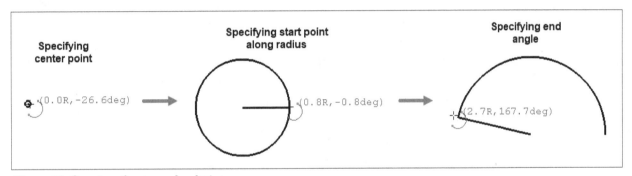

Figure-10. Creation of center and endpoints arc

- Move the cursor to desired location and click to specify starting point of the arc along the radius.
- Click at desired location to specify end point of the arc at a desired angle. The arc will be created.
- Press **ESC** key or press **RMB** to exit the tool.

Creating end points and rim point arc

The **End points and rim point arc** tool is used to create arc using three points (the start point, the end point, and a point on the arc). The procedure to use this tool is discussed next.

- Click on the **End points and rim point arc** tool from **Create arc** drop-down. You will be asked to specify starting point of the arc.
- Click in the 3D view area to specify starting point of the arc; refer to Figure-11. You will be asked to specify end point of the arc.

Figure-11. Creation of End points and rim point arc

- Move the cursor away from the point created and click at desired location to specify end point of the arc. You will be asked to specify rim point of the arc.
- Click at desired location to specify a point on the arc at a desired angle. The arc will be created.
- Press **ESC** key or press **RMB** to exit the tool.

Circle

The tools available in **Create circle** drop-down from **Toolbar** in the **Sketcher** workbench are used to create circle using two and three points. There are two tools in **Create circle** drop-down; **Center and rim point** and **3 rim points**; refer to Figure-12. The procedures to use these tools are discussed next.

Figure-12. Create a circle drop-down

Center and rim point circle

The **Center and rim point** tool is used to create a circle using two points (the center and a point over the circumference of circle). The procedure to use this tool is discussed next.

- Click on **Center and rim point** tool from **Create circle** drop-down or press GC from the Keyboard. You will be asked to specify center point of the circle.
- Click in the 3D view area to specify center point of the circle; refer to Figure-13. You will be asked to specify rim point of the circle.

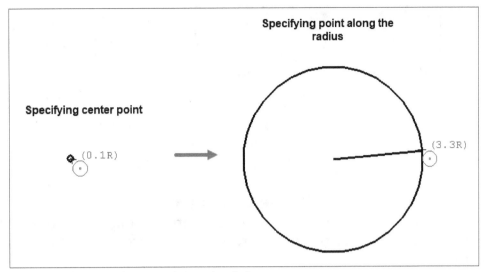

Figure-13. Creation of center and rim point circle

- Click at desired location to specify a point of circumference of the circle. The circle will be created.
- Press **ESC** key or press **RMB** to exit the tool.

3 rim points circle

The **3 rim points** tool is used to create a circle using 3 rim points or circumferential points as reference. The procedure to use this tool is discussed next.

- Click on the **3 rim points** tool from **Create circle** drop-down. You will be asked to specify first rim or circumferential point of a circle.
- Click in the 3D view area to specify first rim or circumferential point of a circle; refer to Figure-14. You will be asked to specify second rim point for circle.

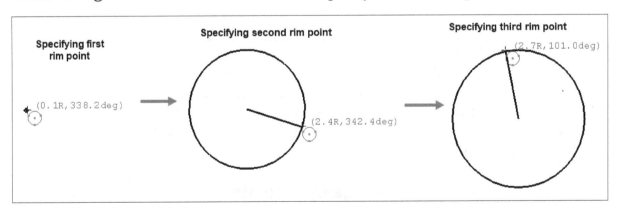

Figure-14. Creation of 3 rim points circle

- Click at desired locations to specify second and then third rim or circumferential point of a circle. The circle will be created.
- Press **ESC** key or press **RMB** to exit the tool.

Conic

There are five tools available in **Create conic** drop-down from **Toolbar** in the **Sketcher** workbench to create ellipses and arcs; **Ellipse by center**, **Ellipse by 3 points**, **Arc of ellipse**, **Arc of hyperbola**, and **Arc of parabola**; refer to Figure-15. The procedures to use these tools are discussed next.

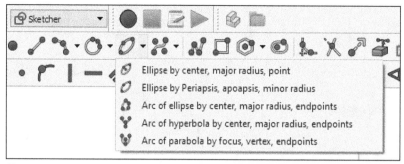

Figure-15. Create conic drop-down

Ellipse by center

The **Ellipse by center** tool is used to create an ellipse using three points (the center, the end of major radius, and the minor radius). The procedure to use this tool is discussed next.

*   Click on **Ellipse by center** tool from **Create conic** drop-down. You will be asked to specify center point of the ellipse.
*   Click in the 3D view area to specify center point of the ellipse; refer to Figure-16.

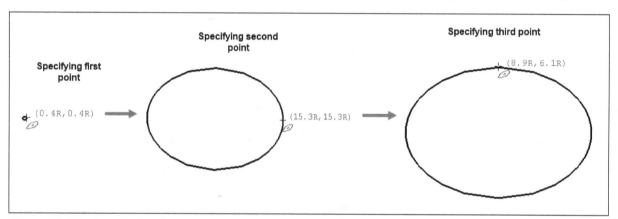

Figure-16. Creation of Ellipse by center

*   Move the cursor away from specified point and click to specify second point of the ellipse.
*   Click at desired location to specify third point of the ellipse. The ellipse will be created.
*   Press **ESC** key or press **RMB** to exit the tool.

Ellipse by 3 points

The **Ellipse by 3 points** tool is used to create an ellipse using three points (the **Periapsis** - first crossing of major diameter with ellipse, the **Apoapsis** - second crossing of major diameter with ellipse, and third point defines the minor radius of ellipse). The procedure to use this tool is discussed next.

*   Click on **Ellipse by 3 points** tool from **Create conic** drop-down. You will be asked to specify first major diameter of ellipse.
*   Click in the 3D view area to specify first point of major diameter of ellipse as shown in Figure-17. You will be asked to specify opposite point for major diameter.

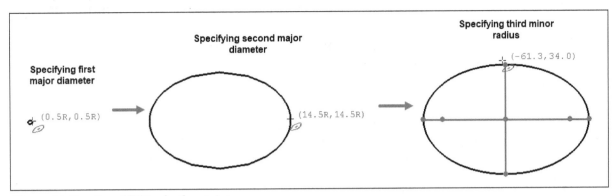

Figure-17. Creation of Ellipse by three points

- Click on the opposite side of the center point to specify second point of major diameter of ellipse. You will be asked to specify a circumferential point for minor diameter of ellipse.
- Click at desired location to specify third minor radius of the ellipse. An ellipse will be created.
- Press **ESC** key or press **RMB** to exit the tool.

Arc of ellipse

The **Arc of ellipse** tool is used to create an arc of ellipse using four points (the center, the end of major radius, the start point, and the end point). The procedure to use this tool is discussed next.

- Click on the **Arc of ellipse** tool from **Create conic** drop-down or press GE from keyboard. You will be asked to specify center point of the arc.
- Click in the 3D view area to specify center point of the arc; refer to Figure-18. You will be asked to specify major radius of ellipse.

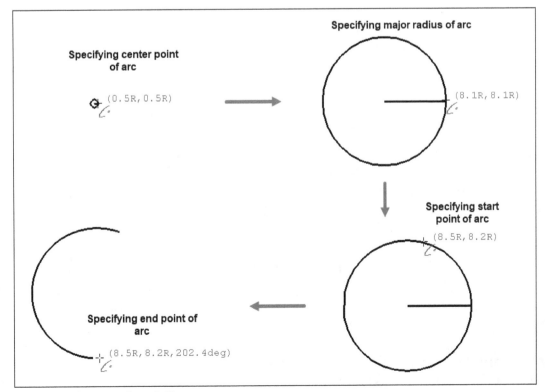

Figure-18. Creation of arc of ellipse

- Move the cursor away from the center point and click to specify major radius of the arc. You will be asked to specify start point of arc.
- Click on desired location to specify starting point and then end point of the arc. The arc of ellipse will be created.
- Press **ESC** key or press **RMB** to exit the tool.

Arc of hyperbola

The **Arc of hyperbola** tool is used to create an arc of hyperbola using four points (the center point, the major radius of arc, the start point, and the end point of arc). The procedure to use this tool is discussed next.

- Click on the **Arc of hyperbola** tool from **Create conic** drop-down or press **GH** from keyboard. You will be asked to specify center point of arc.
- Click in the 3D view area to specify center point of arc; refer to Figure-19. You will be asked to specify major radius of arc.

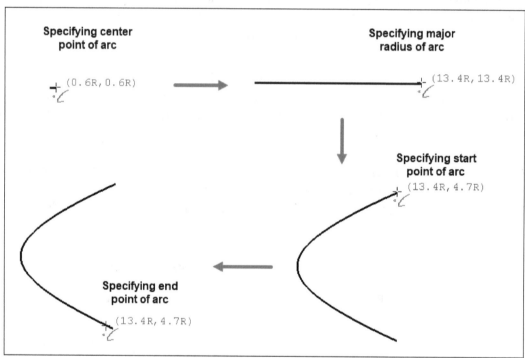

Figure-19. Creation of arc of hyperbola

- Move the cursor away from the center point and click to specify major radius of the arc. You will be asked to specify starting point of arc.
- Click on desired location to specify starting point of arc and then the end point of arc. The arc of hyperbola will be created.
- Press **ESC** key or press **RMB** to exit the tool.

Arc of parabola

The **Arc of parabola** tool is used to create an arc of parabola using four points (the focus of arc, the vertex of arc, the start point, and the end point of arc). The procedure to use this tool is discussed next.

- Click on the **Arc of parabola** tool from **Create conic** drop-down or press **GP** from keyboard. You will be asked to specify focus of the arc.
- Click in the 3D view area to specify focus point of arc; refer to Figure-20.

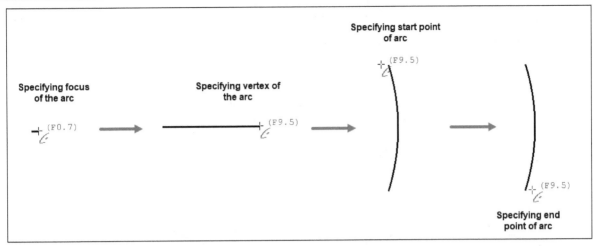

Figure-20. Creation of arc of parabola

- Move the cursor away and click to specify vertex of the arc.
- Click at desired location to specify starting point of arc and then the end point of arc. The arc of parabola will be created.
- Press **ESC** key or press **RMB** to exit the tool.

B-spline

There are four tools available in **Create B-spline** drop-down from **Toolbar** in the **Sketcher** workbench to create B-spline curves; refer to Figure-21. The procedures to use these tools are discussed next.

Figure-21. Create B-spline drop-down

B-spline by control points

The **B-spline by control points** tool is used to create open B-spline curves by tracing its control points. The procedure to use this tool is discussed next.

- Click on the **B-spline by control points** tool from **Create B-spline** drop-down. You will be asked to create the series of points.
- Click in the 3D view area to specify a series of points to create spline; refer to Figure-22.

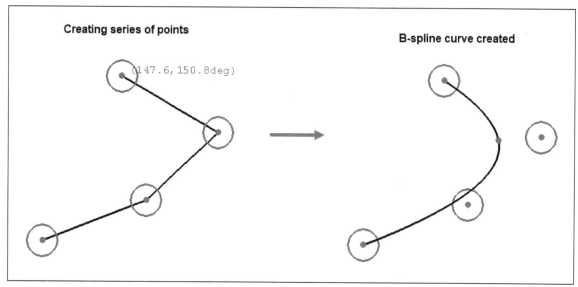

Figure-22. Creation of B spline

Note that if you want to define the degree of the B-spline while creating the B-spline, then press **D** key from keyboard, the **B-Spline Degree** dialog box will be displayed; refer to Figure-23. Specify the value between **1** and **25** as desired and click on the **OK** button.

Figure-23. B Spline Degree dialog box

If you want to delete the last created control point while creating B-spline then press **Backspace** key from keyboard.

* Press **ESC** key or press **RMB** to exit the creation, the B-spline curve will be created.
* Press **ESC** key or press **RMB** again to exit the tool.

Periodic B-spline by control points

The **Periodic B-spline by control points** tool is used to create closed B-spline curves by tracing its control points. The procedure to use this tool is discussed next.

* Click on the **Periodic B-spline by control points** tool from **Create B-spline** drop-down. You will be asked to create series of points.
* Click in the 3D view area to specify a series of points for creating spline; refer to Figure-24. Note that you can specify the B-spline degree value as discussed earlier by pressing **D** key while creating spline.

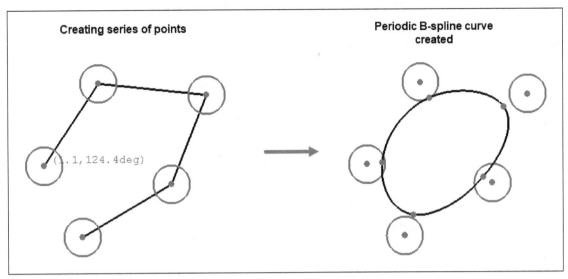

Figure-24. Creation of Periodic B spline

- Press **RMB** to exit the creation, the periodic B-spline curve will be created.
- Press **ESC** key or press **RMB** again to exit the tool.

B-Spline by Knots

The **B-spline by knots** tool is used to create a spline passing through specified points called knots. The procedure to use this tool is discussed next.

- Click on the **B-spline by knots** tool from **Create B-spline** drop-down in Toolbar. You will be asked to specify series of points; refer to Figure-25

Figure-25. B-spline by knots

- Click in the 3D view area to specify a series of points for creating spline; refer to Figure-26

Figure-26. B-spline by knots view

Periodic B-Spline by Knots tool

The **Periodic B-spline by knots** tool is used to create a periodic B-spline curve passing through given points. The procedure to use this tool is discussed next.

- Click on the **Periodic B-spline by knots** tool from the **Create B-spline** drop-down in **Toolbar**; refer to Figure-27. You will be asked to specify series of points.
- Click at desired locations to create the spline as discussed earlier.

Figure-27. Periodic B-Spline by knot

Polyline

The **Create polyline** tool is used to create continuous line and arc segments connected by their vertices. The polyline tool has multiple modes that can be toggled with the **M** key.

For example, you can draw tangent or perpendicular arcs following a line or arc segment. The procedure to use this tool is discussed next.

- Click on the **Create polyline** tool from **Toolbar** in the **Sketcher** workbench; refer to Figure-28. You will be asked to create a line segment.

Figure-28. Polyline tool

- Click in the 3D view area and create a line segment. Press **M** key from the keyboard to toggle different modes of creating polyline; refer to Figure-29.
- After creating a line segment, press **M** key for one time to create a line perpendicular to the previous entity.
- Press **M** key for two times to create a line tangential to the previous entity.
- Press **M** key for three times to create an arc tangential to the previous entity.
- Press **M** key for four times to create an arc perpendicular to the left of the previous entity.
- Press **M** key for five times to create an arc perpendicular to the right of the previous entity.
- Press **M** key for six times to create a line coincident to the previous entity as in the normal state.
- Press **ESC** key or press **RMB** to exit the creation and continue to create a new line.
- Press **ESC** key or press **RMB** again to exit the tool.

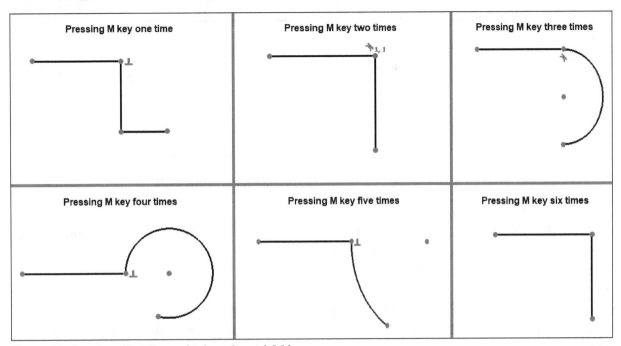

Figure-29. Creation of Polyline multiple modes with M key

Rectangle

There are three tools available in **Create rectangle** drop-down of **Toolbar** in the **Sketcher** workbench to create rectangles; **Rectangle**, **Centered rectangle**, and **Rounded rectangle**; refer to Figure-30. The procedures to use these tools are discussed next.

Figure-30. Create a rectangle drop down

Rectangle

The **Rectangle** tool is used to create rectangle by using two opposite points. The procedure to use this tool is given next.

- Click on **Rectangle** tool from **Create rectangle** drop-down. You will be asked to specify first corner point of the rectangle.
- Click in the 3D view area to specify first corner point of the rectangle; refer to Figure-31.

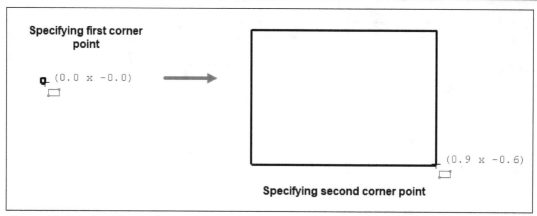

Figure-31. Creation of rectangle

- Move the cursor to desired location and click to specify second corner point of the rectangle. The rectangle will be created.
- Press **ESC** key or press **RMB** to exit the tool.

Centered rectangle

The **Centered rectangle** tool is used to create rectangle by using center point and a corner point. The procedure to use this tool is given next.

- Click on **Centered rectangle** tool from **Create rectangle** drop-down. You will be asked to specify center point of the rectangle.
- Click in the 3D view area to specify center point of the rectangle; refer to Figure-32.

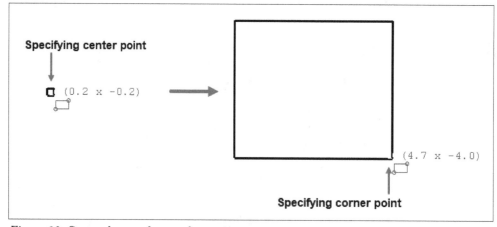

Figure-32. Centered rectangle created

- Move the cursor to desired location and click to specify corner point of the rectangle. The rectangle will be created.
- Press **ESC** key or press **RMB** to exit the tool.

Rounded rectangle

The **Rounded rectangle** tool is used to create rectangle by using two opposite points. The procedure is same as discussed for **Rectangle** tool; refer to Figure-33.

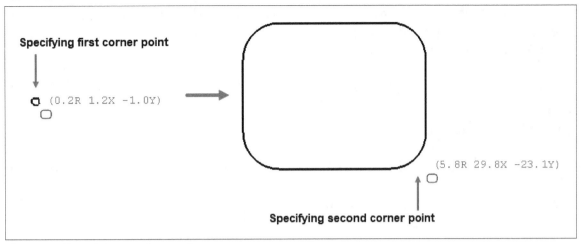

Figure-33. Rounded rectangle created

Polygon

There are seven tools available in **Create regular polygon** drop-down from **Toolbar** in the **Sketcher** workbench to create a polygon; **Triangle**, **Square**, **Pentagon**, **Hexagon**, **Heptagon**, **Octagon**, and **Regular Polygon**; refer to Figure-34. The procedures to use these tools are discussed next.

Figure-34. Create regular polygon drop-down

Triangle

The **Triangle** tool is used to create an equilateral triangle inscribed in a construction geometry circle. The procedure to use this tool is discussed next.

- Click on the **Triangle** tool from the **Create regular polygon** drop-down. You will be asked to specify center of the triangle.
- Click in the 3D view area to specify center of the triangle as shown in Figure-35.

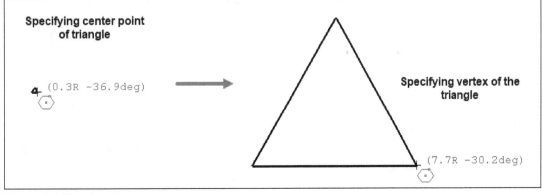

Figure-35. Creation of triangle

- Move the cursor away from the center point and click to specify a vertex of the triangle. The triangle will be created.
- Press **ESC** key or press **RMB** to exit the tool.

Square

The **Square** tool is used to create a square inscribed in a construction geometry circle. The procedure to use this tool is discussed next.

- Click on the **Square** tool from **Create regular polygon** drop-down. You will be asked to specify center point of the square.
- Click in the 3D view area to specify center of the square; refer to Figure-36.

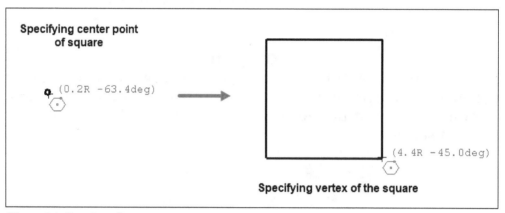

Figure-36. Creation of square

- Move the cursor away from center point and click to specify vertex of the square. The square will be created.
- Press **ESC** key or press **RMB** to exit the tool.

Pentagon

The **Pentagon** tool is used to create a pentagon inscribed in a construction geometry circle. The procedure to use this tool is discussed next.

- Click on the **Pentagon** tool from **Create regular polygon** drop-down. You will be asked to specify center of the pentagon.
- Click in the 3D view area to specify center of the pentagon; refer to Figure-37.

Figure-37. Creation of pentagon

- Move the cursor away from center point and click to specify vertex of pentagon. The pentagon will be created.
- Press **ESC** key or press **RMB** to exit the tool.

Hexagon

The **Hexagon** tool is used to create a hexagon inscribed in a construction geometry circle. The procedure to use this tool is discussed next.

- Click on the **Hexagon** tool from **Create regular polygon** drop-down. You will be asked to specify center of the hexagon.
- Click in the 3D view area to specify center of the hexagon; refer to Figure-38.

Figure-38. Creation of hexagon

- Move the cursor away from center point and click to specify vertex of the hexagon. The hexagon will be created.
- Press **ESC** key or press **RMB** to exit the tool.

Heptagon

The **Heptagon** tool is used to create a heptagon inscribed in a construction geometry circle. The procedure to use this tool is discussed next.

- Click on the **Heptagon** tool from **Create regular polygon** drop-down. You will be asked to specify center of the heptagon.
- Click in the 3D view area to specify center point for heptagon; refer to Figure-39.

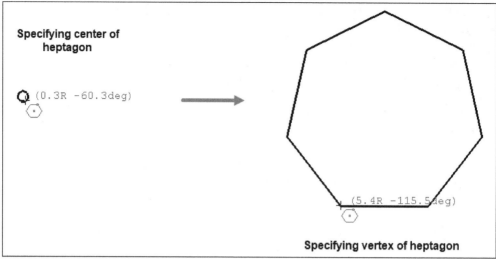

Figure-39. Creation of heptagon

- Move the cursor away from center point and click to specify vertex of heptagon. The heptagon will be created.
- Press **ESC** key or press **RMB** to exit the tool.

Octagon

The **Octagon** tool is used to create an octagon inscribed in a construction geometry circle. The procedure to use this tool is discussed next.

- Click on the **Octagon** tool from **Create regular polygon** drop-down. You will be asked to specify center point of octagon.
- Click in the 3D view area to specify the center of octagon; refer to Figure-40.

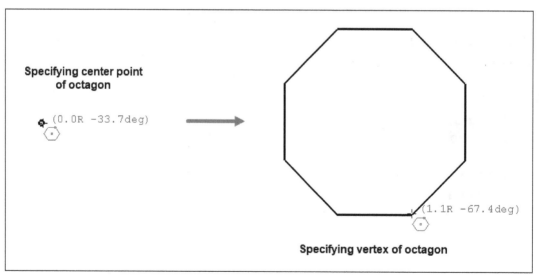

Figure-40. Creation of octagon

- Move the cursor away from center point and click to specify vertex of octagon. The octagon will be created.
- Press **ESC** key or press **RMB** to exit the tool.

Regular Polygon

A **Regular Polygon** tool is used to create a polygon inscribed in a construction geometry circle with specified number of sides. The procedure to use this tool is discussed next.

- Click on the **Regular Polygon** tool from **Create regular polygon** drop-down. The **Create array** dialog box will be displayed asking you to specify the number of sides; refer to Figure-41.

Figure-41. Create array dialog box

- Specify desired number of sides you want to have in polygon in the **Number of Sides** edit box. Click on the **OK** button to close the dialog box. You will be asked to specify center point of polygon.
- Click in the 3D view area to specify the center point for polygon; refer to Figure-42.

Figure-42. Creation of regular polygon

- Move the cursor away from center point and click to specify the vertex of polygon. The regular polygon will be created.
- Press **ESC** key or press **RMB** to exit the tool.

Slot

The **Slot** tool is used to draw a slot by specifying center points of two semicircles at the ends of slot. The procedure to use this tool is discussed next.

- Click on the **Create slot** tool from **Toolbar** in the **Sketcher** workbench; refer to Figure-43. You will be asked to specify center of one semicircle.

Figure-43. Slot tool

- Click in the 3D view area to specify center point of one semicircle; refer to Figure-44.

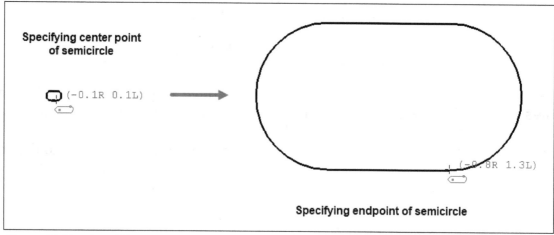

Figure-44. Creation of slot

- Move the cursor away and click at desired location to change the other side of semicircle. The slot will be created.
- Press **ESC** key or press **RMB** to exit the tool.

SKETCH EDITING TOOLS

In **FreeCAD** software, the standard tools to edit sketch entities are available in the **Toolbar** of **Sketcher** workbench; refer to Figure-45.

Figure-45. Sketch editing tools

Fillet

There are two tools available in **Fillet** drop-down from **Toolbar** in the **Sketcher** workbench to create fillets; **Sketch fillet** and **Constraint-preserving sketch fillet**; refer to Figure-46. The procedures to use these tools are discussed next.

Figure-46. Fillet drop down

Sketch fillet

The **Sketch fillet** tool is used to create fillet between two geometries like lines, arcs etc. joined at one point. The procedure to use this tool is discussed next.

- Click on the **Sketch fillet** tool from **Fillet** drop-down. You will be asked to select the intersecting geometries like lines, arcs etc.
- Select the two connecting lines/arcs from the 3D view area or vertex at intersection of two lines. The fillet will be created; refer to Figure-47.

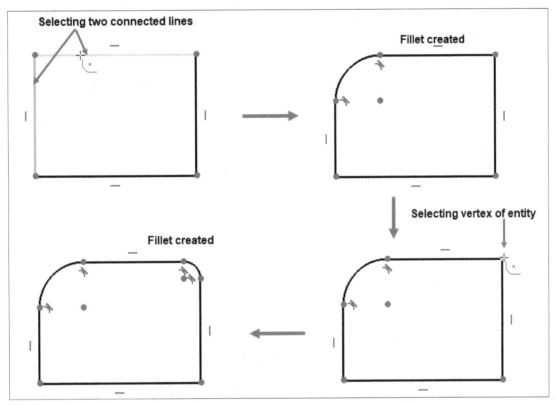

Figure-47. Fillet creation

Note that the distance at which you click from the vertex will specify the fillet radius. You can change the value of fillet radius later.

- Press **ESC** key or press **RMB** to exit the tool.

Constraint-preserving sketch fillet

The **Constraint-preserving sketch fillet** tool is used to create fillet that preserves constraints and intersection point. The procedure is same as discussed for **Sketch fillet** tool.

Trim edge

The **Trim edge** tool is used to trim a curve to the nearest intersection point of other entity. The procedure to use this tool is discussed next.

- Click on the **Trim edge** tool from **Toolbar** in the **Sketcher** workbench; refer to Figure-48. You will be asked to select the curve to be trimmed.

Figure-48. Trim edge tool

- Select an intersecting curve that you want to be trimmed. The curve will be trimmed; refer to Figure-49.

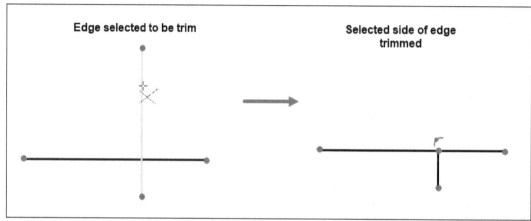

Figure-49. Edge trimmed

- Press **ESC** key or press **RMB** to exit the tool.

Extend edge

The **Extend edge** tool is used to extend a curve to specified location in the sketch or to another curve. The procedure to use this tool is discussed next.

- Click on the **Extend edge** tool from **Toolbar** in the **Sketcher** workbench; refer to Figure-50. You will be asked to select a line or an arc.

Figure-50. Extend edge tool

- Select a line or an arc you want to extend in a specified location or up to selected entity; refer to Figure-51.

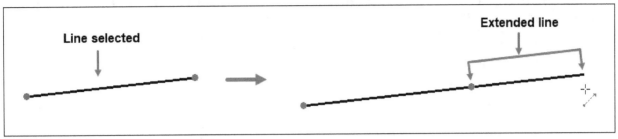

Figure-51. Line extended to an arbitrary location

- To extend a line/arc to another curve, hover the cursor on the curve. When the curve is highlighted and the **Point on object** [r] constraint icon appears then press **LMB** to extend the line. The line will be extended; refer to Figure-52.

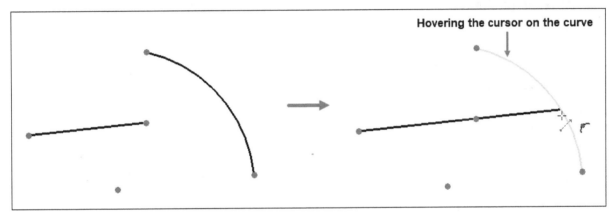

Figure-52. Line extended to an edge

- To extend an edge to a point in the sketch, hover the cursor over the point. When the point is highlighted and the **Coincident** [•] constraint icon appears then press **LMB** to extend the line. The line will be extended; refer to Figure-53.

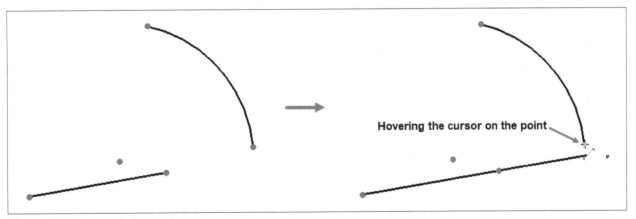

Figure-53. Line extended to a point

Note that only arcs and lines can be extended from this tool.

- Press **ESC** key or **RMB** to exit the tool.

Split edge

The **Split edge** tool is used to divide an edge at selected location. The procedure to use this tool is discussed next.

- Click on the **Split edge** tool from **Toolbar** in the **Sketcher** workbench; refer to Figure-54. You will be asked to select the edge to be split.

Figure-54. Split edge tool

- Select a location on the edge where you want to split the edge. The edge will be split; refer to Figure-55.

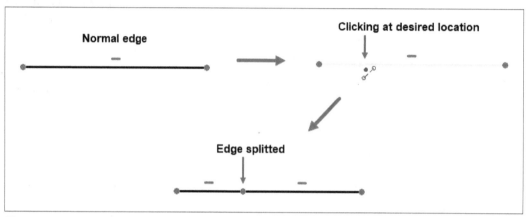

Figure-55. Edge splitted

- Press **ESC** key or **RMB** to exit the tool.

External Geometry

The **External geometry** tool is used to apply a constraint between sketch geometry and geometry external to current sketch. It works by inserting a linked construction geometry in the sketch. The procedure to use this tool is discussed next.

- Click on the **External geometry** tool from **Toolbar** in the **Sketcher** workbench; refer to Figure-56. You will be asked to select a line, arc, or curve.

Figure-56. External Geometry tool

- Select a curve that you want to link to the sketch; refer to Figure-57. The selected curve will become part of current sketch. Now, you can apply constraints between new sketch curves and previous selected sketch curves.

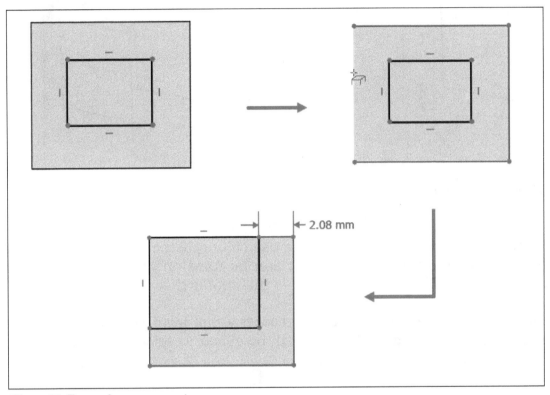

Figure-57. External geometry creation

- Press **ESC** key or press **RMB** to exit the tool.

Create Carbon Copy

The **Create carbon copy** tool copies all the geometry and constraints from another sketch into the active sketch. The procedure to use this tool is discussed next.

- Click on the **Create carbon copy** tool from **Toolbar** in the **Sketcher** workbench; refer to Figure-58. You will be asked to select curves from another sketch.

Figure-58. Create carbon copy tool

- Click on curves from the existing sketch, the geometry elements as well as constraints are copied into the active sketch; refer to Figure-59.
- Press **ESC** key or press **RMB** to exit the tool.

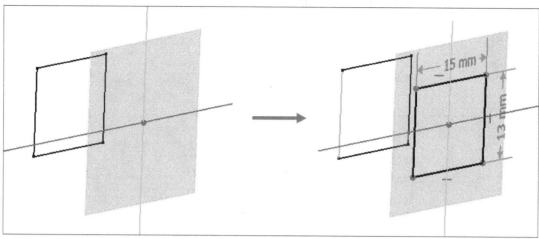

Figure-59. Carbon copy of sketch geometry

Construction Mode

The **Toggle construction geometry** tool is used to toggle the sketch geometry between construction mode and sketch mode. It can be used on any type of geometry like line, arc, and circle. The procedure to use this tool is discussed next.

* Click on the **Toggle construction geometry** tool from **Toolbar** in the **Sketcher** workbench; refer to Figure-60. You will be asked to select the sketch.

Figure-60. Toggle construction geometry tool

* Select one or more sketch geometry in the 3D view to toggle the construction mode; refer to Figure-61. The sketch will be in construction mode.

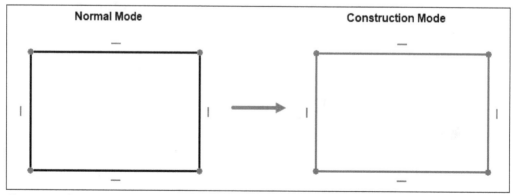

Figure-61. Construction mode sketch

Note that if you have not selected any sketch entity before using this tool and you select **Toggle construction geometry** tool then all the sketches will be created in construction mode until you click the button again.

* Press **ESC** key or press **RMB** to exit the tool.

SKETCHER CONSTRAINTS

Constraints are used to restrict the shape and size of geometric entities.

Geometric constraints

Geometric constraints are not associated with numeric data. The tools to apply geometric constraints are available in the **Toolbar** of **Sketcher** workbench; refer to Figure-62. The procedures to use these tools are discussed next.

Figure-62. Geometric constraints tools

Constrain coincident

The **Constrain coincident** tool is useful when there is a break in the profile, for example: where two lines end near each other and need to be joined. A coincident constraint on their endpoints will close the gap and make two selected points coincident. The procedure to use this tool is discussed next.

- Click on the **Constrain coincident** tool from **Toolbar** in the **Sketcher** workbench; refer to Figure-63. You will be asked to select the endpoints.

Figure-63. Coincident constraint tool

- Select the endpoints of the sketch which you want to join or make coincident to each other. The endpoints of the sketch will become coincident to each other; refer to Figure-64.
- Press **ESC** key or press **RMB** to exit the tool.

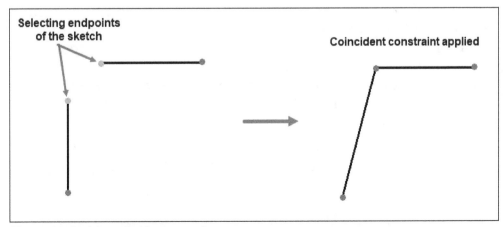

Figure-64. Applying coincident constraint

Point On Object constraint

The **Constrain point onto object** tool affixes the selected point onto another object such as a line, arc, or curve. The procedure to use this tool is discussed next.

- Click on the **Constrain point onto object** tool from **Toolbar** in the **Sketcher** workbench; refer to Figure-65. You will be asked to select the point.

Figure-65. Point On Object constraint tool

- Select the point you want to affix on the line, arc, or sketch curve.
- Select the curve on which you want to affix the point. The point will be affixed on the line and the constraint will be applied; refer to Figure-66.

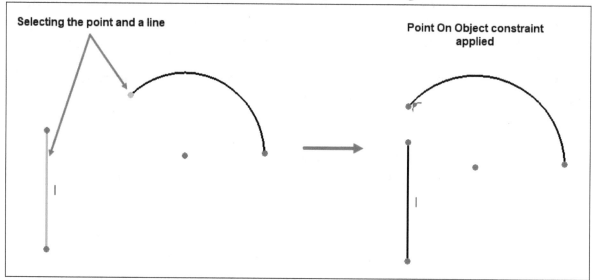

Figure-66. Applying point on object constraint

- Press **ESC** key or press **RMB** to exit the tool.

Vertical constraint

The **Vertical constraint** tool is used to constrain the selected lines or polyline elements to a true vertical orientation. The procedure to use this tool is discussed next. Note that only lines and vertices will be selectable for this tool.

- Click on the **Vertical constraint** tool from **Toolbar** in the **Sketcher** workbench; refer to Figure-67. You will be asked to select the lines or vertices.

Figure-67. Vertical constraint tool

- Select the lines to be constrained vertically. The selected lines will become vertically constrained; refer to Figure-68. You can also select two vertices to make them vertically aligned; refer to Figure-69.
- Press **ESC** key or press **RMB** to exit the tool.

Figure-68. Applying vertical constraint

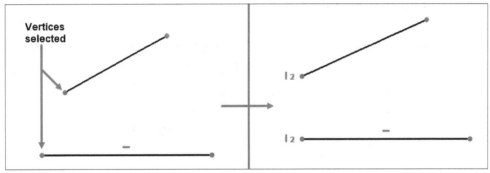

Figure-69. Vertical constrain

Horizontal constraint

The **Horizontal constraint** tool forces a selected line or lines in the sketch to be aligned to the horizontal axis of the sketch. The procedure to use this tool is discussed next.

- Click on the **Horizontal constraint** tool from **Toolbar** in the **Sketcher** workbench; refer to Figure-70. You will be asked to select the lines.

Figure-70. Horizontal constraint tool

- Select the lines which you want to constrain horizontally. The selected lines will become horizontally constrained; refer to Figure-71. You can also align two vertices in horizontal line.

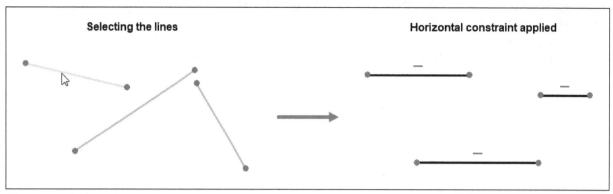

Figure-71. Applying horizontal constraint

• Press **ESC** key or press **RMB** to exit the tool.

Parallel constraint

The **Constrain parallel** tool forces two selected lines to be parallel to each other. The procedure to use this tool is discussed next.

• Click on the **Constrain parallel** tool from **Toolbar** in the **Sketcher** workbench; refer to Figure-72. You will be asked to select the lines.

Figure-72. Parallel constraint tool

• Select the lines or line and axis which you want to make parallel. The selected lines will become parallel to each other; refer to Figure-73.

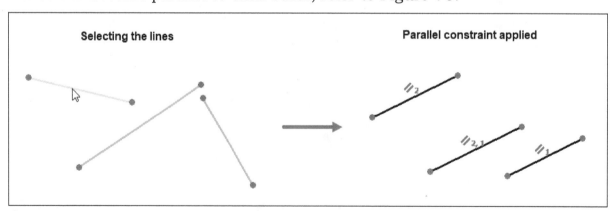

Figure-73. Applying parallel constraint

• Press **ESC** key or **RMB** to exit the tool.

Perpendicular constraint

The **Constrain Perpendicular** tool makes two lines perpendicular to each other or two curves to be perpendicular at their intersection point. The procedure to use this tool is discussed next.

- Click on the **Constrain Perpendicular** tool from **Toolbar** in the **Sketcher** workbench; refer to Figure-74. You will be asked to select desired entities.

Figure-74. Perpendicular constraint tool

- Select the lines, curves, arcs, circles, ellipses, etc. which you want to make perpendicular. The entities selected will become perpendicular to each other; refer to Figure-75.
- Press **ESC** key or **RMB** to exit the tool.

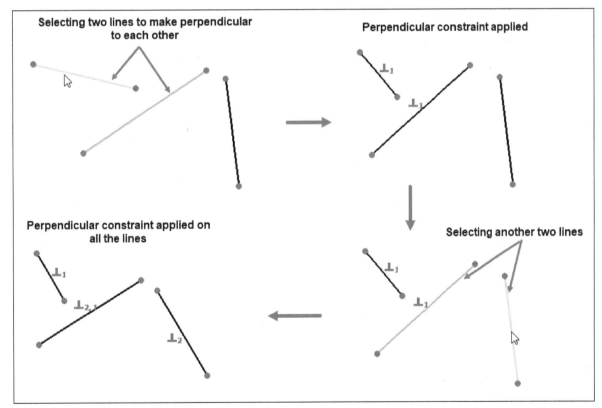

Figure-75. Applying perpendicular constraint

Tangent constraint

The **Constrain tangent** tool applies a tangent constraint between arcs, circles, curves, and lines. If you are applying tangent constraint between two lines then they will become collinear. A line segment does not have to lie directly on an arc or circle to be constrained tangent to that arc or circle. The procedure to use this tool is discussed next.

- Click on the **Constrain tangent** tool from **Toolbar** in the **Sketcher** workbench; refer to Figure-76. You will be asked to select the entities.

Figure-76. Tangent constraint tool

- Select the entities on which you want to apply tangent constraint. The entities will become tangentially constraint to each other; refer to Figure-77.

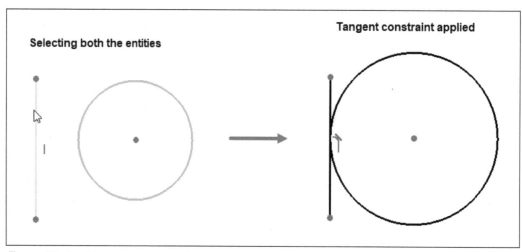

Figure-77. Applying tangent constraint

- Press **ESC** key or **RMB** to exit the tool.

Equal constraint

The **Constraint equal** tool forces two or more sketch segments to have equal length. If applied to arcs or circles, the radii are constrained to be equal. The procedure to use this tool is discussed next. Note that **Constraint equal** cannot be applied to geometric entities which are not of the same type.

- Click on the **Constraint equal** tool from **Toolbar** in the **Sketcher** workbench; refer to Figure-78. You will be asked to select the entities of same geometry type.

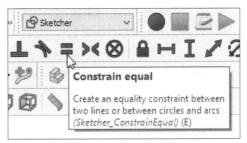

Figure-78. Equal constraint tool

- Select the entities of same geometry type on which you want to apply equal constraint. The selected entities will become equally constrained; refer to Figure-79.
- Press **ESC** key or **RMB** to exit the tool.

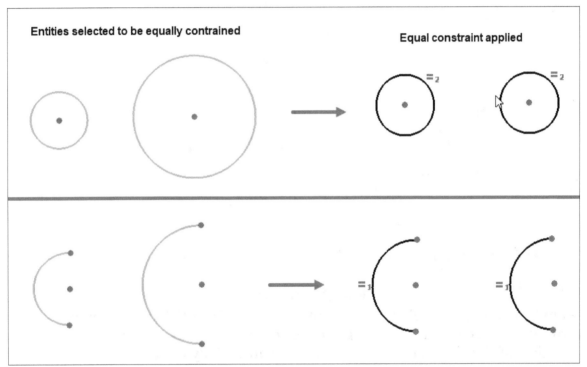

Figure-79. Applying equal constraint

Symmetric constraint

The **Constrain symmetrical** tool is used to constrain the two selected points to be symmetrical about a given line, i.e., both selected points are constrained to lie on a normal to the line through both points and are constrained to be equidistant from the line. The procedure to use this tool is discussed next.

- Click on the **Constrain symmetrical** tool from **Toolbar** in the **Sketcher** workbench; refer to Figure-80. You will be asked to select the points from the sketch.

Figure-80. Symmetric constraint tool

- Select two points and a line from the sketch. The selected entities will be constrained symmetrically; refer to Figure-81.

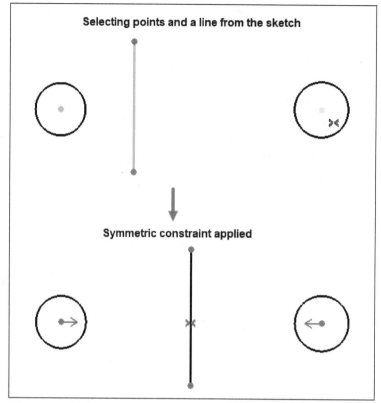

Figure-81. Applying symmetric constraint

- Press **ESC** key or **RMB** to exit the tool.

Block constraint

The **Constrain block** tool is used to block a sketch entity from moving, that is, it prevents its vertices from changing their current positions. It is mainly useful to fix the position of B-splines, otherwise it is difficult to fully constrain it. The procedure to use this tool is discussed next.

- Click on the **Constrain block** tool from **Toolbar** in the **Sketcher** workbench; refer to Figure-82. You will be asked to select the entity.

Figure-82. Block constraint tool

- Select the entity or entities whose position(s) are to be blocked. Now, the position of entity will be blocked; refer to Figure-83.
- Press **ESC** key or **RMB** to exit the tool.

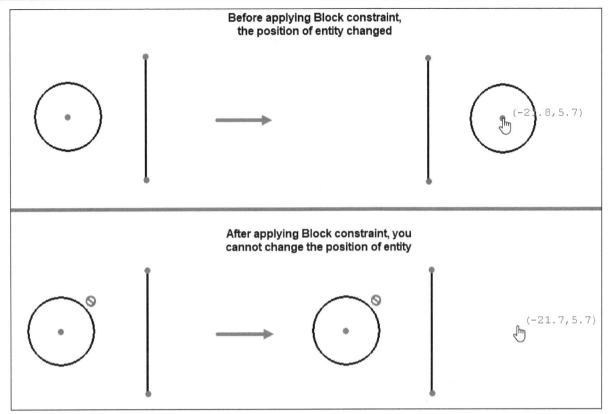

Figure-83. Applying block constraint

Dimensional constraints

Dimensional constraints are associated with numeric data for which you can use the expressions. The tools to apply dimensional constraints are available in the **Toolbar** of **Sketcher** workbench; refer to Figure-84. The procedures to use these tools are discussed next.

Figure-84. Dimensional constraint tools

Lock constraint

Lock constraint tool applies **Horizontal distance** and **Vertical distance** dimensions to selected vertices (points) in the sketch.

If a single vertex is selected, the horizontal and vertical distance constraints will refer to the sketch origin point. If two or more points are selected before activating this tool, horizontal and vertical distance constraints will be added for each pair of points.

The procedure to use this tool is discussed next.

• Click on the **Lock constraint** tool from **Toolbar** in the **Sketcher** workbench; refer to Figure-85. You will be asked to select the vertices.

Figure-85. Lock constraint tool

- Select one or more vertices from the sketch on which you want to apply lock constraint. The lock constraint will be applied to the sketch entity; refer to Figure-86.

Figure-86. Applying lock constraint

- Double-click on the dimension value in the 3D view and modify the value as desired in the edit box displayed.
- Press **ESC** key or press **RMB** to exit the tool.

Horizontal distance constraint

The **Horizontal distance constraint** fixes the horizontal distance between two points or line endpoints. You can also select a line object to apply horizontal length dimension. The procedure to use this tool is discussed next.

- Click on the **Horizontal distance constraint** tool from **Toolbar** in the **Sketcher** workbench; refer to Figure-87. You will be asked to select endpoints or a horizontal line.

Figure-87. Horizontal distance constraint tool

- Select endpoints of a sketch or select a line to fix the horizontal distance or length. The **Insert length** dialog box will be displayed; refer to Figure-88, asking you to specify the horizontal length of entity.

Figure-88. Insert length dialog box

- Specify desired value in the **Length** edit box of the dialog box.
- You can also specify name of the dimension parameter in the **Name** edit box of the **Insert Length** dialog box.
- Select the **Reference** check box to create the reference or constraint dimension.
- Click on **OK** button from the dialog box. The horizontal length or distance of the line/point will be modified; refer to Figure-89.

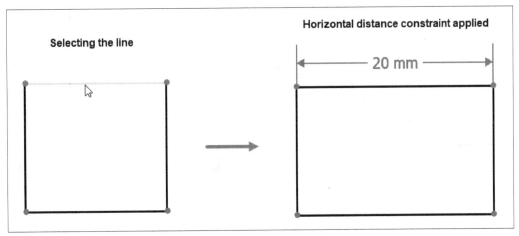

Figure-89. Applying horizontal distance constraint

- Press **ESC** key or press **RMB** to exit the tool.

Vertical distance constraint

The **Vertical distance constraint** tool fixes the vertical distance between two points or line endpoints. The procedure to use this tool is discussed next.

- Click on the **Vertical distance constraint** tool from **Toolbar** in the **Sketcher** workbench; refer to Figure-90. You will be asked to select endpoints or a line.

Figure-90. Vertical distance constraint tool

- Select endpoints of a sketch or select a line to fix the vertical distance. The **Insert length** dialog box will be displayed; refer to Figure-91, asking you to specify the vertical length of entity.

Figure-91. Insert length dialog box

- Specify desired value in the **Length** edit box and click on **OK** button from the dialog box.
- The vertical length or distance of the sketch will be modified; refer to Figure-92.

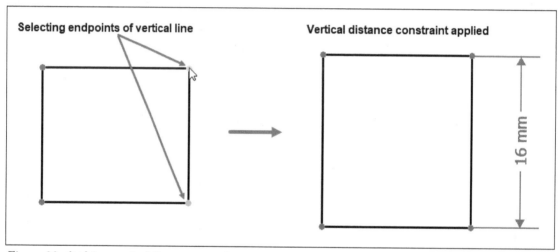

Figure-92. Applying vertical distance constraint

- Press **ESC** key or press **RMB** to exit the tool.

Distance constraint

The **Distance constraint** tool defines the distance of a selected line by constraining its length or defines the distance between two points by constraining the distance between them. Using this tool, you can specify aligned distance of selected line. The procedure to use this tool is discussed next.

- Click on the **Distance constraint** tool from **Toolbar** in the **Sketcher** workbench; refer to Figure-93. You will be asked to select the entities.

Figure-93. Distance constraint tool

- Select two points/a line/a point and a line to apply the distance constraint in the sketch. The **Insert length** dialog box will be displayed as discussed earlier asking you to specify the distance of selected entity.
- Specify desired distance value in the **Length** edit box and click on **OK** button from the dialog box. The length or distance of the selected entity will be modified; refer to Figure-94.

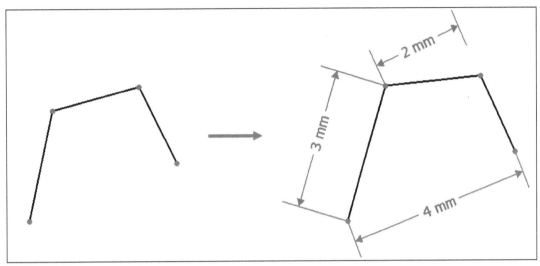

Figure-94. Applying distance constraint

- Press **ESC** key or press **RMB** to exit the tool.

Constrain arc or circle

There are three tools available in **Constrain arc or circle** drop-down from **Toolbar** in the **Sketcher** workbench to apply constraint; **Constrain radius**, **Constrain diameter**, and **Constrain auto radius/diameter**; refer to Figure-95. The procedures to use these tools are discussed next.

Figure-95. Constrain arc or circle drop down

Constrain radius

The **Constrain radius** tool defines the radius of a selected arc or circle by constraining the radius. The procedure to use this tool is discussed next.

- Click on the **Constrain radius** tool from **Constrain arc or circle** drop-down. You will be asked to select a circle or arc.
- Select one or more circles and arcs to define the radius. The **Insert radius** dialog box will be displayed; refer to Figure-96.

Figure-96. Insert radius dialog box

- Specify desired value in **Radius** edit box and click on **OK** button from the dialog box. The radius of selected entity will be modified; refer to Figure-97.

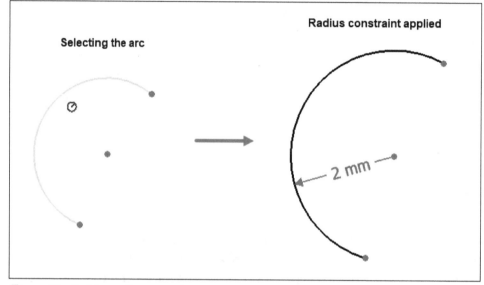

Figure-97. Applying radius constraint

- Press **ESC** key or press **RMB** to exit the tool.

Constrain diameter

The **Constrain diameter** tool defines the diameter of selected arc or circle. The procedure to use this tool is discussed next.

- Click on the **Constrain diameter** tool from **Constrain arc or circle** drop-down. You will be asked to select an arc or circle.
- Select one or more circles or arcs to define the diameter. The **Insert diameter** dialog box will be displayed; refer to Figure-98.

Figure-98. Insert diameter dialog box

- Specify desired value in **Diameter** edit box and click on **OK** button from the dialog box. The diameter of selected entity will be modified; refer to Figure-99.

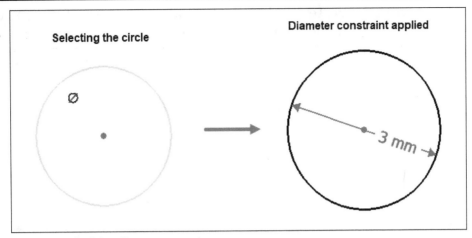

Figure-99. Applying diameter constraint

- Press **ESC** key or press **RMB** to exit the tool.

Constrain auto radius/diameter

The **Constrain auto radius/diameter** tool automatically defines the radius/diameter of a selected arc or circle (weight for a B-spline pole, diameter for a complete circle, and radius for a arc). The procedure to use this tool is same as discussed for previous tools.

Angle constraint

The **Angle constraint** tool defines the internal angle between two selected lines. It is capable of setting slopes of individual lines, angles between lines, angles of intersections of curves, and angle spans of circular arcs. The procedure to use this tool is discussed next.

- Click on the **Angle constraint** tool from **Toolbar** in the **Sketcher** workbench; refer to Figure-100. You will be asked to select one or multiple entities.

Figure-100. Angle constraint tool

- Select the entities from the sketch, the **Insert angle** dialog box will be displayed; refer to Figure-101.

Figure-101. Insert angle dialog box

- Specify desired value in **Angle** edit box and click on **OK** button from the dialog box. The angle between the entities will be modified; refer to Figure-102.

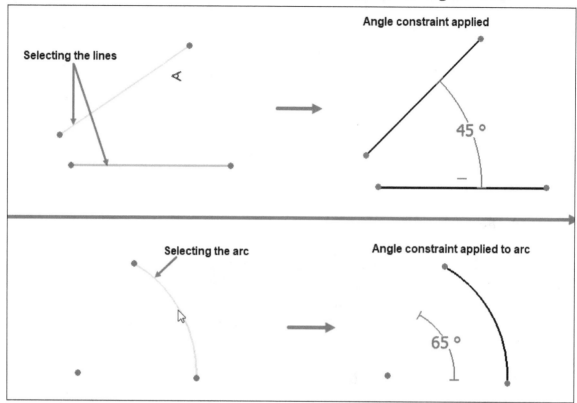

Figure-102. Applying angle constraint

- Press **ESC** key or press **RMB** to exit the tool.

Snell's Law constraint

The **Snell's Law constraint** tool constrains two lines to follow the law of refraction of light as it penetrates through an interface, where two materials of different refraction indices meet. The procedure to use this tool is discussed next.

- You need two lines to follow a beam of light and a curve to act as an interface. The lines should be on different sides of the interface. The interface can be a line, circle/arc, or ellipse/arc of ellipse.
- Select the endpoint of lines and then the interface geometry.
- Click on the **Snell's Law constraint** tool from the **Sketcher constraints** cascading menu in the **Sketch** menu; refer to Figure-103. The **Refractive index ratio** dialog box will be displayed; refer to Figure-104.

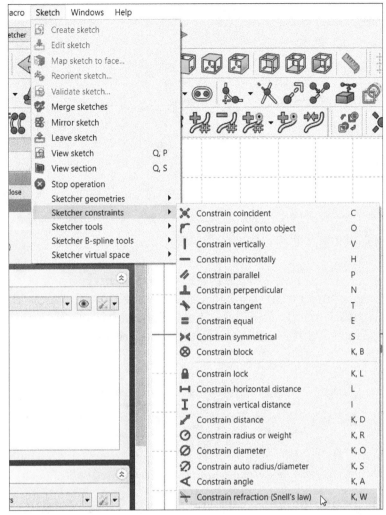

Figure-103. Snell's Law constraint tool

Figure-104. Refractive index ratio dialog box

- Specify desired value in **Ratio n2/n1** edit box and click on **OK** button from the dialog box. The endpoints will be constrained on the interface as per the Snell's law; refer to Figure-105.

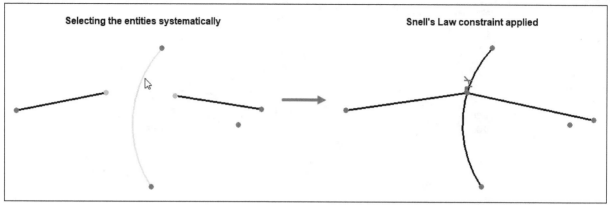

Figure-105. Applying snell's law constraint

- Press **ESC** key or press **RMB** to exit the tool.

Toggle Driving constraint

The **Toggle Driving constraint** tool is used to switch between driving and driven dimension constraints (except **Snell's Law constraint**). The icons in the toolbar turns blue and in place of dimensional constraints, reference dimensions are created; refer to Figure-106.

Figure-106. Dimensional constraints icons turns blue

Reference dimensions do not constrain the sketch. The procedure to use this tool is discussed next.

- Click on the **Toggle driving constraint** tool from **Toolbar** in the **Sketcher** workbench; refer to Figure-107. The dimensional constraint icons turn from red to blue.

Figure-107. Toggle driving constraint tool

- Click on the **Vertical constraint** tool from **Toolbar** in the **Sketcher** workbench. You will be asked to select the entity from sketch to create reference dimension.
- Select the entity from the sketch to apply reference dimension. The reference dimension will be applied to the selected entity; refer to Figure-108.

Figure-108. Applying reference dimension

- Press **ESC** key or press **RMB** to exit the tool.

Toggle Active constraint

The **Toggle Active constraint** tool allows you to activate and deactivate an already placed constraint. The procedure to use this tool is discussed next.

- Select an already placed constraint of an entity which you want to deactivate and click on the **Toggle Active constraint** tool from **Toolbar** in the **Sketcher** workbench; refer to Figure-109. The placed constraint will be deactivated; refer to Figure-110.

Figure-109. Toggle active constraint tool

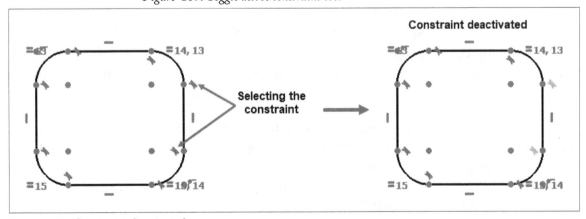

Figure-110. Constraint deactivated

- If you want to reactivate the constraint, then select the deactivated constraint and click on the tool again.

SKETCHER TOOLS

The **Sketcher** tools in **FreeCAD** are available in the **Sketcher** workbench of **toolbar**; refer to Figure-111. You can also access the sketcher tools from the **Sketcher tools** cascading menu in the **Sketch** menu; refer to Figure-112. The procedures to use these tools are discussed next.

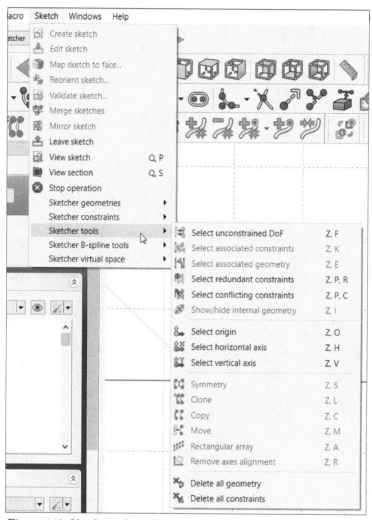

Figure-112. Sketcher tools in sketch menu

Select unconstrained DOFs

The **Select unconstrained DOFs** tool is used to aid in fully constraining a sketch by highlighting the sketch elements with remaining degrees of freedom (DoF) in green color. The procedure to use this tool is discussed next.

- Click on the **Select unconstrained DOFs** tool from the **Sketcher** workbench; refer to Figure-113. The **Solver messages** dialog box will be displayed in the **Tasks** panel of **Combo View** displaying the messages related to constraints and the geometry will be highlighted in green color; refer to Figure-114. Note that in case of lines, only points will be highlighted in green.

Figure-113. Select solver DOFs tool

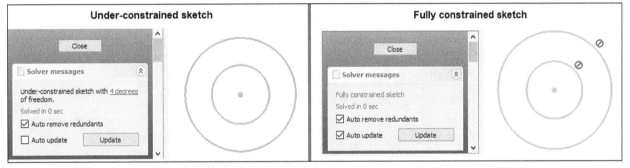

Figure-114. Solver messages dialog box displaying constraint messages

- Press **LMB** to exit the tool.

Select Associated Constraints

The **Select Associated Constraints** tool is used to select the constraints associated to the selected sketcher elements. The procedure to use this tool is discussed next.

- Select the entities of the sketch whose constraints are to be selected.
- Click on the **Select Associated Constraints** tool from **Toolbar** in the **Sketcher** workbench; refer to Figure-115. The constraints associated with selected entities will be selected; refer to Figure-116. Press **LMB** to exit the tool.

Figure-115. Select constraints tool

Figure-116. Selection of constraints

Select Associated constraints

The **Select associated constraints** tool is used to select sketcher elements associated with constraints. The procedure to use this tool is discussed next.

* Select the constraints of those entities which you want to select.
* Click on the **Select associated constraints** tool from **Toolbar** in the **Sketcher** workbench; refer to Figure-117. The entities associated with selected constraints will be selected; refer to Figure-118.

Figure-117. Select associated constraints tool

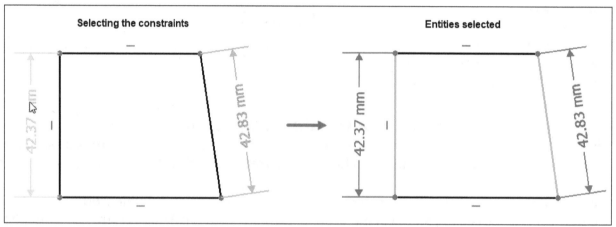

Figure-118. Selection of elements

* Press **LMB** to exit the tool.

Select Associated Geometry

The **Select associated geometry** tool is used to select the sketcher element associated with constraints. The procedure is same as previous topic discussed.

* Click on the **Select associated geometry** tool from the **Toolbar** in the **Sketcher** workbench; refer to figure Figure-119. The associated geometry of sketch will be selected in the 3D view area.

Figure-119. Select associated geometry

* Press **LMB** to exit the tool.

Select Redundant Constraints

The **Select Redundant Constraints** tool is used to select unnecessary constraints of the sketch. The procedure to use this tool is discussed next.

• Select on the **Select Redundant Constraints** tool from the from the **Sketcher tool** cascading menu in the **Sketch** menu; refer to Figure-120. The redundant constraints of the sketch will be selected in the 3D view area; refer to Figure-121.

Figure-120. Select Redundant Constraints tool

Figure-121. Redundant constraints of sketch selected

• Press **LMB** to exit the tool.

Select Conflicting Constraints

The **Select Conflicting Constraints** tool is used to select the conflicting constraints of a sketch. The procedure to use this tool is same as discussed for **Select Redundant Constraints** tool.

Show/Hide internal geometry

The **Show/Hide internal geometry** tool is used to show or hide unnecessary internal geometry or recreates missing geometry of a selected ellipse, arc of ellipse/hyperbola/parabola, or B-spline. The procedure to use this tool is discussed next.

- Select desired entity of sketch from which you want to show or hide the unnecessary geometry.
- Click on the **Show/Hide internal geometry** tool from **Toolbar** in the **Sketcher** workbench; refer to Figure-122. The internal geometry of selected entity will be hidden from the sketch like control points of spline; refer to Figure-123.

Figure-122. Show hide internal geometry tool

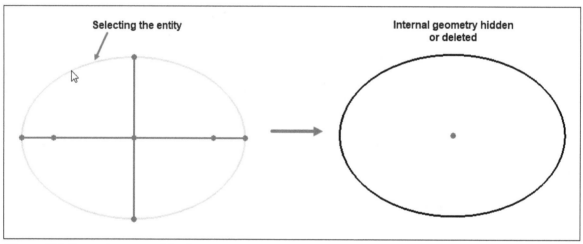

Figure-123. Internal geometry deleted

Symmetry

The **Symmetry** tool is used to mirror sketcher geometry in reference to a chosen line or sketch axis. Note that last selected entity will be used as symmetry axis. The procedure to use this tool is discussed next.

- Select the geometry that you want to copy and then a line or sketch axis to be used as symmetry axis.
- Click on the **Symmetry** tool from **Toolbar** in the **Sketcher** workbench; refer to Figure-124. The selected geometry will be copied symmetrically against the selected line or sketch axis; refer to Figure-125.

Figure-124. Symmetry tool

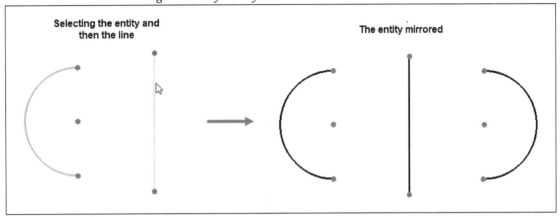
Figure-125. Symmetry entity creation

Creating Clone/Copy

There are three tools available in **Copy** drop-down of **Toolbar** in the **Sketcher** workbench to create a clone of the geometry; **Clone**, **Copy**, and **Move**; refer to Figure-126. The procedures to use these tools are discussed next.

Figure-126. Copy drop down

Clone

The **Clone** tool is used to clone selected sketch elements from one point to another using the last selected point as reference. The procedure to use this tool is discussed next.

- Select the sketch entity which you want to clone.
- Click on the **Clone** tool from a **Copy** drop-down. A line with clone icon will be attached to the cursor asking you to place the clone entity.
- Click at desired location in the 3D view area, the clone of selected entity will be placed; refer to Figure-127.

Figure-127. Creation of clone

Copy

The **Copy** tool is used to copy selected sketch elements from one point to another using the last selected point as reference. The procedure to use this tool is same as discussed for the **Clone** tool.

Move

The **Move** tool moves selected sketch elements from one point to another using the last selected point as reference. The procedure to use this tool is discussed next.

- Select the entity which you want to move from one place to another.
- Click on the **Move** tool from the **Copy** drop-down. A line and a move icon will be attached to the cursor asking you to move the entity.
- Click at desired location in the 3D view area, the selected entity will be moved; refer to Figure-128.

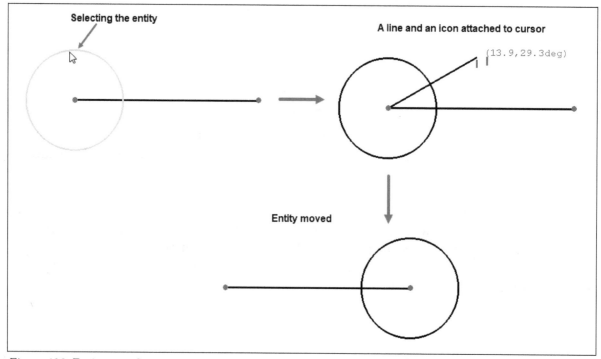

Figure-128. Entity moved

Rectangular Array

The **Rectangular Array** tool creates a rectangular array pattern of the geometry taking last selected point as reference. The procedure to use this tool is discussed next.

- Select the entity using which you want to create a rectangular array pattern.
- Click on the **Rectangular Array** tool from **Toolbar** in the **Sketcher** workbench; refer to Figure-129. The **Create array** dialog box will be displayed; refer to Figure-130, asking you to specify the parameters.

Figure-129. Rectangular array tool

Figure-130. Create array dialog box

- Specify the number of columns and rows in the **Columns** and **Rows** edit boxes, respectively.
- Select the **Equal vertical/horizontal spacing** check box to make the inter-row and inter-column spacing the same.
- Select the **Constrain inter-element separation** check box to constrain each element in the array with respect to the others using construction lines.
- Select the **Clone** check box to substitute dimensional constraint by geometric constraints in the copies, so that a change in the original element is directly reflected on copies.
- After specifying desired parameters, click on **OK** button from the dialog box. A line and an icon will be attached to the cursor.
- Click at desired location in the 3D view area to specify the pattern. The rectangular array pattern will be created; refer to Figure-131.

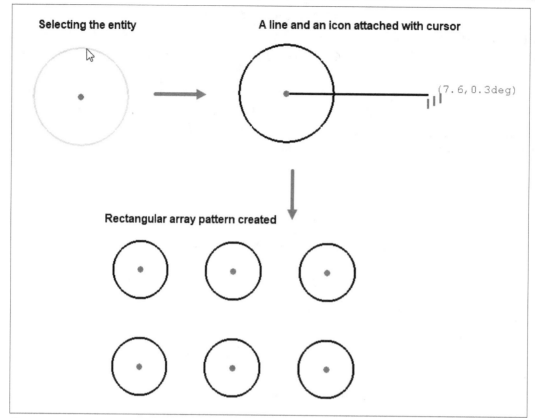

Figure-131. Creation of rectangular array pattern

Remove axes alignment

The **Remove axes alignment** tool modifies constraints to remove axes alignment while trying to preserve the shape of section. The procedure to use this tool is discussed next.

- Select desired geometry from which you want to remove the axes alignment; refer to Figure-132.

Figure-132. Rectangle selected

- Click on the **Remove axes alignment** tool from **Toolbar** in the **Sketcher** workbench; refer to Figure-133. The axes alignment from the geometry will be removed; refer to Figure-134.

Figure-133. Remove axes alignment tool

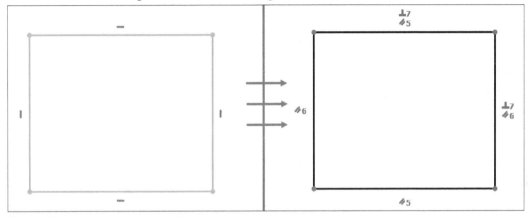

Figure-134. After removing axes alignment

Delete All Geometry

The **Delete All Geometry** tool is used to delete all the geometries of current sketch. The procedure to use this tool is discussed next.

- Click on the **Delete All Geometry** tool from the **Sketcher tools** cascading menu in the **Sketch** menu; refer to Figure-135. The **Delete All Geometry** dialog box will be displayed asking you whether to delete all the constraints or not; refer to Figure-136.

Figure-135. Delete all geometry tool

- Click on the **Yes** button from the dialog box to delete all the geometry from the sketch. The geometries will be deleted.

Figure-136. Delete All Geometry dialog box

Delete All Constraints

The **Delete All Constraints** tool is used to delete all the geometries from the sketch being edited and leaves the geometry intact. The procedure to use this tool is discussed next.

- Click on the **Delete All Constraints** tool from **Toolbar** in the **Sketcher** workbench; refer to Figure-137. The **Delete All Constraints** dialog box will be displayed asking you to whether to delete the constraints or not; refer to Figure-138.

Figure-137. Delete All Constraints tool

Figure-138. Delete All Constraints dialog box

- Click on the **Yes** button from the dialog box to delete all the constraints from the sketch. The constraints will be deleted; refer to Figure-139.

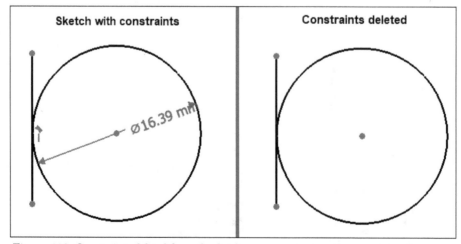

Figure-139. Constraints deleted from the sketch

SKETCHER B-SPLINE TOOLS

The **Sketcher B-spline** tools in **FreeCAD** are available in the **Toolbar** of **Sketcher** workbench; refer to Figure-140. The procedures to use these tools are discussed next.

Figure-140. Sketcher B spline tools

Show/Hide B-spline degree

The **Show/Hide B-spline degree** tool shows or hides the degree of B-spline curve. The procedure to use this tool is discussed next.

• Select the B-spline entity whose degree is to be shown or hidden.
• Click on the **Show/Hide B-spline degree** tool from **Show/hide B-spline information layer** drop-down in the **Toolbar** of **Sketcher** workbench; refer to Figure-141. The degree of B-spline will be displayed as shown in Figure-142.

Figure-141. Show/hide B spline degree tool

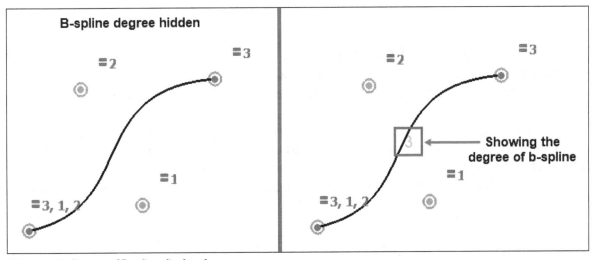

Figure-142. Degree of B spline displayed

Show/Hide B-spline control polygon

The **Show/Hide B-spline control polygon** tool shows or hides the defining polygon of a B-spline curve. The procedure to use this tool is discussed next.

- Select a B-spline entity whose polygon is to be shown or hidden.
- Click on the **Show/Hide B-spline control polygon** tool from **Show/hide B-spline information layer** drop-down in the **Toolbar** of **Sketcher** workbench; refer to Figure-143. The polygon of selected B-spline entity will be shown or hidden; refer to Figure-144.

Figure-143. Show/Hide B spline control polygon tool

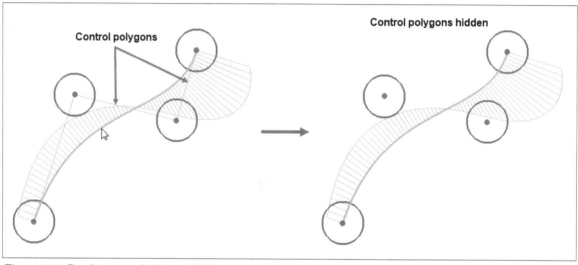

Figure-144. B-spline control polygons hidden

Show/Hide B-spline curvature comb

The **Show/Hide B-spline curvature comb** tool shows or hides the curvature comb of B-spline curve. The procedure to use this tool is discussed next.

- Select the B-spline entity whose curvature comb is to be shown or hidden.
- Click on the **Show/Hide B-spline curvature comb** tool from **Show/hide B-spline information layer** drop-down in the **Toolbar** of **Sketcher** workbench; refer to Figure-145. The curvature comb of selected B-spline entity will be shown or hidden as shown in Figure-146.

Figure-145. Show/hide B spline curvature comb tool

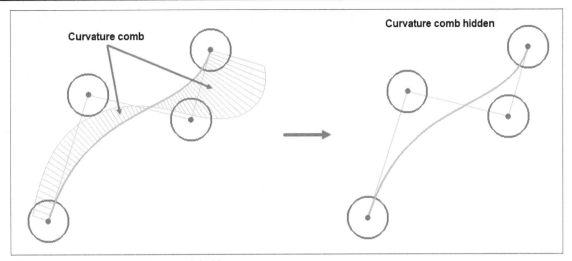

Figure-146. B-spline curvature comb hidden

Show/Hide B-spline knot multiplicity

The **Show/Hide B-spline knot multiplicity** tool shows or hides the knot multiplicity of a B-spline curve. The procedure to use this tool is discussed next.

- Select the B-spline entity whose knot multiplicity is to be shown or hidden.
- Click on the **Show/Hide B-spline knot multiplicity** tool from **Show/hide B-spline information layer** drop-down in the **Toolbar** of **Sketcher** workbench; refer to Figure-147. The knot numbers of spline vertices will be displayed; refer to Figure-148.

Figure-147. Show/hide B spline knot multiplicity tool

Figure-148. Showing knot multiplicity number

Show/hide B-spline control point weight

The **Show/hide B-spline control point weight** tool shows or hides of the weights for the control points of a B-spline curve.

B-splines are basically a combination of Bezier curves'. The Bezier curve is calculated using this formula:

$$\text{Bezier}(n,t) = \sum_{i=0}^{n} \underbrace{\binom{n}{i}}_{\text{polynomial term}} \underbrace{(1-t)^{n-i}t^i}_{\text{polynomial term}} \underbrace{P_i}_{\text{point coordinate}}$$

n is hereby the degree of the curve. So, a bezier curve of degree **n** is a polygon with order **n**. The factors P_i are hereby infact the coordinates of the Bezier curves' control points.

The procedure to use this tool is discussed next.

- Select the B-spline entity whose control point weight is to be shown or hidden.
- Click on the **Show/hide B-spline control point weight** tool from **Show/hide B-spline information layer** drop-down in the **Toolbar** of **Sketcher** workbench; refer to Figure-149. The control point weight of selected B-spline entity will be shown or hidden as shown in Figure-150.

Figure-149. Show/hide B spline control point weight tool

Figure-150. Showing the B spline control point weight

Convert Geometry to B-spline

The **Convert Geometry to B-spline** tool converts compatible geometry, edges, and curves into a B-spline. The procedure to use this tool is discussed next.

- Select the geometry which you want to convert into a B-spline.
- Click on the **Convert Geometry to B-spline** tool from **Toolbar** in the **Sketcher** workbench; refer to Figure-151. The selected geometry will be converted into a B-spline as shown in Figure-152.

Figure-151. Convert geometry to B spline tool

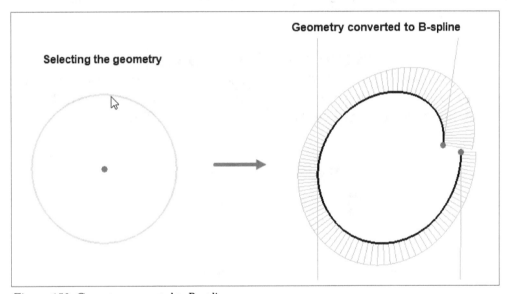

Figure-152. Geometry converted to B-spline

Increase B-spline degree

The **Increase B-spline degree** tool increases the degree of the B-spline. Degrees define the number of vertices of spline that can be modified. The procedure to use this tool is discussed next.

- Select the B-spline entity whose degree is to be increased.
- Click on the **Increase B-spline degree** tool from **Toolbar** in the **Sketcher** workbench; refer to Figure-153. The degree of B-spline entity will be increased; refer to Figure-154.

Figure-153. Increase B spline degree tool

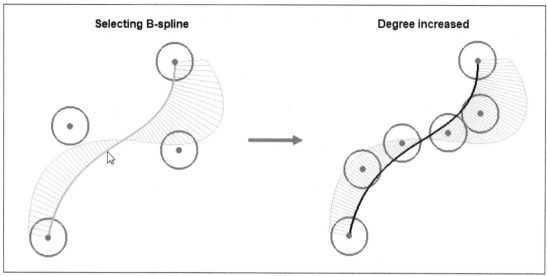

Figure-154. Degree increased of a B spline

Similarly, you can use **Decrease B-spline degree** tool to decrease the degree of the B-spline.

Modify knot multiplicity

There are two tools available in **Modify knot multiplicity** drop-down from **Toolbar** in the **Sketcher** workbench to modify knot multiplicity; **Increase knot multiplicity** and **Decrease knot multiplicity**; refer to Figure-155. The procedures to use these tools are discussed next.

Figure-155. Modify knot multiplicity drop down

Increase knot multiplicity

The **Increase knot multiplicity** tool increases the knot multiplicity of a B-spline curve knot. The number of times a knot value is duplicated is called the knot multiplicity. Knot multiplicity defines the number of times knot appears in the knot sequence of spline. Knot multiplicity higher than degree of spline will cause spline to split in disjoint parts. Select the knot from spline and click on the **Increase knot multiplicity** tool from the drop-down.

Decrease knot multiplicity

The **Decrease knot multiplicity** tool decreases the knot multiplicity of a B-spline curve knot. Select the knot from spline and click on the **Decrease knot multiplicity** tool from the drop-down.

Insert Knot

The **Insert knot** tool is used to insert a knot into an existing B-spline. If a knot already exists at the specified parameter, it increases the multiplicity of that knot by one. The procedure to use this tool is discussed next.

- Select the B-spline into which you want to insert the knot and click on the **Insert knot** tool from **Toolbar** in the **Sketcher** workbench; refer to Figure-156. A knot symbol will be attached to the cursor.

Figure-156. Insert knot tool

- Move the cursor over the spline and click at desired location where you want to insert the knot. The knot will be inserted; refer to Figure-157.

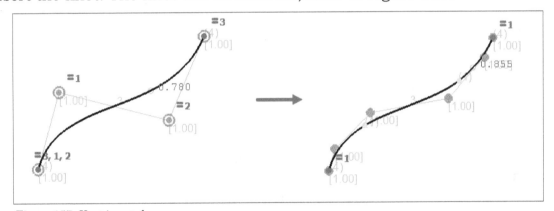

Figure-157. Knot inserted

Join Curves

The **Join Curves** tool is used to join two curves at selected end points. The procedure to use this tool is discussed next.

- Select the end points of two curves to be joined and click on the **Join Curves** tool from **Toolbar** in the **Sketcher workbench**; refer to Figure-158.

Figure-158. Join curves tool

PRACTICAL 1

Create the sketch as shown in Figure-159. Also, dimension the sketch as per the figure.

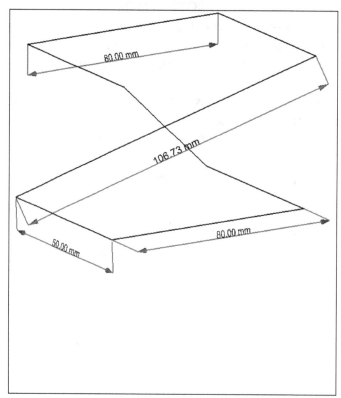

Figure-159. Practical 1

Steps to be performed:
- Start a new part file.
- Select a sketching plane and activate sketching mode.
- Create the sketch using **Line** tool.
- Apply the dimensions using dimension tools.
- Save the file.

Starting a New Sketch

- Start FreeCAD if not started yet.
- Select the **Sketcher** option from the **Switch between workbenches** drop-down in the **Toolbar**. The sketcher environment will be displayed.
- Click on the **Create a new sketch** tool from the **Toolbar** or **Sketch** menu. The **Choose orientation** dialog box will be displayed; refer to Figure-160.

Figure-160. Choose orientation dialog box

- Select the **XY-Plane** tool from the **Sketch orientation** area of dialog box to create sketch on XY plane.
- Click on the **OK** button from the dialog box. The tools to create sketch will become active.

Creating Line Sketch

- Click on the **Polyline** tool from the **Toolbar** or **Sketcher geometries** cascading menu of **Sketch** menu to activate polyline tool. You will be asked to specify start point of line.
- Click at the origin to specify start point of line and create the sketch as shown in Figure-161. Make sure to select the **Snap to grid** check box from **Toggle snap** drop-down in the toolbar of **Sketcher** workbench. The process to set unit to **mm** has been discussed earlier.

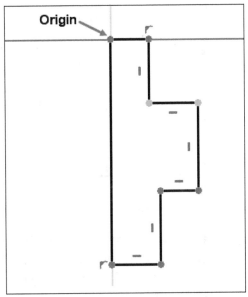

Figure-161. Sketch created for Practical 1

Applying Dimensions

- Click on the **Constrain Vertical distance** tool from the **Toolbar** or **Sketcher constraints** cascading menu of **Sketch** menu. You can also use **SHIFT+V** shortcut key to activate this tool. On activating this tool, you will be asked to select a vertical line.
- Select the vertical base line of sketch. The **Insert Length** dialog box will be displayed with dimension of line; refer to Figure-162.

Figure-162. Insert length dialog box

- Specify the length of vertical line as **135** mm in the **Length** edit box and click on the **OK** button to create the dimension.
- Repeat the steps to create all the vertical dimensions of sketch as shown in Figure-163.

Figure-163. Vertical dimensions of sketch

- Click on the **Constrain Horizontal distance** tool from the **Toolbar** or **Sketcher constraints** cascading menu of **Sketch** menu. You can also use **SHIFT+H** shortcut key to activate this tool. On activating this tool, you will be asked to select a horizontal line.
- Select the horizontal line of sketch. The **Insert Length** dialog box will be displayed with dimension of line; refer to Figure-164.

Figure-164. Creating horizontal dimension of sketch

- Specify the length of horizontal line as **20** mm in the **Length** edit box and click on the **OK** button to create the dimension.
- Repeat the steps to create all the horizontal dimensions of sketch.
- After specifying all the constraints and dimensions, the sketch should be displayed as fully constrained; refer to Figure-165.

Figure-165. Fully constrained sketch

- Click on the **Close** button from **Tasks** panel of **Combo View** to exit sketch mode.
- Click on the **Save** button from the **Toolbar** and save the file at desired location.

PRACTICAL 2

In this practical, we will create the sketch as shown in Figure-166.

Figure-166. Sketch for Practical 2

Steps to be performed:
- Start a new part file.
- Select a sketching plane and activate sketching mode.
- Create the sketch using tools like Line, Circle, Arc, Fillet, and so on.
- Apply the dimensions using dimension tools.
- Save the file.

Starting a New Sketch

- Start FreeCAD if not started yet.
- Select the **Sketcher** option from the **Switch between workbenches** drop-down in the **Toolbar**. The sketcher environment will be displayed.
- Click on the **Create a new sketch** tool from the **Toolbar** or **Sketch** menu. The **Choose orientation** dialog box will be displayed.
- Select the **XY-Plane** tool from the **Sketch orientation** area of dialog box to create sketch on XY plane.
- Click on the **OK** button from the dialog box. The tools to create sketch will become active.

Creating Sketch

- Click on the **Create Line** tool from the **Toolbar** or **Sketcher geometries** cascading menu of **Sketch** menu to activate line tool. You will be asked to specify start point of line.
- Create a horizontal line of length approximately **66** mm starting from origin.
- Select the **Toggle construction geometry** tool from the **Toolbar** or **Sketcher geometries** cascading menu of **Sketch** menu to activate construction mode.
- Create a vertical line of approximately **60** mm length passing through mid-point of horizontal line; refer to Figure-167.

Figure-167. Lines created

- Exit the construction mode by selecting the **Toggle construction geometry** tool again from **Toolbar**.
- Click on the **Create Circle** tool from the **Sketcher geometries** cascading menu of the **Sketch** menu or **Toolbar**. You will be asked to specify location of center of circle.
- Click on the open end point of construction line and create a circle of diameter approximately **13**; refer to Figure-168.

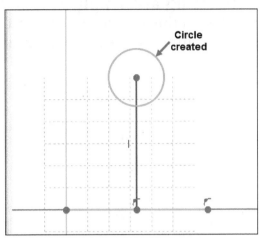

Figure-168. Circle created

- Create a circle of diameter approximately **26** using the same center as previous circle; refer to Figure-169.

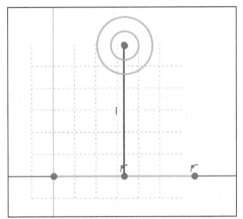

Figure-169. Two circles created

- Press **GL** shortcut key to activate line tool and create line sketch using approximate dimension as per the drawing; refer to Figure-170.

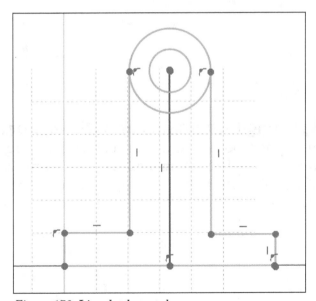

Figure-170. Line sketch created

Trimming and Apply Fillets

- Click on the **Trim edge** tool from the **Sketcher geometries** cascading menu of the **Sketch** menu and click on the inner section of larger circle to trim extra portion; refer to Figure-171.

Figure-171. Trimming circle

- Press **ESC** once to exit the **Trim edge** tool and click on the **Create fillet** tool from the **Sketcher geometries** cascading menu of **Sketch** menu. You will be asked to select two entities between which the fillet will be applied.
- Select the lines in pair near their connection points as shown in Figure-172.

Figure-172. Applying fillets

Applying Constraints and Dimensions

- Select two bottom corner points and centerline passing through the sketch and then select the **Constraint symmetrical** tool from **Sketcher constraints** cascading menu of the **Sketch** menu to keep centerline at the middle of sketch; refer to Figure-173.

Figure-173. Selection for symmetrical constraint

- If there are any disconnection between points that should be coincident then apply the coincident constraint on those points. Note that you may need to zoom on the intersection points to check whether they are actually coincident or not.
- Click on the **Constrain Distance** tool from **Toolbar** or **Sketcher constraints** cascading menu in the **Sketch** menu and apply the dimensions as shown in Figure-174.

Figure-174. Distance dimensions applied

- Click on the **Constraint equal** tool from **Sketcher constraints** cascading menu of **Sketch** menu and one by one make the entities on both sides of centerline equal; refer to Figure-175.

Figure-175. After applying equal constraint

- Select the **Constrain diameter** tool from the **Sketcher constraints** cascading menu of the **Sketch** menu and apply diameter dimensions to arc and circles. The fully defined sketch should display as shown in Figure-176.

Figure-176. Fully defined sketch created

Close the sketcher and save the file as discussed earlier.

PRACTICAL 3

In this practical, we will create the sketch as shown in Figure-177.

Figure-177. Practical 03

Steps to be performed:
- Start a new part file.
- Select a sketching plane and activate sketching mode.
- Create the sketch using various sketching tools.
- Apply the dimensions using dimension tools.
- Save the file.

Starting a New Sketch

- Start FreeCAD if not started yet.
- To change the current unit to inch, click on **Preferences** option from the **Edit** menu. The **Preferences** dialog box will be displayed. Select the **General** option from left area in the dialog box and select **Imperial decimal** option from the **Unit system** drop-down in this page; refer to Figure-178. After setting parameters, click on the **OK** button from the dialog box. The unit system will be changed to inches.

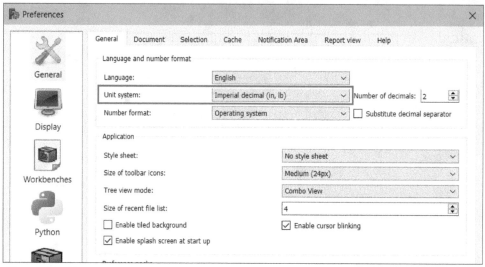

Figure-178. Unit system set to imperial decimal

- Select the **Sketcher** option from the **Switch between workbenches** drop-down in the **Toolbar**. The sketcher environment will be displayed.
- Click on the **Create a new sketch** tool from the **Toolbar** or **Sketch** menu. The **Choose orientation** dialog box will be displayed.
- Select the **XY-Plane** tool from the **Sketch orientation** area of dialog box to create sketch on XY plane.
- Click on the **OK** button from the dialog box. The tools to create sketch will become active.

Creating Sketch

- Click on the **Create circle** tool from the **Toolbar** or **Sketcher geometries** cascading menu of **Sketch** menu to activate circle tool. You will be asked to specify location of center of circle.
- Click at origin in the 3D view area and create concentric circles of diameter approximately **5** and **2.25**; refer to Figure-179.

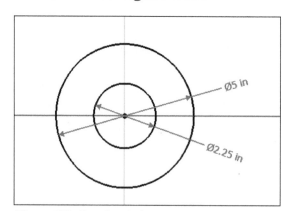

Figure-179. Creating circles

- After creating the circles, apply the dimensions to the circles as per the drawing using **Constrain diameter** tool from **Sketcher constraints** cascading menu of the **Sketch** menu.
- Click on the **Create Line** tool from the **Toolbar** and create line sketch using approximate dimensions as per the drawing; refer to Figure-180.

Figure-180. Creating the sketch using Line tool

- After creating the lines, apply the dimensions to the lines using desired dimension tools from **Sketcher constraints** cascading menu of the **Sketch** menu.
- Create rectangle and line sketches using approximate dimensions as per the drawing using **Create rectangle** and **Create line** tool from the **Toolbar**; refer to Figure-181.

Figure-181. Creating rectangle and line sketches

Creating Arcs

- Click on the **Create arc by three points** tool from **Sketch geometries** cascading menu or **End points and Rim point** tool from **Create arc** drop-down in the **Toolbar** and create the arcs as per the drawing using approximate dimensions.
- After creating the arcs, apply the constraints and dimensions to the arcs using tools from **Sketcher constraints** cascading menu of **Sketch** menu; refer to Figure-182.

Figure-182. Creating arcs

Trimming the entity

- Click on the **Trim edge** tool from the **Sketcher geometries** cascading menu of the **Sketch** menu and click on the edge of both the circles to trim extra portion; refer to Figure-183.

Figure-183. Trimming the extra portions

Close the sketcher and save the file as discussed earlier; refer to Figure-184.

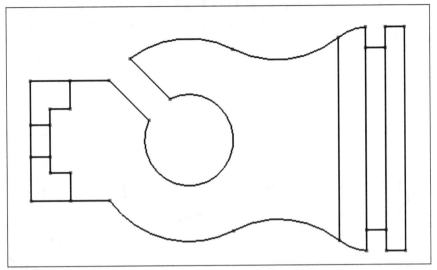

Figure-184. Practical 3 created

PRACTICE 1

In this practice session, you will create a sketch for the drawing shown in Figure-185.

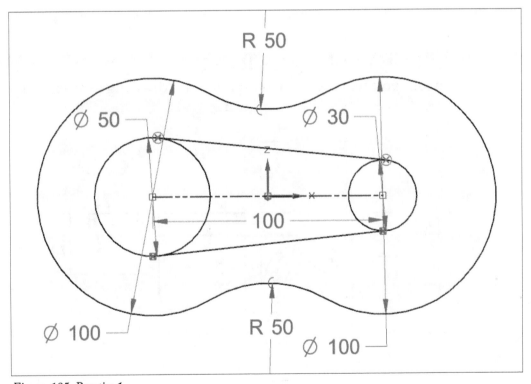

Figure-185. Practice 1

PRACTICE 2

In this practice session, you will create a sketch for drawing shown in Figure-186.

Figure-186. Practice 2

PRACTICE 3

In this practice session, you will create a sketch for the drawing given in Figure-187.

Figure-187. Practice 3

PRACTICE 4

In this practice session, you will create sketches of different views for the drawing given in Figure-188.

Figure-188. Practice 4

FOR STUDENT NOTES

FOR STUDENT NOTES

Chapter 3

Solid Modeling with Part Design

Topics Covered

The major topics covered in this chapter are:

- *Introduction to Solid Modeling*
- *Starting Part Design*
- *Part Design Helper Tools*
- *Part Design Modeling Tools*
- *Datum Tools*
- *Additive Tools*
- *Subtractive Tools*
- *Transformation Tools*
- *Dress-up Tools*
- *Boolean Operation*
- *Involute Gear Tool*
- *Shaft Design Wizard*

INTRODUCTION

Solid Modeling is the most advanced method of geometric modeling in 3D. Solid modeling is the representation of solid objects on your computer. Now, we will learn about various tools and commands used to create solid models.

In FreeCAD, the **Part Design Workbench** provides advanced tools for modeling complex solid parts. It is mostly focused on creating mechanical parts that can be manufactured and assembled into a finished product. Nevertheless, the created solids can be used in general for any other purpose such as architectural design, finite element analysis, or machining and 3D printing.

STARTING PART DESIGN

The **Part Design Workbench** is intrinsically related to the **Sketcher Workbench**. The user normally creates a sketch then uses the **Part Design Pad** tool to extrude it, create a basic solid, and then this solid is further modified.

- To start a new part design file, click on the **New** button from **File** menu and select **Part Design** workbench from **Switch between workbenches** drop-down in the **Toolbar**; refer to Figure-1. The tools to create 3D model will be displayed in the **Toolbar**; refer to Figure-2.

Figure-1. Part Design workbench

Figure-2. Part Design tools

Part Design Helper Tools

The **Part Design Helper** tools help in creating a body and make it active for creating and editing the sketch. You can also map a sketch to desired face using tools of **Part Design Helper** toolbar. These tools are available in the **Toolbar** of **Part Design** workbench; refer to Figure-3. The procedures to use these tools are discussed next.

Figure-3. Part Design Helper tools

Creating a body

A **Part Design** body is the base element to create solid shapes with the **Part Design** workbench. It can contain sketches, datum objects, and part design features that help in building a single contiguous solid. The procedure to use this tool is discussed next.

- Click on the **Create body** tool from **Toolbar** in the **Part Design** workbench; refer to Figure-4. An empty body is created and it automatically becomes active; refer to Figure-5.

Figure-4. Create body tool

Figure-5. A new empty body created

- Now, you can click **Create sketch** tool to create a sketch in the body that can be used with **Pad** tool or other 3D protrusion tools.

Creating a sketch

The **Create sketch** tool creates a new sketch on selected face or plane. If no face is selected while this tool is executed, the user is prompted to select a plane from the **Tasks** panel or from the 3D view area. The procedure to use this tool is discussed next.

- Click on the **Create sketch** tool from **Toolbar** in the **Part Design** workbench; refer to Figure-6. The **Select feature** dialog will be displayed in the **Tasks** panel of **Combo View** to select desired plane; refer to Figure-7. You can also select the plane from 3D view area.

Figure-6. Create sketch tool

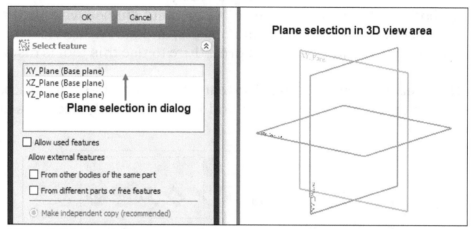

Figure-7. Selecting the plane to create sketch

- Select **Allow used features** check box to reuse earlier used features.
- Select the **From other bodies of the same part** check box from the **Allow external features** section of the dialog to use any elements used in the same body as dimensioning reference for creating sketch.
- Select the **From different parts or free features** check box to use parts and features of other objects in the model as reference for sketch dimensioning.
- Select the **Make independent copy** radio button to create independent copies of selected sketching references so that they do not get changed when original references have been changed.
- Select the **Make dependent copy** radio button to create reference elements dependent on the original features. This is basically use of a Shapebinder.
- Select the **Create cross-reference** radio button to linked the original elements as reference without creating copies. Any changes in original elements are reflected to this sketch.
- After selecting desired plane, click on **OK** button from the dialog. The interface switches to **Sketcher** workbench to create or edit the sketch. You have learned about sketches in previous chapter.

Edit sketch

The **Edit sketch** tool is used to edit an existing sketch. The procedure to use this tool is discussed next.

- Select the sketch that you want to edit from the **Model Tree** or from the 3D view area; refer to Figure-8.

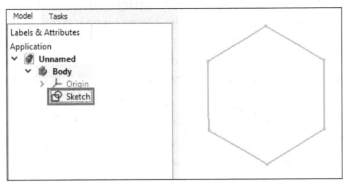

Figure-8. Selecting the sketch to edit

- Click on the **Edit sketch** tool from **Toolbar** in the **Part Design** workbench; refer to Figure-9. The **Sketcher** workbench will be displayed with options to edit the existing sketch.

Figure-9. Edit sketch tool

Map sketch to face

The **Map sketch to face** tool maps an existing sketch on the face of a selected body or a selected plane. Note that this tool is not used to create new sketches. It only maps or remaps existing sketch to the face of a solid/part design feature or a plane. The procedure to use this tool is discussed next.

- Select a face or plane on which you want to map the sketch.
- Click on the **Map sketch to face** tool from **Toolbar** in the **Part Design** workbench; refer to Figure-10. The **Select sketch** dialog box will be displayed; refer to Figure-11, asking you to select the sketch entity to be mapped from the list.

Figure-10. Map sketch to face tool

Figure-11. Select sketch dialog box

- Select desired sketch entity from the list which you want to map and click on **OK** button from the dialog box. The **Sketch attachment** dialog box will be displayed; refer to Figure-12, asking you to select method from the list to be used to attach sketch with selected face. Note that depending on the attachment face/plane selected earlier, you may get different options in the **Sketch attachment** dialog box.

Figure-12. Sketch attachment dialog box

- Select desired method from the list to map the sketch on selected object and click on **OK** button from the dialog box. Generally, it is fine to select the suggested method. The selected sketch entity will be mapped to the part design feature; refer to Figure-13.

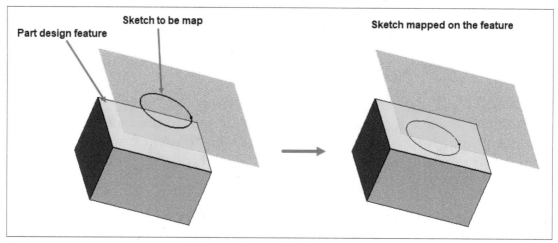

Figure-13. Sketch mapped on the solid feature

Validate Sketch

The **Validate Sketch** tool is used to verify tolerance of different points and adjusts them. The procedure to use this tool is discussed next.

- Click on the **Validate Sketch** tool from **Toolbar** in the **Part Design** workbench; refer to Figure-14 The **Sketcher validation** dialog will be displayed in the **Tasks** panel of **Combo View;** refer to Figure-15.

Figure-14. Validate Sketch tool

Figure-15. Sketcher validation tool

- Click on the **Highlight troublesome vertexes** button from the dialog to highlight vertexes that are overlapping or not connected within tolerance specified in the dialog; refer to Figure-16.

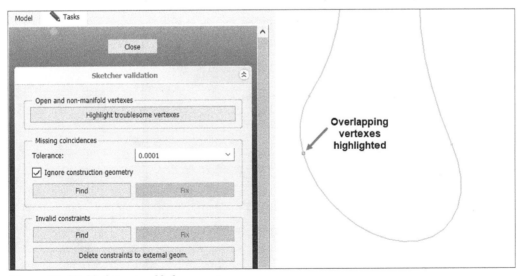

Figure-16. Highlighting troubleshome vertexes

- Clear the **Ignore construction geometry** check box if you do not want to perform validation checks on construction geometries in the sketch.
- Select the **Find** button from the **Missing coincidences** area of the dialog to check end points of curves in sketch which are not connected properly by coincidence constraint.

- Similarly, you can use the other options of the dialog to perform validation checks.
- Click on the **Close** button from the dialog to exit the tool.

Datum tools

The **Datum tools** in FreeCAD are referred to as auxiliary geometrical elements which will not form part of the final shape of the body but can be used as references and supports for sketches and other types of features. The **Datum tools** are available in the **Part Design Helper Toolbar** of **Part Design** workbench; refer to Figure-17. The procedures to use these tools are discussed next.

Figure-17. Datum tools

Create a datum point

The **Create a datum point** tool creates a datum point in the active body which can be used as reference for sketches or other datum geometry. The procedure to use this tool is discussed next.

- Click on the **Create a datum point** tool from **Toolbar** in the **Part Design** workbench; refer to Figure-18. The **Datum Point parameters** dialog will be displayed in the **Tasks** panel of **Combo View**; refer to Figure-19.

Figure-18. Create a datum point tool

Figure-19. Datum Point parameters dialog

- Select the first reference for attachment mode from the 3D view area or from the attachment modes list available in **Attachment mode** area of the dialog.
- Select the **Deactivated** option from the **Attachment mode** area if you want to select desired references for defining datum point without using predefined attachment modes. Select the **Object's origin** option from the area to use origin of selected object as datum point. Select the **Focus1** option from the area to use focal point 1 of selected curve. Select **Focus2** option from the area to use focal point 2 of selected curve. Similarly, use other options in the area to use respective references for creating datum point.
- Specify desired offset values in **Attachment Offset** area of the dialog to move point by specified value.

Figure-20. Datum point offset

- Click on the **Flip sides** check box then the datum point is opposite to the Sketch.
- Click on **OK** button from the dialog. The datum point will be attached to the feature at a specified offset distance; refer to Figure-21.

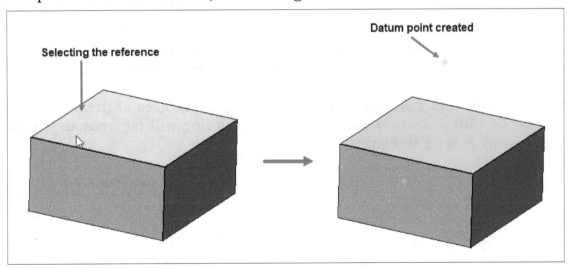

Figure-21. Creation of datum point

Create a datum line

The **Create a datum line** tool creates a datum line which can be used as reference for sketches, other datum geometries, or features. For example, it can be used as revolution axis for **Revolution** and **Groove** features. The procedure to use this tool is discussed next.

- Click on the **Create a datum line** tool from **Part Design Helper Toolbar** in the **Part Design** workbench; refer to Figure-22. The **Datum Line parameters** dialog will be displayed in the **Tasks** panel of **Combo View**; refer to Figure-23.

Figure-22. Create a datum line tool

Figure-23. Datum Line parameters dialog

- Select first reference for attachment mode from the 3D view area or from the attachment modes list available in **Attachment mode** area of the dialog.
- To add an additional reference, click on the **Reference2** button and select desired reference.
- Specify desired offset values in the **Attachment Offset** area of the dialog.
- Click on **OK** button from the dialog. The datum line will be created at specified offset distance; refer to Figure-24.

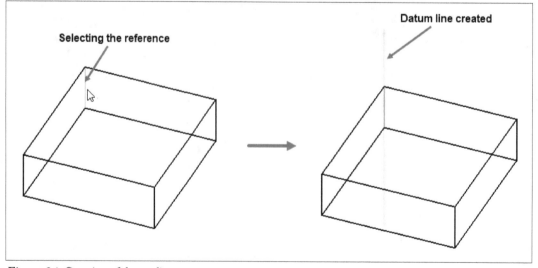

Figure-24. Creation of datum line

Create a datum plane

The **Create a datum plane** tool creates a datum plane which can be used as reference for sketches or other datum geometries. Sketches can also be attached to datum planes. The procedure to use this tool is discussed next.

- Click on the **Create a datum plane** tool from **Part Design Helper Toolbar** in the **Part Design** workbench; refer to Figure-25. The **Datum Plane parameters** dialog will be displayed in the **Tasks** panel of **Combo View**; refer to Figure-26.

Figure-25. Create a datum plane tool

Figure-26. Datum Plane parameters dialog

- Select first reference for attachment mode from the 3D view area or from the attachment modes list available in **Attachment mode** area of the dialog. You can select a point, line, plane, or face as reference. Note that when you hover cursor on an option in the **Attachment mode** area of the dialog box then selection options related to that attachment method.
- Click on the **Deactivated** mode in the attachment mode then the attachment is disabled. The object can be moved by editing its Placement property.
- Click on the **Translate origin** mode in the attachment mode then the origin is aligned to match a vertex. The orientation is still controlled by the Placement property of the attached object.
- Click on the **Object's XY** mode in the attachment mode then the Placement is made equal to the Placement of a linked object.
- Click on the **Object's YZ** mode in the attachment mode then the the X, Y and Z axes are matched with a linked object's local X, Z and -Y axes, respectively.
- Click on the **Object ZX** mode in the attachment mode then the X, Y and Z axes are matched with a linked object's local Y, Z and X axes, respectively.
- Click on the **Plane face** mode in the attachment mode then the plane is aligned to coincide with a planar face.
- Click on the **Tangent to surface** mode in the attachment mode then the plane is made tangent to a face at a vertex.
- Click on the **Normal to edge** mode in the attachment mode then the aligned to be normal to an edge.
- To add an additional reference, click on the **Reference2** button.

- Specify desired offset values in the **Attachment Offset** area of the dialog to move plane by specified value in respective directions.
- Click on **OK** button from the dialog. The datum plane will be created at specified offset distance; refer to Figure-27.

Figure-27. Creation of datum plane

Creating a plane tangent to curved face of body

To create a tangent plane, you will need a curved face and a vertex from which plane will pass. The procedure to create tangent plane is given next.

- Click on the **Create a datum plane** tool from the **Part Design Helper Toolbar**. The **Datum Plane parameters** dialog will be displayed in the **Tasks** tab of **Combo View**.
- Select the **Tangent to surface** option from the **Attachment mode** section of the **Datum Plane parameters** dialog. You will be asked to select references for creating the plane.
- Select the vertex (datum point) and curved face. Preview of tangent plane will be displayed; refer to Figure-28.

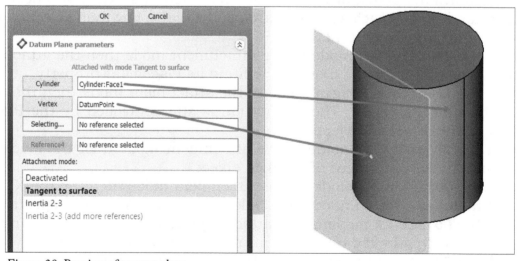

Figure-28. Preview of tangent plane

- Click on the **OK** button from the dialog to create the plane.

Create a local coordinate system

The **Create a local coordinate system** tool creates a local coordinate system which can be used as reference for other datum geometry. It also helps identify the orientation of the referenced datum geometry in 3D space. The procedure to use this tool is discussed next.

- Click on **Create a local coordinate system** tool from **Part Design Helper Toolbar** in the **Part Design** workbench; refer to Figure-29. The **Local Coordinate System parameters** dialog will be displayed in the **Tasks** panel of **Combo View**; refer to Figure-30.

Figure-29. Create a local coordinate system tool

- Select a point, line, face, or plane as first reference for attachment mode from the 3D view area or from the attachment modes list available in **Attachment mode** area of the dialog.
- To add an additional reference, click on the **Reference2** button and select desired entity.
- Specify desired offset values in the **Attachment Offset** area of the dialog.
- Click on **OK** button from the dialog. The local coordinate system will be created at specified offset distance; refer to Figure-31.

Figure-30. Local Coordinate System parameters dialog

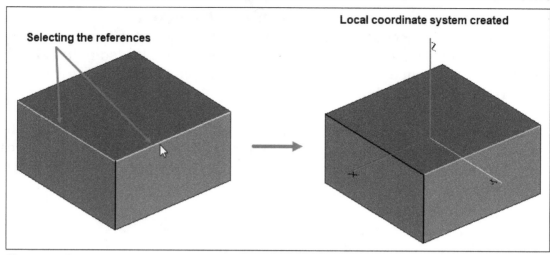

Figure-31. Creation of local coordinate system

Create a shape binder

The **Create a shape binder** tool creates a datum shape binder from selected body inside the active body. A shape binder is a reference object that links edges or faces of another body to active body. It can also be used to link a sketch from one body to another body. The datum shape binder object displays as translucent yellow in the 3D view. The procedure to use this tool is discussed next.

- Click on **Create a shape binder** tool from **Toolbar** in the **Part Design** workbench; refer to Figure-32. The **Datum shape parameters** dialog will be displayed in the **Tasks** panel of **Combo View**; refer to Figure-33.

Figure-32. Create a shape binder tool

Figure-33. Datum shape parameters dialog

- The **Object** button in the dialog will select the whole body and the **Add Geometry** button will select elements of the object (vertex, edge, face). Click on **Object** button and select desired object in the 3D view area or click on **Add Geometry** button and select elements of the object (vertex, edge, face) in the 3D view area. The datum shape binder object will be created; refer to Figure-34.

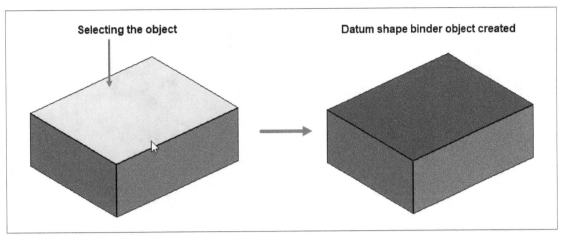

Figure-34. Datum shape binder object created

- To remove selected geometry, click on **Remove Geometry** button from the dialog and select the geometry in the 3D view area.
- To display the datum shape binder object, toggle the visibility of **Pad** feature from **True** to **False** in the **Property editor** of **Combo View** or select the Pad feature from **Model** panel in **Combo View** and press **SPACEBAR** key. The datum shape binder object will be displayed; refer to Figure-35.

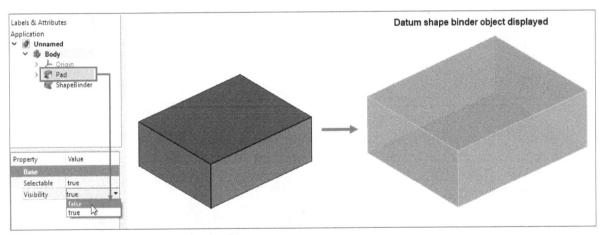

Figure-35. Displaying datum shape binder object

Create a sub-object shape binder

The **Create a sub-object shape binder** tool is used to create a shape binder to a sub-element, like edge or face from another body, while retaining the relative position of that element. Note that although shape binder and sub-object shape binder are created using similar procedure, the shape binder can either reference a single whole object, or sub-elements belonging to a single parent object whereas a sub-object shape binder does not have these restrictions. The procedure to use this tool is discussed next.

- Select the body from the **Model Tree** or from the 3D view area which you want to create shape binder; refer to Figure-36.

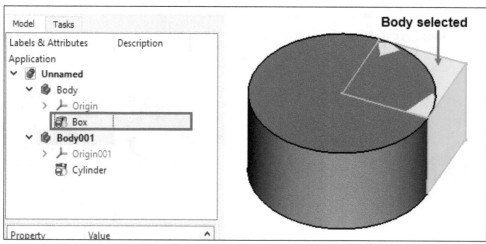

Figure-36. Body selected to create shape binder

- Click on the **Create a sub-object shape binder** tool from **Toolbar** in the **Part Design** workbench; refer to Figure-37. The selected body will be created as a shape binder object; refer to Figure-38.

Figure-37. Create a sub object shape binder tool

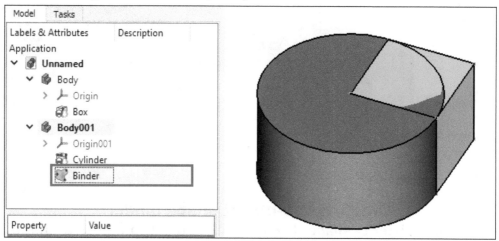

Figure-38. Selected body created as shape binder

An Example of using Shape Binder

- Now, select the shape binder object from the **Model Tree** and click on the **Boolean** tool from **Toolbar** in the **Part Design** workbench (You will learn about this tool later in this chapter). The **Boolean parameters** dialog will be displayed in the **Tasks** panel of **Combo View** and the selected shape binder object will be displayed in the selection area of the dialog; refer to Figure-39.

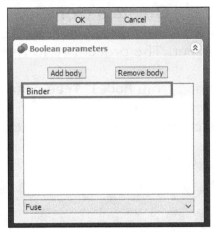

Figure-39. Boolean parameters dialog with shape binder object selected

- Select desired option from the drop-down in the dialog (**Cut** option in this case) and click on the **OK** button from the dialog. The boolean feature will be created; refer to Figure-40.

Figure-40. Boolean feature created

- To display the boolean cut feature, select the first body from the **Model Tree** and select **False** option from **Visibility** drop-down in the **Property editor** dialog or press **Space** button from the keyboard. The boolean cut feature will be displayed; refer to Figure-41.

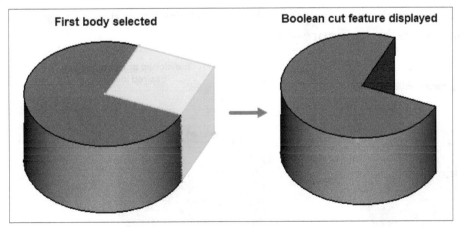

Figure-41. Boolean cut feature displayed

Create a clone

The **Create a clone** tool creates a linked copy of selected object which will follow any future edits to the original object. The procedure to use this tool is discussed next.

- Select the object to be cloned from **Model Tree** or from the 3D view area.
- Click on the **Create a clone** tool from **Toolbar** in the **Part Design** workbench; refer to Figure-42. The selected object will be cloned.

Figure-42. Create a clone tool

- Double-click on cloned object from **Model Tree**. The **Increments** dialog will be displayed in the **Tasks** panel of **Combo View** and a **Placement triad** will be displayed on the object in the 3D view area; refer to Figure-43.

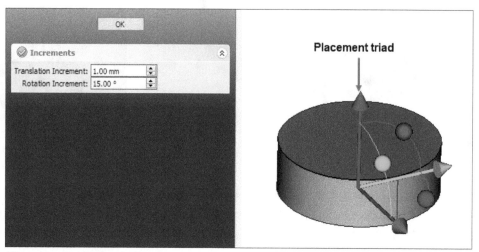

Figure-43. Increments dialog

- Drag the **Placement triad** to place the cloned object in desired location. The object will be placed; refer to Figure-44.

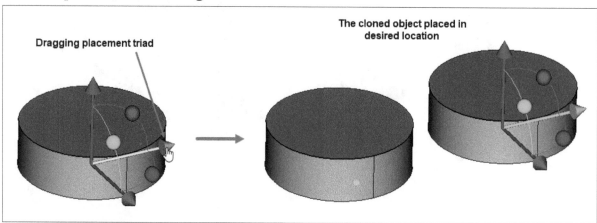

Figure-44. Placement of cloned object

- Click on **OK** button from the dialog to complete the procedure.

PART DESIGN MODELING TOOLS

With the help of Part Design Modeling tools, you can create a part from a conceptual sketch through solid feature-based modeling, as well as build and modify parts through direct intuitive graphical manipulation.

Part Design Modeling tools

The **Part Design Modeling tools** are used for creating base features or adding material to an existing solid body. The **Part Design Modeling tools** are available in the **Toolbar** of **Part Design** workbench; refer to Figure-45. The procedures to use these tools are discussed next.

Figure-45. Part Design Modeling tools

Pad

The **Pad** tool extrudes a sketch to form solid in direction normal to the sketch plane. The procedure to use this tool is discussed next.

- Select desired sketch from the **Model Tree** to create the pad feature; refer to Figure-46. Note that sketch should be a closed loop curve.

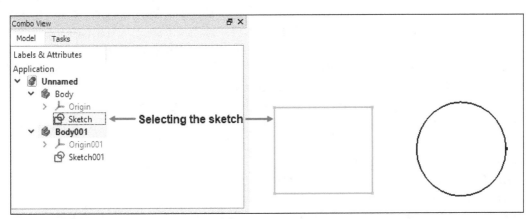

Figure-46. Selecting the sketch to create pad feature

- Click on the **Pad** tool from **Toolbar** in the **Part Design** workbench; refer to Figure-47. The **Pad parameters** dialog will be displayed in the **Tasks** panel of **Combo View** along with the preview of extrude feature; refer to Figure-48.

Figure-47. Pad tool

- Select desired type of extrusion from **Type** drop-down. Select **Dimension** option to extrude selected sketch on either side of the sketch plane at specified length. Select **To last** option to extrude the sketch up to the last face of the base feature. Select **To first** option to extrude the sketch up to the first face of the base feature. Select **Up to face** option to extrude the sketch up to selected face of base feature. Select **Two dimensions** option to extrude the sketch by specifying length for both the sides.
- Specify desired length of extrusion in the **Length** edit box.

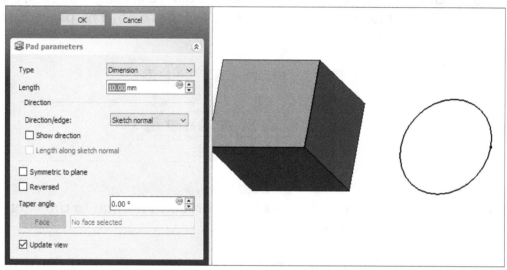

Figure–48. Pad parameters dialog

- Select desired direction of the extrusion from **Direction/edge** drop-down. Select **Sketch normal** option to extrude the sketch or face along its normal. Select the **Select reference** option to extrude the sketch along an edge of the 3D model. Select **Custom direction** option to set the pad length along the custom direction by specifying desired values in **x**, **y**, and **z** edit boxes.
- Select the **Show direction** check box to display **x**, **y**, and **z** edit boxes.
- Select **Length along sketch normal** check box to measure the pad length along the sketch normal.
- Select **Symmetric to plane** check box to extrude the sketch on both the sides by same specified length. This option is available only if the **Dimension** option is selected in the **Length** drop-down.
- Select **Reversed** check box to reverse the direction of extrusion. This option works only if the **Dimension** option is selected in the **Length** drop-down.
- Specify desired value in the **Taper angle** edit box to taper the pad in the extrusion direction by the given angle. This option is available only if the **Dimension** or **Two dimensions** option is selected in the **Length** drop-down.
- Click on **OK** button from the dialog. The selected sketch will be extruded.

Revolution

The **Revolution** tool is used to create a solid by revolving selected sketch or 2D object about given axis. Note that the sketch should be closed loop and revolution axis should not intersect with the sketch. The procedure to use this tool is discussed next.

- Select the sketch to be revolved from **Model** panel in the **Combo View**; refer to Figure-49.

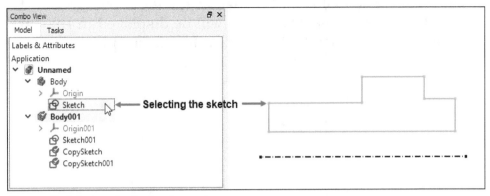

Figure-49. Selecting the sketch to be revolve

- Click on the **Revolution** tool from **Toolbar** in the **Part Design** workbench; refer to Figure-50. The **Revolution parameters** dialog will be displayed in the **Tasks** panel of **Combo View** along with the preview of revolution feature; refer to Figure-51.

Figure-50. Revolution tool

Figure-51. Revolution parameters dialog

- Select desired option from **Axis** drop-down in the dialog about which the sketch is to be revolved. Select **Vertical sketch axis** option to create revolve feature about a vertical sketch axis. Select **Horizontal sketch axis** option to create revolve feature about a horizontal sketch axis. Select **Base X axis**, **Base Y axis**, or **Base Z axis** option to create the revolve feature about **X**, **Y**, or **Z** axis of body's origin, respectively. Select the **Select reference** option to create revolve feature by selecting an edge on the body or a datum line as an axis of revolution.
- Specify the angle of revolution by which the revolution is to be formed in the **Angle** edit box.
- Specify the other parameters as desired and click on **OK** button from the dialog. The revolution feature will be created.

Additive loft

The **Additive loft** tool creates a solid active body by making a transition between two or more sketches. The procedure to use this tool is discussed next.

- Click on **Additive loft** tool from **Toolbar** in the **Part Design** workbench; refer to Figure-52. The **Select feature** dialog will be displayed in the **Tasks** panel of **Combo View** asking you to select the base profile; refer to Figure-53.

Figure-52. Additive loft tool

Figure-53. Select feature dialog

- Select a sketch from the dialog to be used as base profile for additive loft feature and click on **OK** button; refer to Figure-54. The **Loft parameters** dialog will be displayed; refer to Figure-55.

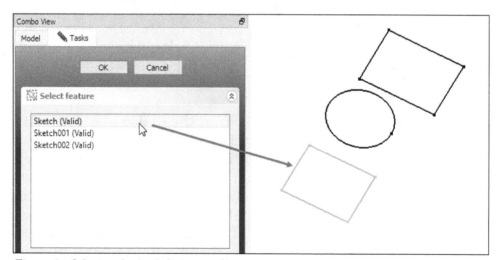

Figure-54. Selecting the sketch for base profile

Figure-55. Loft parameters dialog

- The base profile is preselected in the **Object** selection box of **Profile** area. Click on **Add Section** button and select the second sketch in the 3D view area for the first loft section and select the third sketch for the second loft section. The preview of additive loft feature will be displayed; refer to Figure-56.

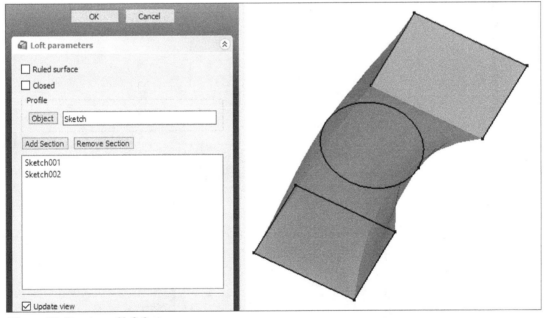

Figure-56. Preview of loft feature

- Select **Ruled surface** check box to make straight transitions between cross-sections.
- Select **Closed** check box to make a transition from the last cross-section to the first to create a loop.
- Specify the parameters as desired and click on **OK** button from the dialog to create the feature.

Additive pipe

The **Additive pipe** tool creates a solid as active body by sweeping one or more sketches along an open or closed path. The procedure to use this tool is discussed next.

- Click on **Additive pipe** tool from **Toolbar** in the **Part Design** workbench; refer to Figure-57. The **Select feature** dialog will be displayed in the **Tasks** panel of **Combo View** as discussed earlier asking you to select the base profile.

Figure-57. Additive pipe tool

- Select desired sketch for base profile of sweep feature from the dialog; refer to Figure-58 and click on **OK** button. The **Pipe parameters** dialog will be displayed in the **Tasks** panel of **Combo View**; refer to Figure-59.

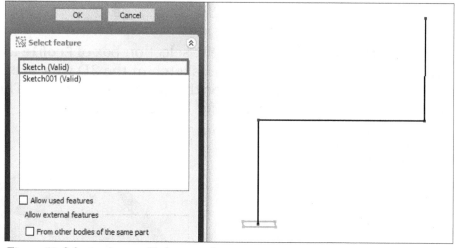

Figure-58. Selecting base profile for sweep feature

Figure-59. Pipe parameters dialog

- The base profile is preselected in the **Object** selection box of the **Profile** area.
- Click on the **Object** button from **Path to sweep along** area and select an edge of the sketch in 3D view area as a path for sweep feature. The whole sketch will be selected as sweep feature path and the preview of sweep feature will be displayed; refer to Figure-60.

Figure-60. Preview of sweep feature

- Click on **Add Edge** button to select single edge of the sketch from the 3D view area as a path for sweep feature.
- To remove an edge of the sketch as a sweep feature path, select an edge from 3D view area and click on **Remove Edge** button from the dialog.
- Select desired option from **Corner Transition** drop-down. The sweep feature will be created as the selected option; refer to Figure-61.

Figure-61. Corner Transition options

- Specify the parameters as desired and click on **OK** button from the dialog to create the sweep feature.

Additive Helix

The **Additive Helix** tool creates a solid by sweeping selected sketch or 2D object along a helix path. The procedure to use this tool is discussed next.

- Create desired sketch using which you want to create a solid sweep along a helical path; refer to Figure-62.

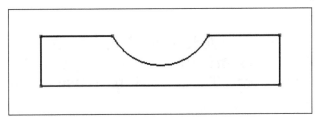

Figure-62. Sketch created for additive helix

- Select the sketch created from the **Model Tree** and click on the **Additive Helix** tool from **Toolbar** in the **Part Design** workbench; refer to Figure-63. The **Helix parameters** dialog will be displayed in the **Tasks** panel of **Combo View** along with the preview of additive helix feature; refer to Figure-64.

Figure-63. Additive Helix tool

Figure-64. Helix parameters dialog along with preview of additive helix feature

- Select desired option from **Axis** drop-down to specify the axis about which the sketch is to be swept.
- Select desired option from **Mode** drop-down to specify what parameters will be used to define the helix. Select the **Pitch-Height** option to define the helix via the height per turn and the overall height. Select the **Pitch-Turns** option to define the helix via the height per turn and the number of turns. Select the **Height-Turns** option to define the helix via the overall height and the number of turns.
- Specify desired value in the **Pitch** edit box to define the distance between turns in the helix.
- Specify desired value in the **Height** edit box to define the height of the helix.
- Specify desired value in the **Turns** edit box to define the number of turns in the helix.
- Specify desired value in the **Cone angle** edit box to define the rate at which the radius of the helix increase along the axis.
- Select **Left handed** check box to reverse the turning direction of helix from clockwise to counter clockwise.
- Select the **Reversed** check box to reverse the axis direction of helix from default.
- After specifying desired parameters in the dialog, click on the **OK** button to create the additive helix feature.

Additive Box

The **Additive Box** tool inserts a primitive box in the active body as the first feature or fuses it to the existing feature. The procedure to use this tool is discussed next.

- Click on the **Additive Box** tool from **Create an additive primitive** drop-down in the **Toolbar** of **Part Design** workbench; refer to Figure-65. The **Primitive parameters** dialog will be displayed in the **Tasks** panel of **Combo View** along with the preview of additive box feature; refer to Figure-66.

Figure-65. Additive Box tool

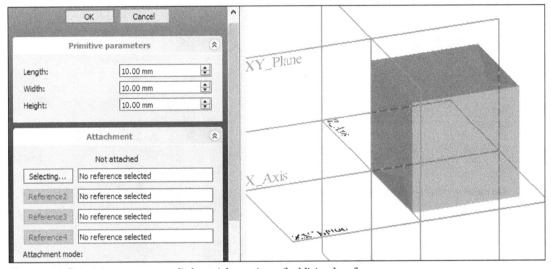

Figure-66. Primitive parameters dialog with preview of additive box feature

- Specify the length, width, and height of the feature in the **Length**, **Width**, and **Height** edit boxes of the **Primitive parameters** dialog, respectively.
- Specify the other parameters in the dialog as discussed earlier.
- Click on **OK** button from the dialog. The additive box feature will be created; refer to Figure-67.

Figure-67. Additive box feature created

Additive Cylinder

The **Additive Cylinder** tool inserts a primitive cylinder in the active body as the first feature or fuses it to the existing feature. The procedure to use this tool is discussed next.

- Click on the **Additive Cylinder** tool from **Create an additive primitive** drop-down in the **Toolbar** of **Part Design** workbench; refer to Figure-68. The **Primitive parameters** dialog will be displayed in the **Tasks** panel of **Combo View** along with the preview of additive cylinder feature; refer to Figure-69.

Figure-68. Additive Cylinder tool

Figure-69. Primitive parameters dialog with preview of additive cylinder feature

- Specify the radius and height of the feature in the **Radius** and **Height** edit boxes, respectively.
- Specify desired angle of first and second direction in the **Angle in first direction** and **Angle in second direction** edit boxes to define revolution in clockwise and counter clockwise directions, respectively.
- Specify the angle of revolution of the cross-section in the **Rotation angle** edit box of the dialog.
- Specify the other parameters in the dialog as discussed earlier.
- Click on **OK** button from the dialog. The additive cylinder feature will be created; refer to Figure-70.

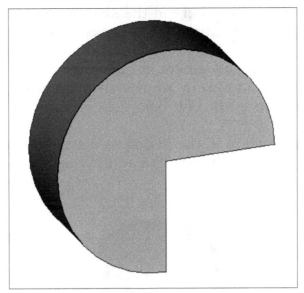

Figure-70. Additive cylinder feature created

Additive Sphere

The **Additive Sphere** tool inserts a primitive sphere in the active body as the first feature or fuses it to the existing feature. The procedure to use this tool is discussed next.

- Click on the **Additive Sphere** tool from **Create an additive primitive** drop-down in the **Toolbar** of **Part Design** workbench; refer to Figure-71. The **Primitive parameters** dialog will be displayed in the **Tasks** panel of **Combo View** along with the preview of additive sphere feature; refer to Figure-72.

Figure-71. Additive Sphere tool

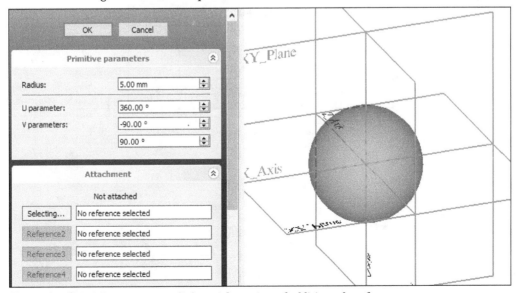

Figure-72. Primitive parameters dialog with preview of additive sphere feature

- Specify desired radius of the sphere in the **Radius** edit box of the dialog.
- Specify desired angle of revolution of the cross section in the **U parameter** edit box of the dialog.
- Specify lower truncation of the sphere parallel to the circular cross section in the **V parameters** edit box and specify upper truncation of the ellipsoid parallel to the circular cross section in the edit box just below to the **V parameters** edit box of the dialog.
- Specify the other parameters in the dialog as discussed earlier.
- Click on **OK** button from the dialog. The additive sphere feature will be created; refer to Figure-73.

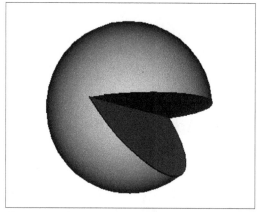

Figure-73. Additive sphere feature created

Additive Cone

The **Additive Cone** tool inserts a primitive cone in the active body as the first feature or fuses it to the existing feature. The procedure to use this tool is discussed next.

- Click on the **Additive Cone** tool from **Create an additive primitive** drop-down in the **Toolbar** of **Part Design** workbench; refer to Figure-74. The **Primitive parameters** dialog will be displayed in the **Tasks** panel of **Combo View** along with the preview of additive cone feature; refer to Figure-75.

Figure-74. Additive Cone tool

Figure-75. Primitive parameters dialog with preview of additive cone feature

- Specify desired radius value for base of the cone in the **Radius 1** edit box and specify desired radius value for top of the cone in the **Radius 2** edit box of the dialog.
- Specify the height of the cone along its axis in the **Height** edit box of the dialog.
- Specify the angle of revolution of the cross section in the **Angle** edit box of the dialog.
- Specify the other parameters in the dialog as discussed earlier.
- Click on **OK** button from the dialog. The additive cone feature will be created; refer to Figure-76.

Figure-76. Additive cone feature created

Additive Ellipsoid

The **Additive Ellipsoid** tool inserts a primitive ellipsoid in the active body as the first feature or fuses it to the existing feature. The procedure to use this tool is discussed next.

- Click on the **Additive Ellipsoid** tool from **Create an additive primitive** drop-down in the **Toolbar** of **Part Design** workbench; refer to Figure-77. The **Primitive parameters** dialog will be displayed in the **Tasks** panel of **Combo View** along with the preview of additive ellipsoid feature; refer to Figure-78.

Figure-77. Additive Ellipsoid tool

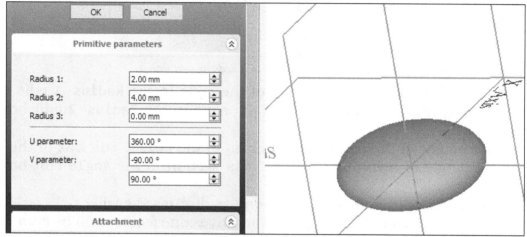

Figure-78. Primitive parameters dialog with preview of additive ellipsoid feature

- Specify desired radius value along the ellipsoid's vertical axis, along the ellipsoid's length, and along the ellipsoid's width in the **Radius 1**, **Radius 2**, and **Radius 3** edit boxes, respectively.
- Specify desired angle of revolution of the elliptical cross section in the **U parameter** edit box of the dialog.
- Specify the lower truncation of the ellipsoid parallel to the circular cross section in the **V parameter** edit box and specify upper truncation of the ellipsoid parallel to the circular cross section in the edit box just below to the **V parameter** edit box.
- Specify other parameters in the dialog as discussed earlier.
- Click on **OK** button from the dialog. The additive ellipsoid feature will be created; refer to Figure-79.

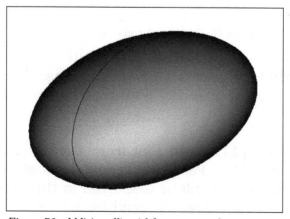

Figure-79. Additive ellipsoid feature created

Additive Torus

The **Additive Torus** tool inserts a primitive torus in the active body as the first feature or fuses it to the existing feature. The procedure to use this tool is discussed next.

- Click on the **Additive Torus** tool from **Create an additive primitive** drop-down in the **Toolbar** of **Part Design** workbench; refer to Figure-80. The **Primitive parameters** dialog will be displayed in the **Tasks** panel of **Combo View** along with the preview of additive ellipsoid feature; refer to Figure-81.

Figure-80. Additive Torus tool

- Specify desired radius of the imaginary orbit around which the circular cross section revolves in the **Radius 1** edit box and specify desired radius of circular cross section defining the form of the torus in **Radius 2** edit box of the dialog.

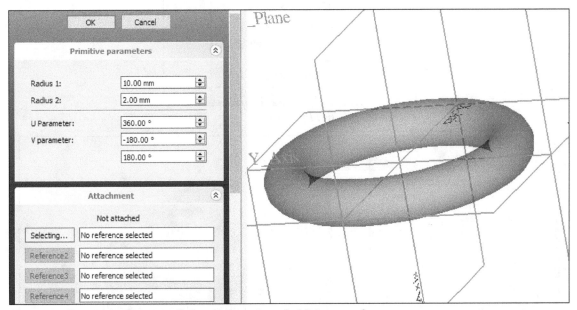

Figure-81. Primitive parameters dialog with preview of additive torus feature

- Specify other parameters in the dialog as discussed earlier.
- Click on **OK** button from the dialog. The additive torus feature will be created; refer to Figure-82.

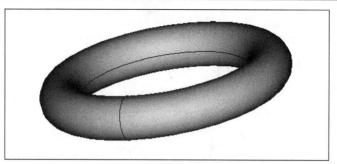

Figure-82. Additive torus feature created

Additive Prism

The **Additive Prism** tool inserts a primitive prism in the active body as the first feature or fuses it to the existing feature. The procedure to use this tool is discussed next.

- Click on the **Additive Prism** tool from **Create an additive primitive** drop-down in the **Toolbar** of **Part Design** workbench; refer to Figure-83. The **Primitive parameters** dialog will be displayed in the **Tasks** panel of **Combo View** along with the preview of additive prism feature; refer to Figure-84.

Figure-83. Additive Prism tool

Figure-84. Primitive parameters dialog with preview of additive prism feature

- Specify desired number of sides in the polygon cross section of the prism in the **Polygon** edit box of the dialog.
- Specify the circumscribed radius of the polygon cross section of the prism in the **Circumradius** edit box of the dialog.
- Specify desired height of the prism in the **Height** edit box of the dialog.
- Specify desired angle value for the first direction and second direction in the **Angle in first direction** and **Angle in second direction** edit boxes of the dialog, respectively.
- Specify other parameters in the dialog as discussed earlier.
- Click on **OK** button from the dialog. The additive prism feature will be created; refer to Figure-85.

Figure-85. Additive prism feature created

Additive Wedge

The **Additive Wedge** tool inserts a primitive wedge in the active body as the first feature or fuses it to the existing feature. The procedure to use this tool is discussed next.

- Click on the **Additive Wedge** tool from **Create an additive primitive** drop-down in the **Toolbar** of **Part Design** workbench; refer to Figure-86. The **Primitive parameters** dialog will be displayed in the **Tasks** panel of **Combo View** along with the preview of additive wedge feature; refer to Figure-87.

Figure-86. Additive Wedge tool

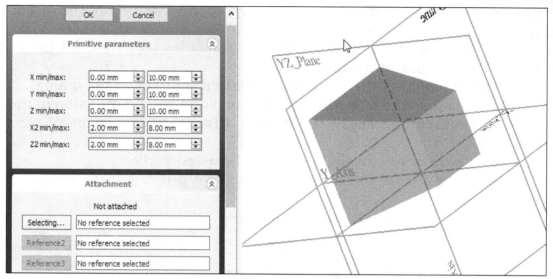

Figure-87. Primitive parameters dialog with preview of additive wedge feature

- Specify the minimum/maximum value for base face of the feature along X axis in the **X min/max** edit boxes, respectively.
- Specify desired height of the wedge in the **Y min/max** edit boxes, respectively.
- Specify the minimum/maximum value for base face of the feature along Z axis in the **Z min/max** edit boxes, respectively.
- Specify the minimum/maximum value for top face of the feature along X axis in the **X2 min/max** edit boxes, respectively.
- Specify the minimum/maximum value for top face of the feature along Z axis in the **Z2 min/max** edit boxes, respectively.
- Specify the other parameters in the dialog as discussed earlier.
- Click on **OK** button from the dialog. The additive wedge feature will be created; refer to Figure-88.

Figure-88. Additive wedge feature created

Subtractive tools

The **Subtractive tools** are used for subtracting materials from an existing body. The **Subtractive tools** are available in the **Toolbar** of **Part Design** workbench; refer to Figure-89. The procedures to use these tools are discussed next.

Figure-89. Subtractive tools in Part Design workbench

Pocket

The **Pocket** tool cuts out a solid by extruding a sketch (or a face of the solid) in a straight path and subtracting it from the solid. The procedure to use this tool is discussed next.

- Click on the **Pocket** tool from **Toolbar** in the **Part Design** workbench; refer to Figure-90. The **Pocket parameters** dialog will be displayed in the **Tasks** panel of **Combo View** along with the preview of pocket feature; refer to Figure-91.

Figure-90. Pocket tool

- Select desired option from **Type** drop-down in the dialog. Select the **Dimension** option to extrude cut the feature to the defined sketch plane at a specified length. Select the **Through all** option to extrude cut the feature through all material in the extrusion direction. Select the **To first** option to extrude cut the feature up to the first face of the solid entity in the extrusion direction. Select the **Up to face** option to extrude cut the feature up to the selected face. Select the **Two dimensions** option to extrude cut the feature in opposite direction by specifying second length of the extrusion.

Figure-91. Pocket parameters dialog with preview of pocket feature

- Specify desired length of pocket in the **Length** edit box.
- Select desired direction of the extrusion from **Direction/edge** drop-down. Select **Sketch normal** option to extrude cut the sketch or face along its normal. Select the **Select reference** option to extrude cut the sketch along an edge of the 3D model. Select **Custom direction** option to set the pocket length along the custom direction by specifying desired values in **x**, **y**, and **z** edit boxes.

- Select the **Show direction** check box to display **x, y,** and **z** edit boxes.
- Select **Length along sketch normal** check box to measure the pocket length along the sketch normal.
- Select **Symmetric to plane** check box to extrude cut the feature in both directions with respect to sketch plane.
- Select **Reversed** check box to reverse the direction of extrusion.
- Specify desired value in the **Taper angle** edit box to taper the pocket in the extrusion direction by the given angle. This option is available only if the **Dimension** or **Two dimensions** option is selected in the **Length** drop-down.
- Click on **OK** button from the dialog. The pocket feature will be created; refer to Figure-92.

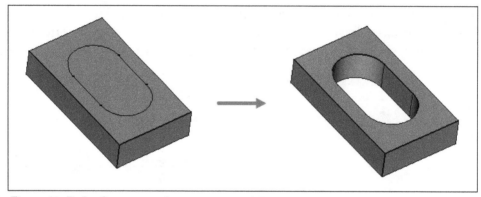

Figure-92. Pocket feature created

Hole

The **Hole** tool creates one or more holes from selected sketch. The sketch must contain one or multiple circles. The procedure to use this tool is discussed next.

- Click on the **Hole** tool from **Toolbar** in the **Part Design** workbench; refer to Figure-93. The **Hole parameters** dialog will be displayed in the **Tasks** panel of **Combo View** along with the preview of hole feature; refer to Figure-94.

Figure-93. Hole tool

- Select desired option from **Profile** drop-down in the **Threading and size** area of the dialog to specify the type of thread.
- Select the **Threaded** check box to add threading data to the hole feature and the hole minor diameter is used.
- Select the **Model Thread** check box to create a modelled thread on the component.
- Select the **Custom Thread** check box to customize the thread clearance by specifying desired value in the **Clearance** edit box.
- The radio buttons in **Direction** area will be activate when the **Threaded** check box will be selected. Select desired radio button to specify the thread direction.

Figure-94. Hole parameters dialog with preview of hole feature

- Select desired size of thread from **Size** drop down in the **Threading and size** area of the dialog.
- Select desired option from **Clearance** drop-down to define standard, close, or wide fit for threaded profiles.
- Specify desired tolerance class for hole in the **Class** drop-down.
- Specify desired hole diameter in the **Diameter** edit box.
- Select **Dimension** option from **Depth** drop-down to specify the depth value in the edit box and select **Through All** option from drop-down to cut the hole through the whole body.
- Select desired option from **Thread Depth** drop-down and specify desired depth for the thread in the respective edit box.
- Specify desired type of hole from **Type** drop-down in the **Hole cut** area of the dialog.
- Select **Custom values** check box to override the values predefined by the type of hole cut.
- Specify the major diameter for all hole cut types in the **Diameter** edit box from **Hole cut** area of the dialog.
- Specify the depth of hole cut type from the sketch plane in the **Depth** edit box.
- Specify the angle of conical hole cut in the **Countersink angle** edit box.
- Select **Flat** radio button in the **Type** option from **Drill point** area to create the flat bottom of the hole and select **Angled** radio button to create the conical point of the hole.
- Select **Take into account for depth** check box to subtract the conical height from the dimension.
- Select the **Tapered** check box from **Misc** area of the dialog to specify taper angle of the hole in the edit box.

- Select the **Reversed** check box to reverse the direction of hole extrusion.
- Click on **OK** button from the dialog. The hole feature will be created; refer to Figure-95.

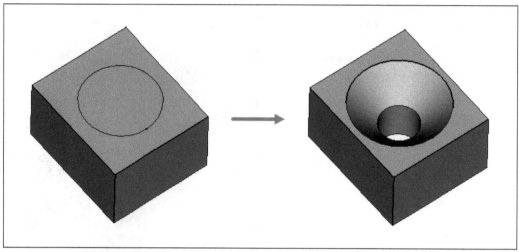

Figure-95. Hole feature created

Groove

The **Groove** tool revolves selected sketch or profile about a given axis, cutting out material from the solid model. The procedure to use this tool is discussed next.

- Select the closed loop sketch to be revolved and click on the **Groove** tool from **Toolbar** in the **Part Design** workbench; refer to Figure-96. The **Revolution parameters** dialog will be displayed in the **Tasks** panel of **Combo View**; refer to Figure-97.

Figure-96. Groove tool

Figure-97. Revolution parameters dialog

- Select the sketch of axis from **Model** panel in the **Combo View** and press **Space** key to toggle the visibility of axis of revolution; refer to Figure-98.

Figure-98. Toggle the visibility of sketch axis

- Select the **Select Reference** option from **Axis** drop-down in the dialog and select the sketch of axis in the 3D view area. The preview of groove feature will be displayed; refer to Figure-99.

Figure-99. Preview of groove feature

- Specify the angle of revolution in the **Angle** edit box through which the groove is to be formed.
- Specify the other parameters in the dialog as discussed earlier.
- Click on **OK** button from the dialog. The groove feature will be created; refer to Figure-100.

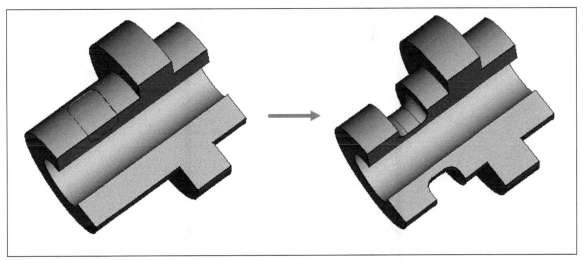

Figure-100. Groove feature created

Subtractive Loft

The **Subtractive Loft** tool removes material from the active body by making a transition between two or more sketches. The procedure to use this tool is discussed next.

- Click on the **Subtractive Loft** tool from **Toolbar** in the **Part Design** workbench; refer to Figure-101. If no object is selected before selecting the tool then the **Select feature** dialog will be displayed in the **Tasks** panel of **Combo View** as discussed earlier asking you to select the base profile.

Figure-101. Subtractive loft tool

- Select desired sketch from the dialog as base profile for subtractive loft feature; refer to Figure-103 and click on **OK** button. The **Loft parameters** dialog will be displayed in the **Tasks** panel of **Combo View**; refer to Figure-103.

Figure-102. Selecting base profile for subtractive loft feature

Figure-103. Loft parameters dialog

- The base profile is preselected in the **Object** selection box of **Profile** area. Click on **Add Section** button and select the first, second, and third sketches for loft section.
- Select the **Ruled surface** check box to make straight transitions between cross-sections.
- Select the **Closed** check box to make a transition from the last cross-section to the first by creating a loop.
- Click on **OK** button from the dialog. The subtractive loft feature will be created; refer to Figure-104

Figure-104. Subtractive loft feature created

Subtractive Pipe

The **Subtractive Pipe** tool creates a subtractive solid in the active body by sweeping one or more sketches along an open or closed path and subtracts it from the active body. The procedure to use this tool is discussed next.

- Click on the **Subtractive Pipe** tool from **Toolbar** in the **Part Design** workbench; refer to Figure-105. The **Pipe parameters**, **Section orientation**, **and Section transformation** dialogs will be displayed in the **Tasks** panel of **Combo View**; refer to Figure-106.

Figure-105. Subtractive pipe tool

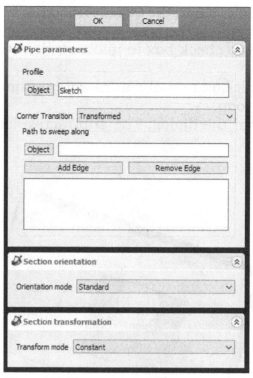

Figure-106. Pipe parameters dialog

- The first cross-section is preselected in the **Object** selection box in the **Profile** area of dialog.
- Click on the **Object** button from **Path to sweep along** area and select an edge of the sketch to be used as path for sweep feature in the 3D view area. The preview of subtractive pipe feature will be displayed; refer to Figure-107.

Figure-107. Preview of subtractive pipe feature

- Select desired transition mode from **Corner Transition** drop-down in the dialog.
- Select desired orientation mode from **Orientation mode** drop-down in the **Section orientation** dialog.
- Select desired transformation mode from **Transform mode** drop-down in the **Section transformation** dialog.
- Specify the other parameters in the **Pipe parameters** dialog as desired and click on **OK** button. The subtractive pipe feature will be created; refer to Figure-108.

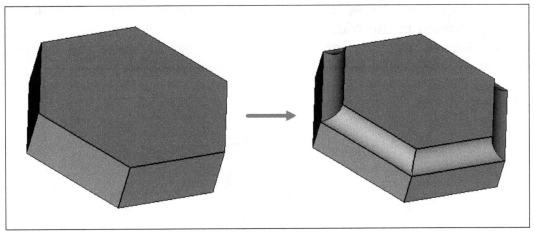

Figure-108. Subtractive pipe feature created

Subtractive Helix

The **Subtractive Helix** tool creates a solid shape by sweeping a sketch along a helix and subtracts it from the active body. The procedure to use this tool is discussed next.

• Create and select the sketch to be swept into a helix; refer to Figure-109.
• Click on the **Subtractive Helix** tool from **Toolbar** in the **Part Design** workbench; refer to Figure-110. The **Helix parameters** dialog will be displayed in the **Tasks** panel of **Combo View** along with the preview of subtractive helix feature; refer to Figure-111.

Figure-110. Subtractive Helix tool

Figure-109. Sketch selected to be swept

Figure-111. Helix parameters dialog along with preview of subtractive helix feature

- All the parameters in the **Helix parameters** dialog of **Subtractive Helix** tool is same as discussed in the **Additive Helix** tool.
- Select the **Remove outside of profile** check box to create the feature by the intersection of the swept profile and the pre-existing body.
- Click on the **OK** button from the dialog. The subtractive helix feature will be created; refer to Figure-112.

Figure-112. Subtractive helix feature created

Subtractive Box

The **Subtractive Box** tool inserts a subtractive box in the active body. Its shape is subtracted from the existing solid. The procedure to use this tool is discussed next. Note that an existing solid body will be required for creating a subtractive box feature.

- Click on the **Subtractive Box** tool from **Create a subtractive primitive** dropdown in the **Toolbar** of **Part Design** workbench; refer to Figure-113. The **Primitive parameters** dialog will be displayed in the **Tasks** panel of **Combo View** along with the preview of subtractive box feature; refer to Figure-114.

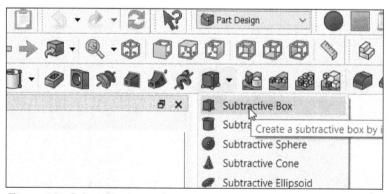

Figure-113. Subtractive box tool

- Specify the length, width, and height of the feature in the **Length**, **Width**, and **Height** edit boxes, respectively.
- Specify the other parameters in the dialog as discussed earlier.
- Click on **OK** button from the dialog. The subtractive box feature will be created; refer to Figure-115.

Figure-114. Primitive parameters dialog with preview of subtractive box feature

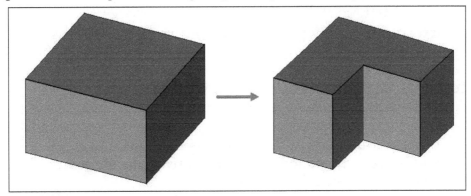

Figure-115. Subtractive box feature created

Subtractive Cylinder

The **Subtractive Cylinder** tool inserts a subtractive cylinder in the active body. Its shape is subtracted from the existing solid. The procedure to use this tool is discussed next.

- Click on the **Subtractive Cylinder** tool from **Create a subtractive primitive** drop-down in the **Toolbar** of **Part Design** workbench; refer to Figure-116. The **Primitive parameters** dialog will be displayed in the **Tasks** panel of **Combo View** along with the preview of subtractive cylinder feature; refer to Figure-117.

Figure-116. Subtractive cylinder tool

Figure-117. Primitive parameters dialog with preview of subtractive cylinder feature

- Specify desired radius and height of the feature in the **Radius** and **Height** edit boxes, respectively.
- Specify desired angle of first and second direction in the **Angle in first direction** and **Angle in second direction** edit boxes, respectively.
- Specify desired value of angle of revolution of the cross-section in the **Angle** edit box.
- Specify the other parameters in the dialog as discussed earlier.
- Click on **OK** button from the dialog. The subtractive cylinder feature will be created; refer to Figure-118.

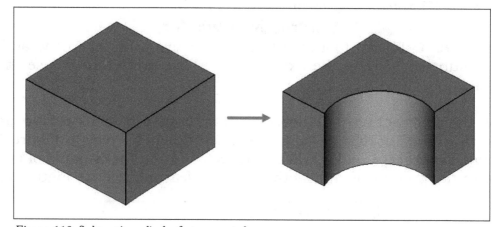

Figure-118. Subtractive cylinder feature created

Subtractive Sphere

The **Subtractive Sphere** tool inserts a subtractive sphere in the active body. Its shape is subtracted from the existing solid. The procedure to use this tool is discussed next.

- Click on the **Subtractive Sphere** tool from **Create a subtractive primitive** drop-down in the **Toolbar** of **Part Design** workbench; refer to Figure-119. The **Primitive parameters** dialog will be displayed in the **Tasks** panel of **Combo View** along with the preview of subtractive sphere feature; refer to Figure-120.

Figure-119. Subtractive sphere tool

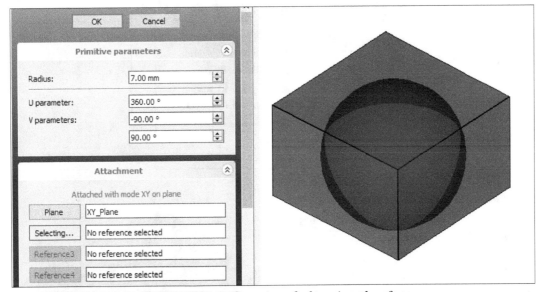

Figure-120. Primitive parameters dialog with preview of subtractive sphere feature

- Specify desired radius of the sphere in the **Radius** edit box.
- Specify desired angle of revolution of the cross section in the **U parameter** edit box.
- Specify lower truncation of the sphere parallel to the circular cross-section in the **V parameters** edit box and specify upper truncation of the ellipsoid parallel to the circular cross-section in the edit box just below to the **V parameters** edit box of the dialog.
- Specify other parameters in the dialog as discussed earlier.
- Click on **OK** button from the dialog. The subtractive sphere feature will be created; refer to Figure-121.

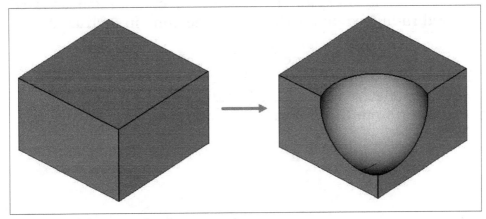

Figure-121. Subtractive sphere feature created

Subtractive Cone

The **Subtractive Cone** tool inserts a subtractive cone in the active body. Its shape is subtracted from the existing solid. The procedure to use this tool is discussed next.

- Click on the **Subtractive Cone** tool from **Create a subtractive primitive** drop-down in the **Toolbar** of **Part Design** workbench; refer to Figure-122. The **Primitive parameters** dialog will be displayed in the **Tasks** panel of **Combo View** along with the preview of subtractive cone feature; refer to Figure-123.

Figure-122. Subtractive cone tool

Figure-123. Primitive parameters dialog with preview of subtractive cone feature

- Specify desired radius value for the base of the cone in **Radius 1** edit box and specify desired radius value for the top of the cone in **Radius 2** edit box of the dialog.
- Specify desired height of the cone in the **Height** edit box.
- Specify desired angle of revolution of the cross section in the **Angle** edit box.
- Specify other parameters in the dialog as discussed earlier.
- Click on **OK** button from the dialog. The subtractive cone feature will be created; refer to Figure-124.

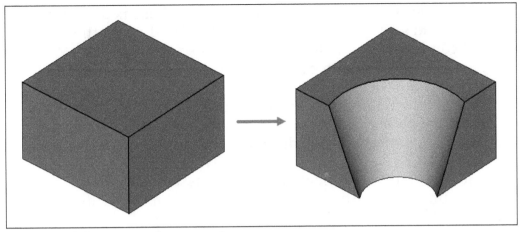

Figure-124. Subtractive cone feature created

Subtractive Ellipsoid

The **Subtractive Ellipsoid** tool inserts a subtractive ellipsoid in the active body. Its shape is subtracted from the existing solid. The procedure to use this tool is discussed next.

- Click on the **Subtractive Ellipsoid** tool from **Create a subtractive primitive** drop-down in the **Toolbar** of **Part Design** workbench; refer to Figure-125. The **Primitive parameters** dialog will be displayed in the **Tasks** panel of **Combo View** along with the preview of subtractive ellipsoid feature; refer to Figure-126.

Figure-125. Subtractive ellipsoid tool

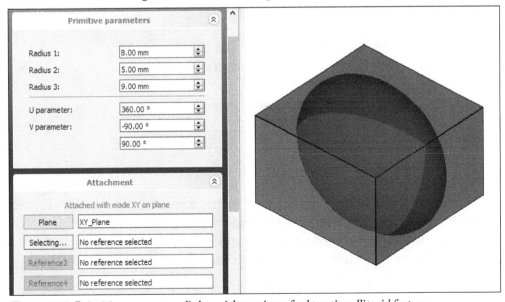

Figure-126. Primitive parameters dialog with preview of subtractive ellipsoid feature

- Specify desired radius value along the ellipsoid's vertical axis in **Radius 1** edit box, along the ellipsoid's length in **Radius 2** edit box, and along the ellipsoid's width in **Radius 3** edit box of the dialog.
- Specify the angle of revolution of the elliptical cross section in the **U parameter** edit box.
- Specify lower truncation of the ellipsoid parallel to the circular cross-section in the **V parameter** edit box and specify upper truncation of the ellipsoid parallel to the circular cross section in the edit box just below to the **V parameter** edit box of the dialog.
- Specify other parameters in the dialog as discussed earlier.
- Click on **OK** button from the dialog. The subtractive ellipsoid feature will be created; refer to Figure-127.

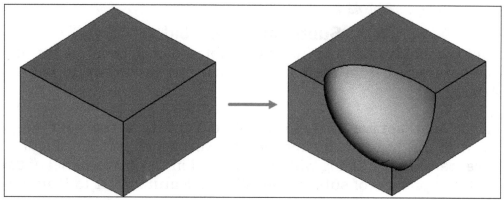

Figure-127. Subtractive ellipsoid feature created

Subtractive Torus

The **Subtractive Torus** tool inserts a subtractive torus in the active body. Its shape is subtracted from the existing solid. The procedure to use this tool is discussed next.

- Click on the **Subtractive Torus** tool from **Create a subtractive primitive** drop-down in the **Toolbar** of **Part Design** workbench; refer to Figure-128. The **Primitive parameters** dialog will be displayed in the **Tasks** panel of **Combo View** along with the preview of subtractive torus feature; refer to Figure-129.

Figure-128. Subtractive torus tool

- Specify desired radius of the imaginary orbit around which the circular cross-section revolves in **Radius 1** edit box and specify the radius of circular cross-section defining the form of the torus in the **Radius 2** edit box of the dialog.
- Specify the angle of revolution of the circular cross section in the **U parameter** edit box.

- Specify lower truncation of the torus parallel to the circular cross-section in the **V parameter** edit box and specify upper truncation of the ellipsoid parallel to the circular cross section in the edit box just below to the **V parameter** edit box of the dialog.

Figure-129. Primitive parameters dialog with preview of subtractive torus feature

- Specify other parameters in the dialog as discussed earlier.
- Click on **OK** button from the dialog. The subtractive torus feature will be created; refer to Figure-130.

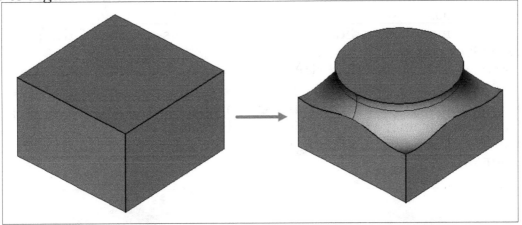

Figure-130. Subtractive torus feature created

Subtractive Prism

The **Subtractive Prism** tool inserts a subtractive prism in the active body. Its shape is subtracted from the existing solid. The procedure to use this tool is discussed next.

- Click on the **Subtractive Prism** tool from **Create a subtractive primitive** drop-down in the **Toolbar** of **Part Design** workbench; refer to Figure-131. The **Primitive parameters** dialog will be displayed in the **Tasks** panel of **Combo View** along with the preview of subtractive prism feature; refer to Figure-132.

Figure-131. Subtractive prism tool

Figure-132. Primitive parameters dialog with preview of subtractive prism feature

- Specify desired number of sides for polygon cross-section of the prism in the **Polygon** edit box.
- Specify circumscribed radius of the polygon cross-section of the prism in the **Circumradius** edit box.
- Specify desired height of the prism in the **Height** edit box of the dialog.
- Specify desired angle value for the first and second directions in the **Angle in first direction** and **Angle in second direction** edit boxes of the dialog, respectively.
- Specify other parameters in the dialog as discussed earlier.
- Click on **OK** button from the dialog. The subtractive prism feature will be created; refer to Figure-133.

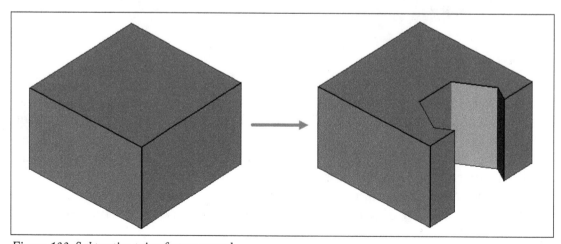

Figure-133. Subtractive prism feature created

Subtractive Wedge

The **Subtractive Wedge** tool inserts a subtractive wedge in the active body. Its shape is subtracted from the existing solid. The procedure to use this tool is discussed next.

- Click on the **Subtractive Wedge** tool from **Create a subtractive primitive** drop-down in the **Toolbar** of **Part Design** workbench; refer to Figure-134. The **Primitive parameters** dialog will be displayed in the **Tasks** panel of **Combo View** along with the preview of subtractive wedge feature; refer to Figure-135.
- Specify the minimum/maximum value for base face of the feature along X axis in the **X min/max** edit boxes, respectively. In other words, you can specify the length of base feature in both positive and negative directions along the X axis.
- Specify desired height of the wedge in the **Y min/max** edit boxes, respectively.
- Specify the minimum/maximum value for base face of the feature along Z axis in the **Z min/max** edit boxes, respectively.
- Specify the minimum/maximum value for top face of the feature along X axis in the **X2 min/max** edit boxes, respectively.
- Specify the minimum/maximum value for top face of the feature along Z axis in the **Z2 min/max** edit boxes, respectively.

Figure-134. Subtractive wedge tool

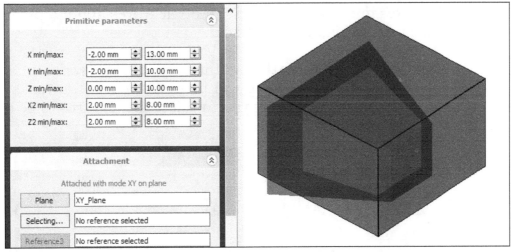

Figure-135. Primitive parameters dialog with preview of subtractive wedge feature

- Specify other parameters in the dialog as discussed earlier.
- Click on **OK** button from the dialog. The subtractive wedge feature will be created; refer to Figure-136.

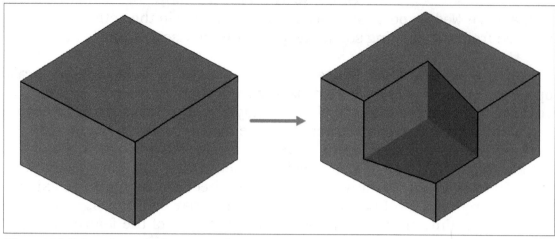

Figure-136. Subtractive wedge feature created

Transformation Tools

The **Transformation** tools are used for transforming existing features using mirror and pattern operations. They will allow you to choose which features to transform. The **Transformation** tools are available in the **Part Design Helper Toolbar** of **Part Design** workbench; refer to Figure-137. The procedures to use these tools are discussed next.

Figure-137. Transformation tools in part design workbench

Mirrored

The **Mirrored** tool creates mirror copy of one or more features about a plane or face. The procedure to use this tool is discussed next.

- Click on the **Mirrored** tool from **Toolbar** in the **Part Design** workbench; refer to Figure-138. The **Select feature** dialog will be displayed in the **Tasks** panel of **Combo View**; refer to Figure-139, asking you to select the features to be mirrored.

Figure-138. Mirrored tool

Figure-139. Select feature dialog

- Select the feature from the dialog which you want to mirror and click on **OK** button. The **Mirrored parameters** dialog will be displayed in the **Tasks** panel of **Combo View** along with the preview of mirrored feature; refer to Figure-140.

Figure-140. Mirrored parameters dialog with preview of mirrored feature

- To add more features to be mirrored, click on the **Add feature** button from the dialog and select the feature in the 3D view.
- To remove the feature to be mirrored, select the feature from the list and click on **Remove feature** button.
- Select desired plane or face to be used as mirror axis from **Plane** drop-down in the dialog.
- Click on **OK** button from the dialog. The mirrored feature will be created; refer to Figure-141.

Figure-141. Mirrored feature created

Linear Pattern

The **Linear Pattern** tool creates evenly spaced copies of a feature in linear direction. The procedure to use this tool is discussed next.

- Click on the **Linear Pattern** tool from **Toolbar** in the **Part Design** workbench; refer to Figure-142. The **Select feature** dialog will be displayed in the **Tasks** panel of **Combo View** as discussed earlier asking you to select the feature to be linear patterned.

Figure-142. Linear pattern tool

- Select the feature from the dialog to be linear patterned and click on **OK** button. The **LinearPattern parameters** dialog will be displayed in the **Tasks** panel of **Combo View** along with the preview of linear pattern of the feature; refer to Figure-143.

Figure-143. Linear Pattern parameters dialog with preview of linear patterned feature

- To add more features to be patterned, click on **Add feature** button from the dialog and select the feature from **Model Tree** or from the 3D view.
- To remove a feature from the list of features being patterned, select the feature from the list and click on **Remove feature** button.
- Specify desired direction of pattern from the **Direction** drop-down.
- Select the **Reverse direction** check box to reverse the direction of linear pattern.
- Specify the distance between patterned feature and the original feature in the **Length** edit box.
- Specify desired number of occurrences of linear pattern in the **Occurrences** edit box.
- Click on **OK** button from the dialog. The linear pattern of the feature will be created; refer to Figure-144.

Figure-144. Linear pattern of the feature created

Polar Pattern

The **Polar Pattern** tool creates a set of copies of selected features rotated around a chosen axis. The procedure to use this tool is discussed next.

- Click on the **Polar Pattern** tool from **Toolbar** in the **Part Design** workbench; refer to Figure-145. The **Select feature** dialog will be displayed in the **Tasks** panel of **Combo View** as discussed earlier asking you to select the feature.

Figure-145. Polar pattern tool

- Select the feature to be polar patterned from the dialog and click on **OK** button. The **PolarPattern parameters** dialog will be displayed in the **Tasks** panel of **Combo View** along with the preview of polar pattern of the feature; refer to Figure-146.

Figure-146. Polar Pattern parameters dialog with preview of polar patterned feature

- Specify desired axis of rotation for polar pattern from **Axis** drop-down in the dialog.
- Select the **Reverse direction** check box to reverse the direction of polar pattern.
- Specify the angle between patterned feature and the original feature in the **Angle** edit box.
- Specify desired number of occurrences of polar pattern in the **Occurrences** edit box.
- Click on **OK** button from the dialog. The polar pattern of the feature will be created; refer to Figure-147.

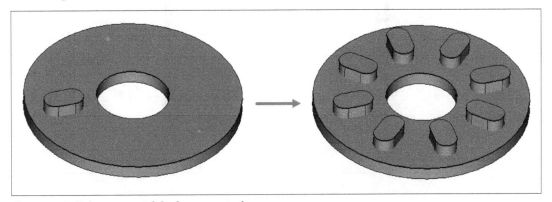

Figure-147. Polar pattern of the feature created

Create Multi Transform

The **Multi Transform** tool takes one or a set of part features as its input and allows the user to apply multiple transformations to that feature or a set of features progressively in sequence - creating a combined or compound transformation. The procedure to use this tool is discussed next.

- Click on the **Multi transform** tool from **Toolbar** in the **Part Design** workbench; refer to Figure-148. The **Select feature** dialog will be displayed in the **Tasks** panel of **Combo View** as discussed earlier asking you to select the feature.
- Select the feature you want to be multi transformed and click on **OK** button from the dialog. The **MultiTransform parameters** dialog will be displayed in the **Tasks** panel of **Combo View**; refer to Figure-149.

Figure-148. Multi transform tool

Figure-149. MultiTransform parameters dialog

- If you want to include additional features for the transformations, click on the **Add feature** button from the dialog and select the feature you want to add from the **Model Tree** or from the 3D view.
- Click **RMB** in the **Transformations** list view in the dialog. A shortcut menu will be displayed; refer to Figure-150.

Figure-150. Shortcut menu

- Select desired transformation option from the shortcut menu. The parameters related to the selected transformation will be displayed in the dialog along with the preview of linear pattern of the feature; refer to Figure-151.

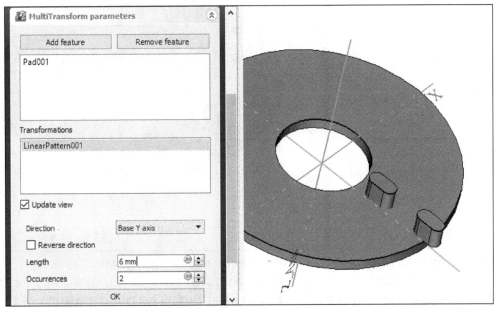

Figure-151. Linear pattern type parameters displayed along with the preview of linear pattern of the feature

- Specify all the parameters related to transformation as discussed earlier.
- Click on **OK** button below **Transformations** list view area. The linear pattern of the feature will be created; refer to Figure-152.
- If you want to add more transformation to the feature then click **RMB** in the **Transformations** list view in the dialog and select desired pattern type from the shortcut menu displayed.
- If you have selected **Add polar pattern** option then the parameters related to selected polar pattern type will be displayed in the dialog along with the preview of polar pattern of the feature; refer to Figure-153.

Figure-152. Linear pattern of the feature created

Figure-153. Polar pattern type parameters displayed along with the preview of polar pattern of the feature

- Specify all the parameters related to polar pattern type as discussed earlier.
- Click on **OK** button below **Transformations** list box. The polar pattern of the feature will be created; refer to Figure-154.
- Click on **OK** button to close the dialog.

Figure-154. Polar pattern of the feature created

Dress-up tools

The **Dress-up** tools are used to apply treatment to the selected edges or faces. These tools are available in the **Part Design Helper Toolbar** of **Part Design** workbench; refer to Figure-155. The procedures to use these tools are discussed next.

Figure-155. Dress up tools in part design workbench

Fillet

The **Fillet** tool creates fillets (rounds) on the selected edges of an object. The procedure to use this tool is discussed next.

- Select an edge or multiple edges or face of an object on which you want to apply fillet; refer to Figure-156.

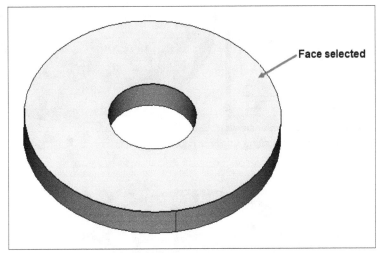

Figure-156. Face selected to create fillet

- Click on the **Fillet** tool from **Toolbar** in the **Part Design** workbench; refer to Figure-157. The **Fillet parameters** dialog will be displayed in the **Tasks** panel of **Combo View** along with the preview of fillets created on the object; refer to Figure-158.

Figure-157. Fillet tool

- Specify desired radius of the fillet in the **Radius** edit box.
- If you want to add more edges or faces to be fillet then click on the **Add** button from the dialog and select the edges or faces from the 3D view area. If you want to add all the edges of the model to apply fillet then press **RMB** in the list area of the dialog and select **Add all edges** option from the options displayed.
- If you want to remove the fillet then select the face or an edge from the list view area in the dialog and click on the **Remove** button.
- Select **Use All Edges** check box to select all edges of the feature to apply fillet.
- Click on **OK** button from the dialog. The fillets will be created; refer to Figure-159.

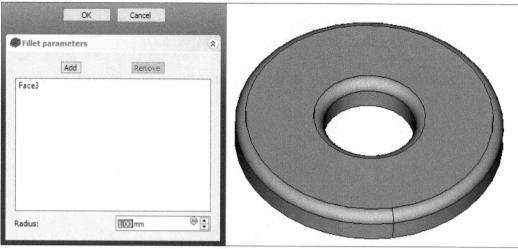

Figure-158. Fillet parameters dialog along with the preview of fillets

Figure-159. Fillets created on the object

Chamfer

The **Chamfer** tool creates chamfers on the selected edges of an object. The procedure to use this tool is discussed next.

- Select an edge or multiple edges or face of an object on which you want to apply chamfer as discussed for fillet tool.
- Click on the **Chamfer** tool from **Toolbar** in the **Part Design** workbench; refer to Figure-160. The **Chamfer parameters** dialog will be displayed in the **Tasks** panel of **Combo View** along with the preview of chamfers created on the object; refer to Figure-161.

Figure-160. Chamfer tool

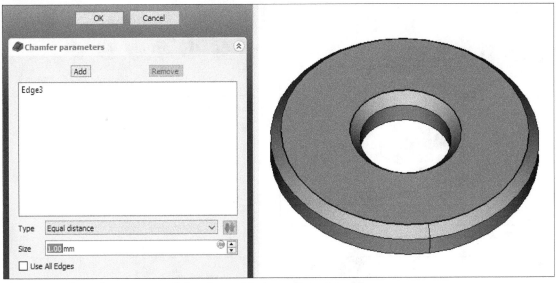

Figure-161. Chamfer parameters dialog along with the preview of chamfers

- Select desired chamfer type from **Type** drop-down in the dialog. Select the **Equal distance** option to equally distance the chamfer edges from the body edge. Select the **Two distances** option to specify the distances of the chamfer edge to the body edge. Select the **Distance and angle** option to specify one distance of the chamfer edge to the body edge. The second chamfer edge is defined by the angle of the chamfer.
- Specify desired size of the chamfer in the **Size** edit box.
- The other parameters of the dialog is same as discussed for **Fillet** tool.
- Click on **OK** button from the dialog. The chamfers will be created; refer to Figure-162.

Figure-162. Chamfers created on the object

Draft

The **Draft** tool creates angular draft on the selected faces of an object. The procedure to use this tool is discussed next.

- Select one or multiple faces of an object on which you want to apply an angular draft; refer to Figure-163.
- Click on the **Draft** tool from **Toolbar** in the **Part Design** workbench; refer to Figure-164. The **Draft parameters** dialog will be displayed in the **Tasks** panel of **Combo View** along with the preview of draft applied; refer to Figure-165.

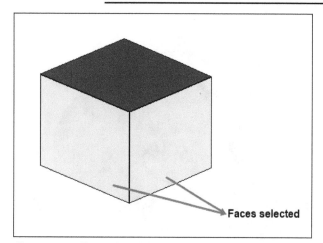

Figure-163. Faces selected for angular draft

Figure-164. Draft tool

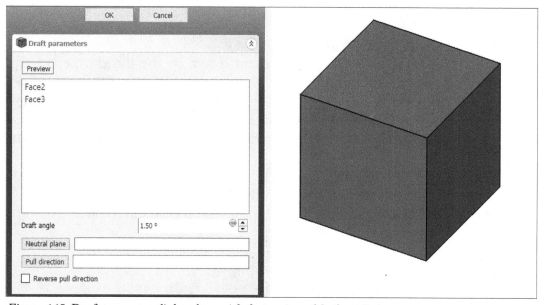

Figure-165. Draft parameters dialog along with the preview of draft applied

- If you want to add more faces to apply the draft then click on the **Add face** button from the dialog and select the faces of an object in the 3D view area.
- If you want to remove the faces then select the faces from the list view area in the dialog and press **DELETE** key from keyboard.
- Specify desired angle of draft in the **Draft angle** edit box.
- Click on the **Neutral plane** button and select the face to be used as reference for defining draft angle.
- Click on **Pull direction** button and select desired edge of an object to define direction in which draft will be applied.
- Select the **Reverse pull direction** check box to reverse the direction of draft pulling.
- Click on **OK** button from the dialog. The draft will be created on the object; refer to Figure-166.

Figure-166. Draft created on the object

Thickness

The **Thickness** tool works on a solid body and transforms it into a thick-walled hollow object with at least one open face. The procedure to use this tool is discussed next.

- Select the face or multiple faces of an object to be hollowed; refer to Figure-167.

Figure-167. Faces selected to be hollowed

- Click on the **Thickness** tool from **Toolbar** in the **Part Design** workbench; refer to Figure-168. The **Thickness parameters** dialog will be displayed in the **Tasks** panel of **Combo View** along with the preview of hollowed faces; refer to Figure-169.

Figure-168. Thickness tool

Figure-169. Thickness parameters dialog along with the preview of opened faces

- If you want to add more faces to be hollowed then click on the **Add face** button from the dialog and select the faces of the object from the 3D view area.
- If you want to remove the faces then select the faces from the list of the dialog and right click on the mouse and click on the remove button.
- Specify desired wall thickness value in the **Thickness** edit box.
- Specify desired mode of thickness from the **Mode** drop-down. Select the **Skin** option if you want to get an item like a vase, headless but with the bottom. Select the **Pipe** option if you want to get an object like a pipe, headless, and bottomless.
- Select desired option from the **Join Type** drop-down. Select the **Arc** option if you want to remove the outer edges and create a fillet with radius equal to the defined thickness. Select the **Intersection** option when faces are offset outward, sharp edges are kept between faces.
- Select the **Make thickness inwards** check box if you want to offset the faces inwards.
- Click on **OK** button from the dialog. The thick-walled hollow object with opened faces will be created; refer to Figure-170.

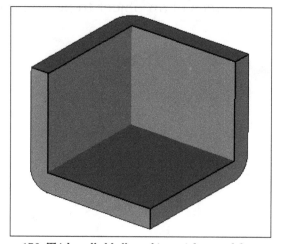

Figure-170. Thick walled hollow object with opened faces created

Boolean operation

The **Boolean** tool applies a boolean operation (fuse, cut, or common) to the Part Design bodies. The procedure to use this tool is discussed next.

- Activate the body on which you want to apply the boolean feature. To activate the body, right-click on desired body, and select the **Toggle active body** option from the shortcut menu. The body will be activated.
- Click on the **Boolean** tool from **Toolbar** in the **Part Design** workbench; refer to Figure-171. The **Boolean** parameters dialog will be displayed in the **Tasks** panel of **Combo View**; refer to Figure-172.

Figure-171. Boolean tool

Figure-172. Boolean parameters dialog

- Click on the **Add body** button from the dialog, the active body temporarily disappears from the 3D view area; refer to Figure-173.

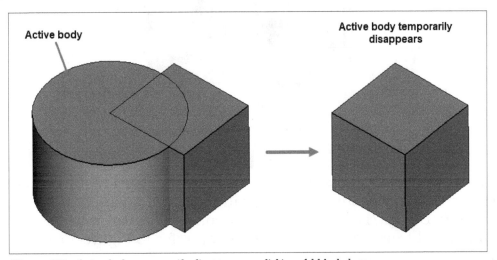

Figure-173. Active body temporarily disappears on clicking Add body button

- Select the body or multiple bodies which you want to use for creating the boolean feature from the 3D view area.
- Select desired type of boolean operation from the drop-down in the dialog. The preview of cut boolean feature will be displayed; refer to Figure-174.

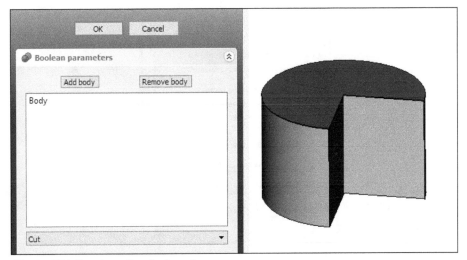

Figure-174. Preview of cut boolean feature

- Select the **Fuse** option from drop-down to merge the tool body or bodies to the active body. Select the **Cut** option to subtract the tool body or bodies from the active body. Select the **Common** option to extract the intersection from the selected body or bodies with the active body.
- If you want to remove the body to be boolean then select the body from the 3D view area and click on **Remove body** button from the dialog.
- Click on **OK** button from the dialog. The cut boolean feature will be created; refer to Figure-175.

Figure-175. Cut boolean feature created

Migrate Tool

The **Migrate** tool is used to migrate files created with older FreeCAD versions. If the file is pure PartDesign feature-based, migration should succeed. If the file contains mixed Part/Part Design/Draft objects then conversion will most likely fail. Click on the **Migrate** tool from the **Part Design** menu; refer to Figure-176 to perform the operation.

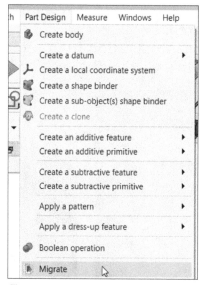

Figure-176. Migrate tool

Creating Sprocket

The **Sprocket** tool allows you to create a 2D profile of a sprocket. It can be padded with the **Part Design Pad** feature. The procedure to use this tool is discussed next.

- Click on the **Sprocket** tool from the **Part Design** menu; refer to Figure-177. The profile of sprocket will be displayed with **Sprocket parameter** dialog in the **Tasks** panel of **Combo View**; refer to Figure-178.

Figure-177. Sprocket tool

Figure-178. Sprocket parameter dialog with sprocket profile

- Specify desired number of teeth for the sprocket in the **Number of teeth** edit box.
- Select desired sprocket type from **Sprocket Reference** drop-down.
- Specify desired distance between the two teeth in the **Chain Pitch** edit box of the dialog.
- Specify desired diameter of the chain rollers in the **Roller Diameter** edit box.
- Specify desired thickness of sprocket in the **Thickness** edit box.
- Click on the **OK** button from the dialog. The sprocket will be created.
- Now, if you want to pad the sprocket feature then select the sprocket created from the **Model Tree**; refer to Figure-179, and click on the **Pad** tool from the **Toolbar** as discussed earlier in this chapter. The pad feature of sprocket profile will be created; refer to Figure-180.

Figure-179. Sprocket selected

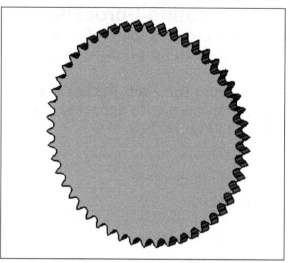

Figure-180. Pad feature of sprocket profile created

Creating Involute Gear

The **Involute gear** tool is used to generate profile for creating involute gear. You can later use this profile for **Pad** tool to create solid gear feature. The procedure to use this tool is given next.

- Click on the **Involute gear** tool from the **Part Design** menu; refer to Figure-181. The profile of involute gear will be displayed with **Involute parameter** dialog in the **Tasks** panel of **Combo View**; refer to Figure-182.

Figure-181. Involute gear tool

*Figure-182. Involute parameter
dialog with involute gear profile*

- Specify the number of teethes to be created in profile in the **Number of teeth** edit box of dialog.

- Specify the other parameters like module of gear, pressure angle, and so on in respective edit boxes of the dialog and then click on the **OK** button to create the profile.

Select the profile created and click on the **Pad** tool from **Toolbar** to create solid gear.

Shaft Design Wizard

The **Shaft design wizard** tool is used to design shaft based on specified parameters. The procedure to use this tool is given next.

- Click on the **Shaft design wizard** tool from **Part Design** menu; refer to Figure-183. A dialog will be displayed in the **Tasks** panel of **Combo View** to specify parameters for creating shaft; refer to Figure-184.

Figure-183. Shaft design wizard tool

Figure-184. Dialog for shaft design

- Specify length, diameter, inner diameter, and constraint type for shaft sections in respective fields of dialog.
- If you want to add more sections to the shaft then right-click in the dialog and select **Add column** option; refer to Figure-185.

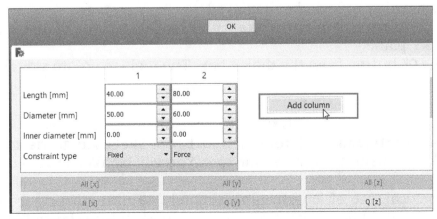

Figure-185. Add column option

• You can check various parameters of shaft using the buttons at the bottom in the dialog.

PRACTICAL

Create the sketch as shown in Figure-186.

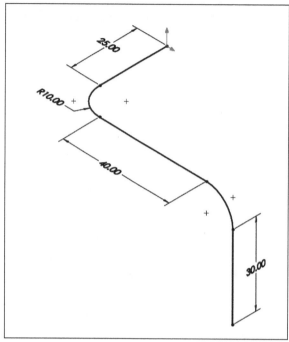

Figure-186. Practical 1

Steps to be performed:
• Start a new part file.
• Select a sketching plane and activate sketching mode.
• Create the sketch using **Line** tool.
• Apply the dimensions using dimension tools.
• Save the file.

Starting a New Sketch

• Start FreeCAD if not started yet.
• Select the **Sketcher** option from the **Switch between workbenches** drop-down in the **Toolbar**. The sketcher environment will be displayed.
• Click on the **Create sketch** tool from the **Toolbar** or **Sketch** menu. The **Choose orientation** dialog box will be displayed.
• Select the **XY-Plane** tool from the **Sketch orientation** area of dialog box to create sketch on XY plane.
• Click on the **OK** button from the dialog box. The tools to create sketch will become active.

Creating Sketch

• Click on the **Create line** tool from the **Toolbar** or **Sketcher geometries** cascading menu of **Sketch** menu to activate line tool. You will be asked to specify start point of line.
• Create a vertical line of length approximately **40** mm starting from origin and create a horizontal line of approximately **25** mm length; refer to Figure-187.

Figure-187. Creating lines

- Click on the **Sketch fillet** tool from the **Toolbar** and apply the fillet of radius **10** mm between the lines; refer to Figure-188.

Figure-188. Fillet applied

- Apply the dimensions as per the drawing. Click on the **Close** button from the **Tasks** panel or press **ESC** key to exit the sketching mode.

Creating Second Sketch

- Click on the **Create a new datum plane** tool from the **Part Design** workbench and create a datum plane on the **XZ** plane as shown in Figure-189.

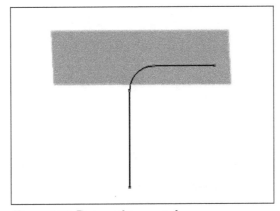

Figure-189. Datum plane created

- Select the datum plane recently created and click on the **Create sketch** tool from the **Toolbar** and create the sketch as shown in Figure-190.

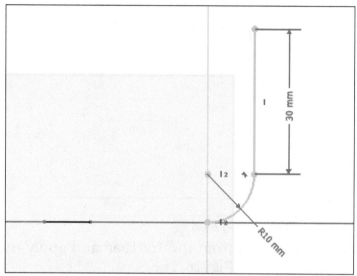

Figure-190. Creating second sketch

- Click on the **Close** button from the **Tasks** panel or press **Esc** key to exit the sketching mode. The 3D sketch will be created.

Chapter 4

Solid Modeling
with Part Workbench

Topics Covered

The major topics covered in this chapter are:

- *Introduction to Solid Modeling*
- *Starting Part Workbench*
- *Primitive Tools*
- *Modifying Object Tools*
- *Offset Tools*
- *Compound Tools*
- *Join Features Tools*
- *Splitting Tools*
- *Measure Tools*

INTRODUCTION

The solid modeling capabilities of FreeCAD are based on the **OpenCASCADE Technology** (OCCT) kernel, a professional-grade CAD system that features advanced 3D geometry creation and manipulation. The Part Workbench is a layer sitting on top of the OCCT libraries that gives the user access to OCCT geometric primitives and functions. Essentially, all 2D and 3D drawing functions in every workbench (Draft, Sketcher, PartDesign, etc.) are based on these functions exposed by the Part Workbench. Therefore, the **Part Workbench** is considered the core component of the modeling capabilities of FreeCAD.

Part objects are more complex than mesh objects created with the **Mesh Workbench** as they permit more advanced operations like coherent boolean operations, modifications history, and parametric behavior.

The **Part Workbench** is the basic layer that exposes the OCCT drawing functions to all workbenches in FreeCAD; refer to Figure-1.

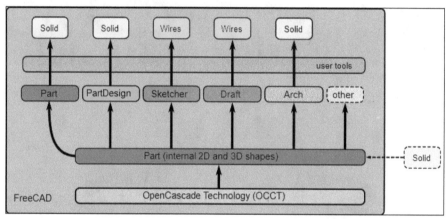

Figure–1. OpenCascade Technology

STARTING PART WORKBENCH

The objects created with the **Part Workbench** are relatively simple; they are intended to be used with boolean operations (unions and cuts) in order to build more complex shapes. This modeling paradigm is known as the constructive solid geometry (CSG) workflow and it was the traditional methodology used in early CAD systems.

- To start a new part file, click on the **New** button from **File** menu and select **Part** workbench from **Switch between workbenches** drop-down in the **Toolbar**; refer to Figure-2. The tools to create 3D models will be displayed in the **Toolbar**; refer to Figure-3.

Figure–2. Part workbench

Figure-3. Part workbench tools

Primitive tools

Primitives are the tools for creating primitive objects. These tools are available in the **Toolbar** of **Part** workbench; refer to Figure-4. The procedures to use these tools are discussed next.

Figure-4. Primitive tools

Cube

The **Cube** tool inserts a parametric, rectangular cuboid, using geometric primitive into the active document. The procedure to use this tool is discussed next.

• Click on the **Cube** tool from **Toolbar** in the **Part** workbench; refer to Figure-5. A rectangular cuboid primitive will be inserted in the 3D view area; refer to Figure-6.

Figure-5. Box tool

Figure-6. Rectangular cuboid primitive inserted

• If you want to edit the properties of inserted cuboid then select the cube created from the **Model Tree**; refer to Figure-7. The **Property editor** dialog will be displayed in the **Model** tab of **Combo View**; refer to Figure-8.

Figure-7. Selecting the object

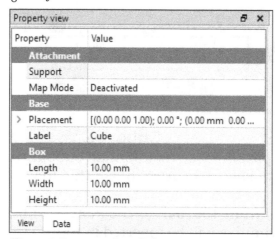

Figure-8. Property editor dialog

- Click on the **Placement** node from **Base** section of the dialog. The parameters related to placement will be displayed to specify the orientation and position of the cube in the 3D view area.
- Specify desired label for the cube in the **Label** edit box.
- Specify desired length, width, and height of the cube in the **Length**, **Width**, and **Height** edit boxes, respectively from **Box** section of the dialog. The properties of the rectangular cuboid will be edited.

Cylinder

The **Cylinder** tool creates a simple parametric cylinder with position, angle, radius, and height parameters. The procedure to use this tool is discussed next.

- Click on the **Cylinder** tool from **Toolbar** in the **Part** workbench; refer to Figure-9. A cylinder primitive will be inserted in the 3D view area; refer to Figure-10.

Figure-9. Cylinder tool

Figure-10. Cylinder primitive inserted

- The procedure to edit the properties of cylinder is same as discussed for **Box** tool.

Sphere

The **Sphere** tool creates a simple parametric sphere with position, angle1, angle2, angle3, and radius parameters. The procedure to use this tool is discussed next.

- Click on the **Sphere** tool from **Toolbar** in the **Part** workbench; refer to Figure-11. The sphere primitive will be inserted in the 3D view area; refer to Figure-12.

Figure-11. Sphere tool

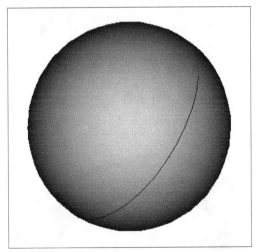

Figure-12. Sphere primitive inserted

- The procedure to edit the properties of sphere is same as discussed earlier.

Cone

The **Cone** tool creates a simple parametric truncated cone. The procedure to use this tool is discussed next.

- Click on the **Cone** tool from **Toolbar** in the **Part** workbench; refer to Figure-13. The cone primitive will be inserted in the 3D view area; refer to Figure-14.

Figure-13. Cone tool

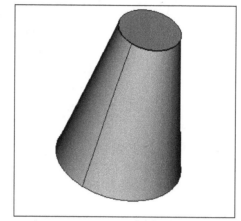

Figure-14. Cone primitive inserted

- The procedure to edit the properties of cone is same as discussed earlier.

Torus

The **Torus** tool creates a simple parametric torus with position, angle1, angle2, angle3, radius1, and radius2 as parameters. The procedure to use this tool is discussed next.

- Click on the **Torus** tool from **Toolbar** in the **Part workbench**; refer to Figure-15. The torus primitive will be inserted in the 3D view area; refer to Figure-16.

Figure-15. Torus tool

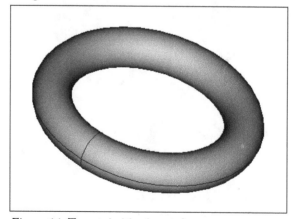

Figure-16. Torus primitive inserted

- The procedure to edit the properties of torus is same as discussed earlier.

Tube

The **Tube** tool inserts a tube into the active document. The tube is geometrically treated as a cut of a smaller cylinder into a larger one. The procedure to use this tool is discussed next.

- Click on the **Tube** tool from **Toolbar** in the **Part workbench**; refer to Figure-17. The **Tube** dialog will be displayed in the **Tasks** panel of **Combo View** along with the tube primitive inserted in the 3D view area; refer to Figure-18.

Figure-17. Tube tool

Figure-18. Tube dialog displayed along with the tube primitive

- Specify desired outer and inner radius of the tube in the **Outer radius** and **Inner radius** edit boxes of the dialog, respectively.
- Specify desired height of the tube in the **Height** edit box and click on the **OK** button from the dialog to create the feature.
- The procedure to edit the other properties of tube is same as discussed earlier.

Create primitives

The **Create primitives** tool is used to create any of the parametric geometric primitives defined in the part workbench. The procedure to use this tool is discussed next.

- Click on the **Create primitives** tool from **Toolbar** in the **Part** workbench; refer to Figure-19. The **Geometric Primitives** dialog will be displayed in the **Tasks** panel of **Combo View**; refer to Figure-20.

Figure-19. Create primitives tool

Figure-20. Geometric Primitives dialog

- Click on the **Primitive type** drop-down in the dialog. The list of available parametric geometric primitives will be displayed; refer to Figure-21.

Figure-21. List of available geometric primitives

- Select desired primitive type from the drop-down to be created. The parameters related to the selected primitive type will be displayed in the dialog.
- Specify the parameters as desired and click on **Create** button from the dialog. The selected geometric primitive will be created; refer to Figure-22.

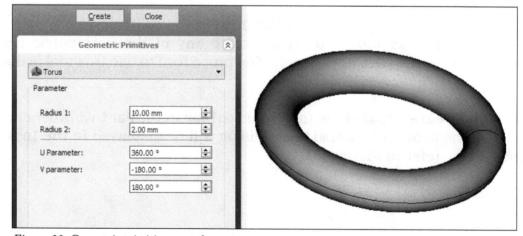

Figure-22. Geometric primitive created

- Specify all the parameters as desired in the dialog and click on **Close** button to close the dialog.

Shape builder

The **Shape builder** tool is used to create more complex shapes from various parametric geometric primitives. The procedure to use this tool is discussed next.

- Click on the **Shape builder** tool from **Toolbar** in the **Part** workbench; refer to Figure-23. The **Create shape** dialog will be displayed in the **Tasks** panel of **Combo View**; refer to Figure-24.

Figure-23. Builder tool

Figure-24. Create shape dialog

- Select desired radio button from **Create shape** area of the dialog for the shape to be created. The conditions to create the selected shape will be displayed at the bottom of the dialog; refer to Figure-24.
- Select the objects as required for the shape and click on **Create** button from the dialog. The shape will be created; refer to Figure-25.

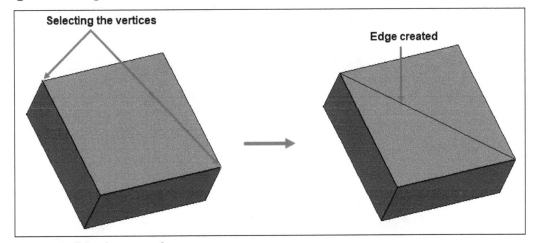

Figure-25. Edge shape created

- Click on **Close** button to close the dialog.

Modifying Object tools

Modifying object tools are used for modifying existing objects. They will allow you to choose which object to modify. These tools are available in the **Toolbar** of **Part** workbench; refer to Figure-26. The procedures to use these tools are discussed next.

Figure-26. Modifying object tools

Extrude

The **Extrude** tool extends a shape by a specified distance in a specified direction. The output shape type will vary depending on the input shape type and the options selected. The procedure to use this tool is discussed next.

- Select desired shape from the 3D view area or from the **Model Tree** which you want to extrude; refer to Figure-27.

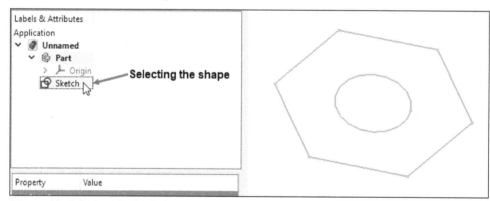

Figure-27. Selecting the sketch for extrusion

- Click on the **Extrude** tool from **Toolbar** in the **Part** workbench; refer to Figure-28. The **Extrude** dialog will be displayed in the **Tasks** panel of **Combo View**; refer to Figure-29.

Figure-28. Extrude tool

Figure-29. Extrude dialog

- Select the **Along normal** radio button from **Direction** area of the dialog to create the extrusion in a normal direction.
- Select the **Along edge** radio button and click on **Select** button below the radio button to select the edge along the direction of which the extrusion will be created.
- Select the **Reversed** check box next to the **Along edge** radio button to reverse the direction of extrusion along an edge.
- Select the **Custom direction** radio button and specify desired value in **X**, **Y**, and **Z** edit boxes to specify the extrusion direction along x, y, and z axis, respectively.
- Specify desired length of extrusion along normal direction in the **Along** edit box from **Length** area of the dialog.
- Specify the length of extrusion along reverse direction in the **Against** edit box.
- Select the **Symmetric** check box to create the extrusion in both the directions.
- Specify desired value in **Taper angle along** edit box to apply taper angle along the extrusion direction and specify desired value in **Taper angle against** edit box to apply taper angle against the extrusion direction.
- Select the **Create solid** check box to extrude the closed wire or edge as a solid model. The selected shape to be extruded will display in the **Shape list** area of the dialog. If you want to select more shapes to extrude then select desired shape from the **Shape list** area.
- Click on **Apply** button from the dialog. The preview of extrusion will be created; refer to Figure-30.

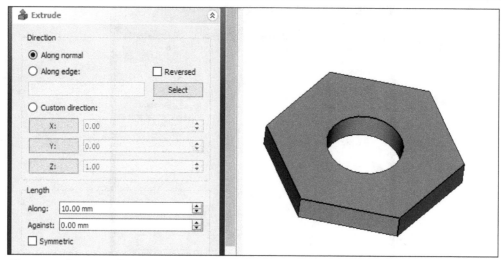

Figure-30. Preview of extrusion

- Click on **OK** button to close the dialog. The extrusion will be created.

Revolve

The **Revolve** tool revolves selected object about an axis. The procedure to use this tool is discussed next.

- Select desired shape from 3D view area or from the **Model Tree** which you want to revolve; refer to Figure-31.
- Click on the **Revolve** tool from **Toolbar** in the **Part** workbench; refer to Figure-32. The **Revolve** dialog will be displayed in the **Tasks** panel of **Combo View**; refer to Figure-33.

Figure-31. Selecting the shape to revolve

Figure-32. Revolve tool

Figure-33. Revolve dialog

- The selected shape will display in the **Shape list** area of the dialog. If you want to select more shapes then select the shape from shape list area.
- Specify the parameters as desired from **Revolution axis** area of the dialog.
- Click on the **Select reference** button from **Revolution axis** area. You are asked to select the line or an arc.
- Select the line or an arc from 3D view area which you want to make reference for the revolution of shape; refer to Figure-34.

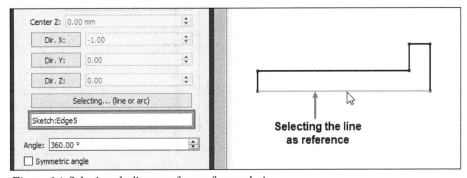

Figure-34. Selecting the line as reference for revolution

- Specify desired angle of revolution in the **Angle** edit box.
- Select the **Symmetric angle** check box to revolve the shape in both the direction at the given angle.
- Select the **Create solid** check box to revolve the shape as a solid model.
- Click on **OK** button from the dialog. The revolved model of the shape will be created; refer to Figure-35.

Figure-35. Revolved model of the shape created

Mirror

The **Mirror** tool creates a new object (image) which is a reflection of the original object (source). The image object is created on the other side of mirror plane. The procedure to use this tool is discussed next.

• Click on the **Mirror** tool from **Toolbar** in the **Part** workbench; refer to Figure-36. The **Mirroring** dialog will be displayed in the **Tasks** panel of **Combo View**; refer to Figure-37.

Figure-36. Mirror tool

Figure-37. Mirroring dialog

• Select desired shape which you want to mirror from **Shapes list** area of the dialog.
• Select desired mirror plane from the **Mirror plane** drop-down.
• Specify desired value in **x**, **y**, and **z** edit boxes from the **Base point** area of the dialog to move the mirror plane parallel to the selected standard mirror plane.
• Click on **OK** button from the dialog. The mirrored entity of the shape will be created; refer to Figure-38.

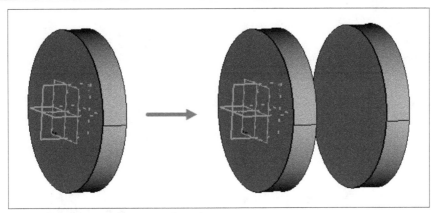

Figure-38. The mirrored entity of the shape

Fillet

The **Fillet** tool creates a fillet (round) on the selected edges of an object. The procedure to use this tool is discussed next.

- Click on the **Fillet** tool from **Toolbar** in the **Part** workbench; refer to Figure-39. The **Fillet Edges** dialog will be displayed in the **Tasks** panel of **Combo View**; refer to Figure-40.

Figure-39. Fillet tool

Figure-40. Fillet Edges dialog

- Click on the **Selected shape** drop-down from **Shape** area of the dialog and select the shape from the drop-down list you want to be fillet. All the edges of the selected shape will be displayed in the **Shapes list** area of the dialog.
- If you want to select the edges of the shape then select the **Select edges** radio button from **Selection** area in the **Fillet Parameter** section of the dialog and if you want to select the faces of the shape then select the **Select faces** radio button from the **Selection** area.
- Click on the **All** button from the **Selection** area to select all the edges available in the **Shapes list** area of the dialog and click on **None** button to deselect all the edges from the **Shapes** list area; refer to Figure-41.

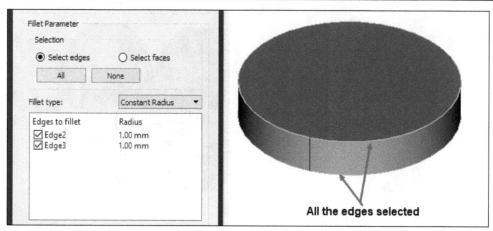

Figure-41. All the edges selected to be fillet

- Select desired type of fillet from **Fillet type** drop-down in the **Fillet Parameter** section. Select **Constant Radius** option to specify same fillet radius on both ends of the edges. Select **Variable Radius** option to specify different radius values at start and end points of the edges.
- Click on **OK** button from the dialog. The fillets will be created on the edges; refer to Figure-42.

Figure-42. Fillets created

Chamfer

The **Chamfer** tool chamfers selected edges of an object. The procedure to use this tool is discussed next.

- Click on the **Chamfer** tool from **Toolbar** in the **Part** workbench; refer to Figure-43. The **Chamfer Edges** dialog will be displayed in the **Tasks** panel of **Combo View**; refer to Figure-44.

Figure-43. Chamfer tool

- The procedure to create the chamfer is same as discussed for **Fillet** tool.
- Specify the parameters and click on **OK** button from the dialog. The chamfers will be created on the edges; refer to Figure-45.

Figure-44. Chamfer Edges dialog

Figure-45. Chamfer created

Make face from wires tool

The **Make face from wires** tool is used to create face from closed wireframe (sketching) curves in custom shape. The procedure to use this tool is discussed next.

- Select desired closed sketch from the **Model Tree** or from the 3D view area that you want to make a face; refer to Figure-46.

Figure-46. Sketch selected to make face

- Click on the **Make face from wires** tool from **Toolbar** in the **Part** workbench; refer to Figure-47. The face will be created; refer to Figure-48.

Figure-47. Make face from wires tool

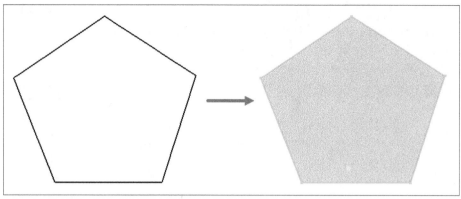

Figure-48. Face created

Ruled Surface

The **Ruled Surface** tool is used to create a ruled surface using two edges or sketch curves. The procedure to use this tool is discussed next.

• Select the two edges or two wires from the 3D view area; refer to Figure-49.

Figure-49. Selecting the wires

• Click on the **Ruled Surface** tool from **Toolbar** in the **Part** workbench; refer to Figure-50. The ruled surface will be created; refer to Figure-51.

Figure-50. Ruled surface tool

Figure-51. Ruled surface created

Loft

The **Loft** tool is used to create a face, shell, or a solid shape from two or more profiles. The profiles can be a point (vertex), line (Edge), wire, or face. Edges and wires may be either open or closed. The procedure to use this tool is discussed next.

*   Click on the **Loft** tool from **Toolbar** in the **Part** workbench; refer to Figure-52. The **Loft** dialog will be displayed in the **Tasks** panel of **Combo View**; refer to Figure-53.

Figure-52. Loft tool

Figure-53. Loft dialog

*   The list of profiles which are available to be loft will display in the **Available profiles** list area in the dialog.
*   To create the loft feature, add the two or more available profiles one by one in the **Selected profiles** list area by clicking on **Add** button.
*   Select desired profile from the **Selected profiles** list area as a base profile for loft; refer to Figure-54.

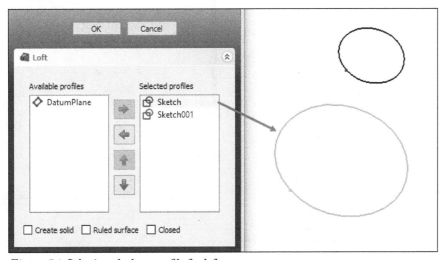

Figure-54. Selecting the base profile for loft

- Select the **Create solid** check box to create solid model of the profiles if the profiles are of closed geometry. If the profiles are open geometry then it creates a face or shell of the profiles.
- Select the **Ruled surface** check box to create a face or ruled surface using the profiles.
- Select the **Closed** check box to loft the last profile to the first profile creating a closed body.
- Click on **OK** button from the dialog. The loft feature will be created; refer to Figure-55.

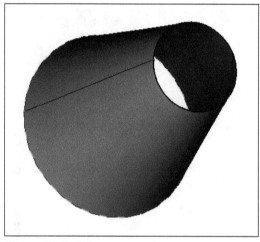

Figure-55. Loft feature created

Sweep

The **Sweep** tool is used to create a face, a shell, or a solid shape from one or more profiles (cross-sections) projected along a path. The procedure to use this tool is discussed next.

- Click on the **Sweep** tool from **Toolbar** in the **Part** workbench; refer to Figure-56. The **Sweep** dialog will be displayed in the **Tasks** panel of **Combo View**; refer to Figure-57.

Figure-56. Sweep tool

Figure-57. Sweep dialog

- Select the profile to be used as a base profile for sweep feature from the **Available profiles** list area of the dialog and add it in the **Selected profiles** list area by clicking on **Add** ➡ button.
- Select the profile from the **Selected profiles** list area and click on **Sweep Path** button. You are asked to select the profile for sweep path from the 3D view area.
- Select the profile from the 3D view area to be used as path for creating a sweep feature; refer to Figure-58.

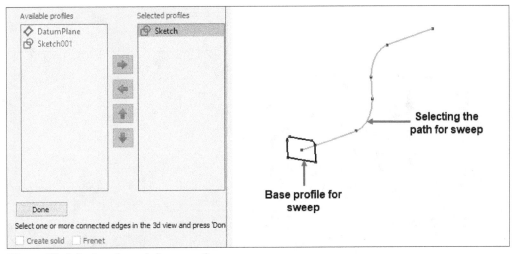

Figure-58. Selecting the path for sweep feature

- Click on **Done** button from the dialog.
- Select the **Create solid** check box to create the solid model of the profiles, if the profiles are of closed geometry. If the profiles are of open geometry, it creates a face or shell of the profiles.
- Select the **Frenet** check box to compute the profile orientation basing on local curvature and tangency vectors of the path.
- Click on **OK** button from the dialog. The sweep feature will be created; refer to Figure-59.

Figure-59. Sweep feature created

Section

The **Section** tool is used to create a section by intersecting an object with a section plane. The procedure to use this tool is discussed next.

- Select the model and section plane from the **Combo View** and then click on the **Section** tool from the **Toolbar** in the **Part** workbench; refer to Figure-60. The section view will be created; refer to Figure-61.

Figure-60. Section tool

Figure-61. Section created

Cross-Sections

The **Cross-sections** tool creates one or more cross-sections of selected body. The procedure to use this tool is discussed next.

- Select the shape from which you want to create cross-sections from the **Model Tree** or from the 3D view area; refer to Figure-62.

Figure-62. Selecting the shape to cross-section

- Click on the **Cross-sections** tool from **Toolbar** in the **Part** workbench; refer to Figure-63. The **Cross sections** dialog will be displayed in the **Tasks** panel of **Combo View** along with the preview of cross section; refer to Figure-64.

Figure-63. Cross sections tool

Figure-64. Cross sections dialog with preview of cross section

- Select desired guiding plane of creating the cross-section from the **Guiding plane** section of the dialog. Select the **XY**, **XZ**, or **YZ** radio button to create the cross-section along xy, xz, or yz plane, respectively.
- Specify the position of cross section in the **Position** edit box.
- Select the **Sections** check box to edit the parameters of cross-section.
- Select **On both sides** check box to create the cross-section on both sides of the selected guiding plane.
- Specify the number of cross sections to be created in the **Count** edit box.
- Specify distance between the cross sections in the **Distance** edit box of the dialog.
- Click on the **Apply** button. The cross sections of the shape will be created; refer to Figure-65.

Figure-65. Cross sections of the shape created

- Click on **OK** button to close the dialog.

Offset tools

The **Offset** tools are used to create 2D as well as 3D copies of selected shapes. The tools available to create 2D and 3D offsets are discussed next.

3D Offset

The **3D Offset** tool creates parallel copies of selected shape at a certain distance from the base shape, giving a new object. The procedure to use this tool is discussed next.

- Select the object from the **Model Tree** or from the 3D view area of which you want to create the offset; refer to Figure-66.

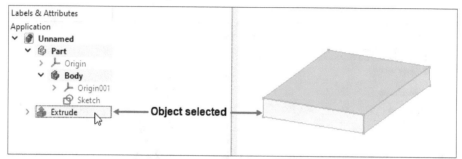

Figure-66. Object selected to create the 3D offset

- Click on the **3D Offset** tool from **Offset** drop-down in the **Toolbar** of **Part** workbench; refer to Figure-67. The **Offset** dialog will be displayed in the **Tasks** panel of **Combo View** along with the preview of offset creation; refer to Figure-68.

Figure-67. 3D Offset tool

Figure-68. 3D Offset dialog with the preview of offset creation

- Select the mode of offset creation from the **Mode** drop-down. Select the **Skin** option to create a new shape around the source shape.
- Select desired option from the **Join type** drop-down. Select the **Arc** and **Tangent** option to create the offset of rounded corners. Select the **Intersection** option to create the offset of sharp corners by linear extension of the edges.
- Select the **Intersection** check box to allow the offsets pointing inwards to over flood the gap by intersecting the resulting shape until opposite faces are reached.
- Select the **Self-intersection** check box to allow creation of self-intersecting body after apply offset operation.
- Select the **Fill offset** check box while using the 2 dimensional shape in which the gap in between the two shapes gets filled.

- Click on **OK** button from the dialog. The 3D offset of the original object will be created; refer to Figure-69.

Figure-69. 3D offset of original object created

2D Offset

The **2D Offset** tool constructs a wire parallel to the original wire at a certain distance from it or enlarges/shrinks a planar face. The procedure to use this tool is discussed next.

- Select the object from the **Model Tree** or from the 3D view area which you want to offset; refer to Figure-70.

Figure-70. Object selected to create the 2D offset

- Click on the **2D Offset** tool from the **Offset** drop-down in the **Toolbar** of **Part** workbench; refer to Figure-71. The **Offset** dialog will be displayed in the **Tasks** panel of **Combo View** along with the preview of offset creation; refer to Figure-72.

Figure-71. 2D Offset tool

Figure-72. 2D Offset dialog with the preview of offset creation

- The parameters in the **2D Offset** dialog are same as discussed for **3D Offset** tool.
- Specify the parameters and click on **OK** button from the dialog. The 2D offset of the original object will be created; refer to Figure-73.

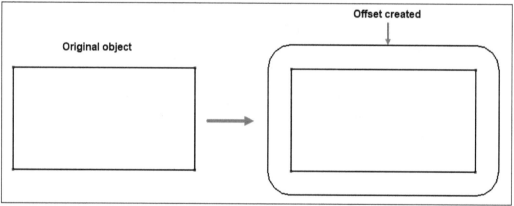

Figure-73. 2D offset of original object created

Thickness

The **Thickness** tool works on a solid shape and transforms it into a hollow object giving each of its faces a defined thickness. The procedure to use this tool is discussed next.

- Select one or more faces of the solid object from the 3D view area which you want to remove after applying thickness; refer to Figure-74.

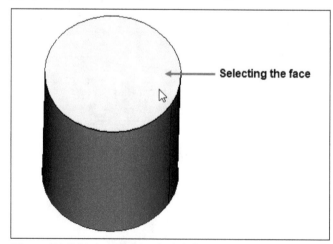

Figure-74. Selecting the face of solid object

- Click on the **Thickness** tool from **Toolbar** in the **Part** workbench; refer to Figure-75. The **Thickness** dialog will be displayed in the **Tasks** panel of **Combo View** along with the preview of thickness defined to the object; refer to Figure-76.

Figure-75. Thickness tool

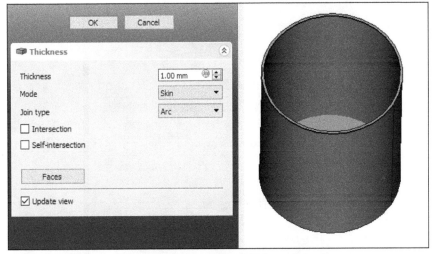

Figure-76. Thickness dialog with the preview of thickness defined

- The parameters of **Thickness** dialog are same as discussed for **Offset** tools.
- Specify the parameters and click on **OK** button from the dialog. The thickness will be defined to the object; refer to Figure-77.

Figure-77. Thickness defined to the object

Projection On Surface

The **Projection On Surface** tool is used to project a shape on top of a face from another object. This can be used to project a logo or textual object onto different surfaces to create interesting effects. The procedure to use this tool is discussed next.

- Click on the **Projection On Surface** tool from **Toolbar** in the **Part** workbench; refer to Figure-78. The **Projection on surface** dialog will be displayed in the **Tasks** panel of **Combo View**; refer to Figure-79, asking you to select the projection surface.

Figure-78. Projection On Surface tool

Figure-79. Projection on surface dialog

- Click on the **Select projection surface** button from the dialog and select the surface of the model on which you want to project the shape; refer to Figure-80. The other options in the dialog will be activated.

Figure-80. Selecting the surface to project the shape

- Click on the **Add face** button from the dialog and keep the surface parallel to the screen.
- Select the faces of the shape to project on the surface. The preview of projection will be displayed on the surface in real time; refer to Figure-81.

Figure-81. Preview of projection in real time

- Click on the **Add wire** button to select the source edge. This button will select the entire wire to which the edges belongs. For example, by choosing a single edge of a polygon, it will project the entire polygon.
- Click on the **Add edge** button to select the source edge. This button will only project the selected edge.
- Select the **Show all** radio button to show all types of closed wires and edges on the target surface.
- Select the **Show faces** radio button to show a preview of a filled face on the target surface.
- Select the **Show Edges** radio button to show a preview of the edges on the target surface.
- Specify desired height of the projected shape in the **Extrude height** edit box of the dialog.
- Specify desired depth of the projected face along the projection direction in the **Solid depth** edit box of the dialog.
- Click on the **Get current camera direction** button from **Direction** area of the dialog to change the projection direction. Alternatively, click on the **X:**, **Y:**, or **Z:** buttons to set the projection direction to the main global axes, **+X**, **-X**, **+Y**, **-Y**, **+Z**, or **-Z**.
- After specifying desired parameters, click on the **OK** button from the dialog to create the projection; refer to Figure-82.

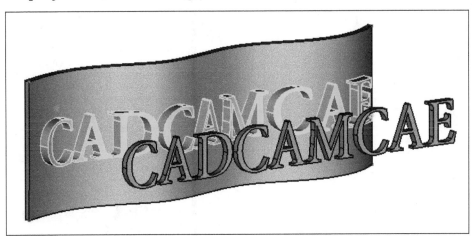

Figure-82. Projection created

Attachment

The **Attachment** tool is a utility to attach an object to another one. The attached object is linked to the other object which means that if the latter's placement is changed afterwards, the attached object will update to its new position. The procedure to use this tool is discussed next.

- Select the object which you want to attach from the **Model Tree** or from the 3D view area; refer to Figure-83.
- Click on the **Attachment** tool from **Part** menu in the **Ribbon**; refer to Figure-84. The **Attachment** dialog will be displayed in the **Tasks** panel of **Combo View**; refer to Figure-85, asking you to select the vertex, edge, face/plane as a reference.

Figure-83. Object selected to attach

Figure-84. Attachment tool

Figure-85. Attachment dialog

- Select desired reference on the object on which you want to attach the selected object; refer to Figure-86.

Figure-86. Face selected as reference

- Select desired attachment mode from **Attachment mode** area of the dialog; refer to Figure-87.

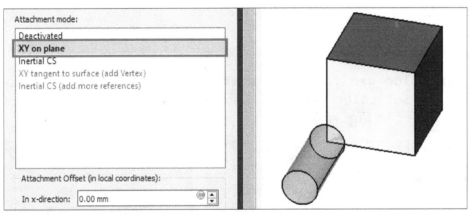

Figure-87. Attachment mode selected

- Specify desired offset values relative to the local coordinate system in the edit boxes from **Attachment Offset (in local coordinates)** area of the dialog.
- Select **Flip sides** check box to reverse the attached object from its **XY** plane.
- After specifying desired parameters, click on the **OK** button from the dialog. The object will be attached; refer to Figure-88.

Figure-88. Object attached

Color Per Face

The **Color per face** tool allows you to define a color for each face or surface of an object. This way you can assign multiple colors to one part. The procedure to use this tool is discussed next.

- Select the object from the **Model Tree** on which you want to define the color; refer to Figure-89.

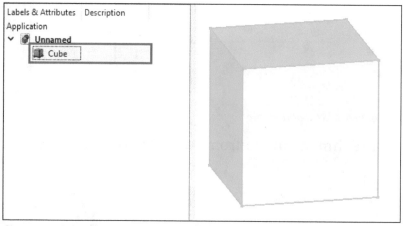

Figure-89. Selecting the object to define the color

- Click on the **Color per face** tool from **Toolbar** in the **Part** workbench; refer to Figure-90. The **Set color per face** dialog will be displayed in the **Tasks** panel of **Combo View**; refer to Figure-91.

Figure-90. Face Colors tool

Figure-91. Set color per face dialog

- Select desired face of the object in the 3D view area to define the color; refer to Figure-92.

Figure-92. Selecting face of the object to define the color

- Click on the [_____] button in the **Set color per face** dialog. The **Select Color** dialog will be displayed; refer to Figure-93.

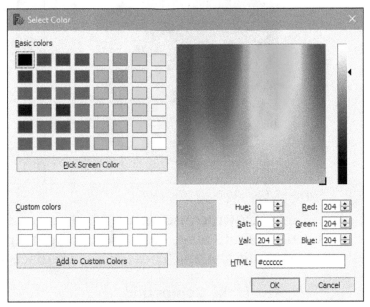

Figure-93. Select Color dialog

- Select desired color from the dialog to apply to the face and click on **OK** button from the **Select Color** dialog.
- Click **RMB** in the 3D view area and the selected color will be defined to the selected face; refer to Figure-94.

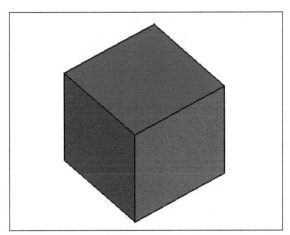

Figure-94. Color defined to the face

- Repeat the procedure to define colors to the other faces of the object.
- Click on the **Set to default** button in the dialog to set the color of all the faces of the object to the default color.
- If you want to select multiple faces of an object, then click on the **Box selection** button and press & hold **LMB** and drag with the mouse a selection rectangle in the object.
- Click on the **OK** button from the dialog to complete the procedure. The colors will be defined to the faces of the object; refer to Figure-95.

Figure-95. Colors defined to the object

Compound tools

The **Compound** tools are used to create a set of shapes, to split up the shapes, or to extract individual pieces from split shapes. These tools are discussed next.

Make Compound

The **Make Compound** tool creates a compound of any kind of topological shapes. These can be solids, meshes, or any other kind of topological shapes. A compound is a set of shapes grouped into one object. The procedure to use this tool is discussed next.

*   Select the topological shapes from the **Model Tree** to be added to the compound; refer to Figure-96.

Figure-96. Selecting the topological shapes

*   Click on the **Make compound** tool from **Compound** drop-down in the **Toolbar** of **Part** workbench; refer to Figure-97. The compound of the selected topological shapes will be created; refer to Figure-98.

Figure-97. Make compound tool

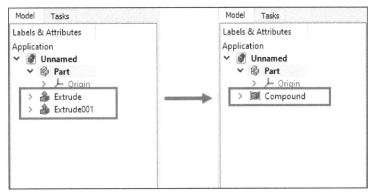

Figure-98. Compound of the shapes created

Explode Compound

The **Explode compound** tool is used to split up compounds of shapes to make each contained shape available as a separate object in Model tree view. The procedure to use this tool is discussed next.

- Select the compound body to be split into individual objects from the **Model Tree**; refer to Figure-99.

Figure-99. Compound shape selected

- Click on the **Explode compound** tool from **Compound** drop-down in the **Toolbar** of **Part** workbench; refer to Figure-100. The Compound shape will explode to generate separate objects; refer to Figure-101.

Figure-100. Explode compound tool

Figure-101. Compound shape exploded

Compound Filter

The **Compound Filter** tool can be used to extract individual pieces of the result of a **Part Slice** operation with which you have split an object. The procedure to use this tool is discussed next.

• Select the sliced object from **Model Tree** or from the 3D view area to extract the individual pieces; refer to Figure-102.

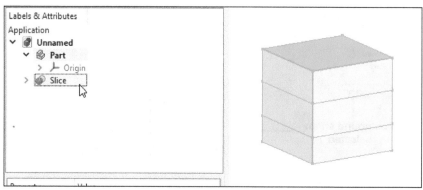

Figure-102. Selecting the sliced object

• Click on the **Compound Filter** tool from **Compound** drop-down in the **Toolbar** of **Part** workbench; refer to Figure-103. The compound filter object will be created; refer to Figure-104.

Figure-103. Compound Filter tool

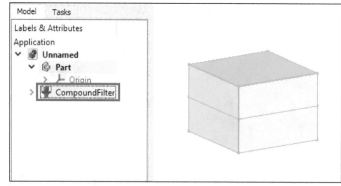

Figure-104. Compound filter object created

• Select the compound filter object from the **Model Tree**. The **Property editor** dialog will be displayed in the **Model** tab of **Combo View** as discussed earlier.

- Select the **specific items** option from **Filter Type** drop-down in the **Compound Filter** section of the dialog.
- Specify the number of elements you want to extract in the **items** edit box of the dialog. For a single piece, this is a number starting with 0, i.e. if you want to extract the first element, enter 0 in this edit box, 1 for the next element.
- After specifying the parameters, click **LMB** in the 3D view area. The element will be extracted; refer to Figure-105.

You can separate desired sliced object from the other sliced objects using **Property editor**.

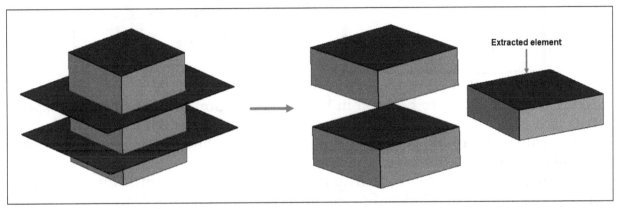

Figure-105. Element extracted

Boolean

The **Boolean** tool is a generic all-in-one boolean tool. It allows you to specify the objects and operation to perform via a single dialog. The procedure to use this tool is discussed next.

- Click on the **Boolean** tool from **Toolbar** in the **Part** workbench; refer to Figure-106. The **Boolean Operation** dialog will be displayed in the **Tasks** panel of **Combo View**; refer to Figure-107.

Figure-106. Boolean tool

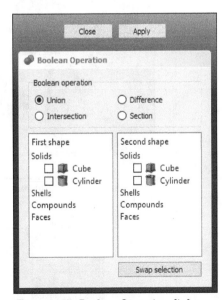

Figure-107. Boolean Operation dialog

- The objects available to perform boolean operation will display in the both **First shape** list and **Second shape** list area of the dialog; refer to Figure-108.

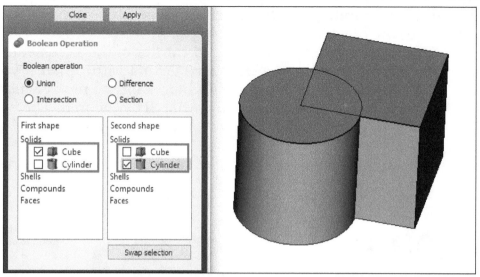

Figure-108. Objects to perform boolean operation

- Select the first object to be used in boolean operation from the **First shape** list area and select the second object from the **Second shape** list area.
- If you want to swap the selection of objects in the shape lists area then click on the **Swap selection** button at the bottom of the dialog.
- Select the type of boolean operation you want to perform from the **Boolean operation** area of the dialog. Select the **Union** radio button if you want to unite the selected part objects into one. Select the **Difference** radio button if you want to subtract second object from first object. Select the **Intersection** radio button if you want to extract the common part between selected part objects. Select the **Section** radio button if you want to extract a section from the intersection of two selected shapes.
- Click on **Apply** button from the dialog. The boolean operation will be performed; refer to Figure-109.

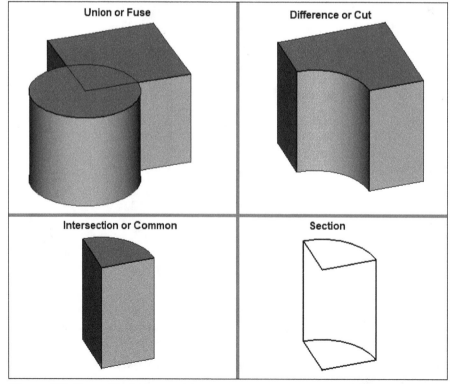

Figure-109. Types of boolean operation

- Click on **Close** button to close the dialog.

If you want the quicker access to these operations, use the **Fuse** ⬤ tool, **Cut** ◉ tool, **Common** ◉ tool, and **Section** ▣ tool from **Toolbar** in the **Part** workbench.

Join Objects tools

The **Join objects** tools are used to connect, embed, and cutout the walled objects. These tools are discussed next.

Connect objects

The **Connect objects** tool connects the interiors of two walled objects (e.g., pipes). It can also join shells and wires. The procedure to use this tool is discussed next.

- Select two or more objects from the **Model Tree** or from the 3D view area which you want to connect; refer to Figure-110.

Figure-110. Selecting the objects to connect

- Click on the **Connect objects** tool from **Join objects** drop-down in the **Toolbar** of **Part** workbench; refer to Figure-111. A connected parametric object will be created; refer to Figure-112.

Figure-111. Connect objects tool

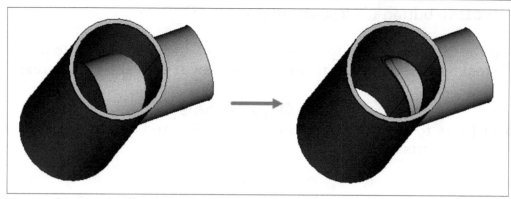

Figure-112. Connect objects created

Embed object

The **Embed object** tool embeds a walled object (e.g., a pipe) into another walled object. The procedure to use this tool is discussed next.

*   Select the base object first and then select the object to be embedded from the **Model Tree** or from the 3D view area; refer to Figure-113.

Figure-113. Selection of objects to embed

*   Click on the **Embed object** tool from **Join Objects** drop-down in the **Toolbar** of **Part** workbench; refer to Figure-114. The embedded object will be created; refer to Figure-115.

Figure-114. Embed object tool

Figure-115. Embed object created

Cutout for object

The **Cutout for object** tool creates a cutout in a walled object (e.g., a pipe) to fit another walled object. The procedure to use this tool is discussed next.

- Select the base object first and then select the object to define the cutout from the **Model Tree** or from the 3D view area; refer to Figure-116.

Figure-116. Selection of objects to be cutout

- Click on the **Cutout for object** tool from **Join Objects** drop-down in the **Toolbar** of **Part** workbench; refer to Figure-117. The cutout object will be created; refer to Figure-118.

Figure-117. Cutout for object tool

Figure-118. Cutout object created

Splitting tools

The **Splitting** tools are used to compute boolean fragments and split features. These tools are discussed next.

Boolean Fragments

The **Boolean fragments** tool is used to compute all fragments that can result from applying boolean operations between input shapes. The procedure to use this tool is discussed next.

*   Select the objects to create boolean fragments from the **Model Tree** or from the 3D view area; refer to Figure-119.

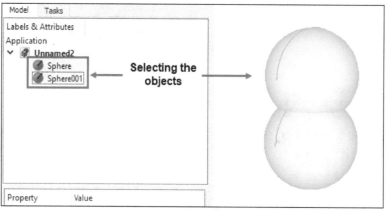

Figure-119. Selecting the objects to create boolean fragments

*   Click on the **Boolean fragments** tool from **Split objects** drop-down in the **Toolbar** of **Part** workbench; refer to Figure-120. The boolean fragments of the objects will be created; refer to Figure-121.

Figure-120. Boolean fragments tool

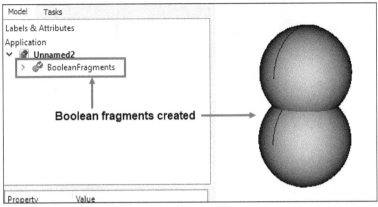

Figure-121. Boolean fragments created

- The individual pieces of the boolean fragments will not be displayed immediately as the pieces remain grouped together.
- If you want to extract the individual pieces of the object then explode the object using **Downgrade** tool.
- Select both the objects from 3D view area and click on the **Downgrade** tool from **Draft modifications tools** in the **Draft** workbench; refer to Figure-122. The exploded pieces of the boolean fragments will be displayed in the **Model Tree**; refer to Figure-123.

Figure-122. Downgrade tool

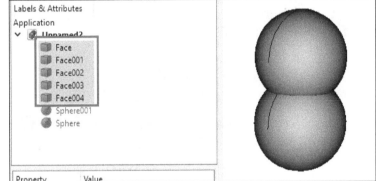

Figure-123. Exploded pieces of the boolean fragments

- Select the pieces of the object one by one from 3D view area and move apart manually using the **Property editor**. The exploded view of the boolean fragments will be displayed; refer to Figure-124.

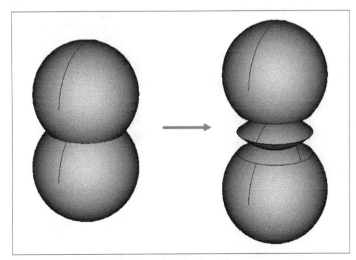

Figure-124. Exploded view of the boolean fragments

Slice Apart

The **Slice apart** tool is used to split shapes by intersection with other shapes. The procedure to use this tool is discussed next.

- Select the base object first which is to be sliced and then select the other object to slice with from the **Model Tree** or from the 3D view area; refer to Figure-125.

Figure-125. Selection of objects to slice apart

- Click on the **Slice apart** tool from **Split objects** drop-down in the **Toolbar** of **Part** workbench; refer to Figure-126. The sliced part of base object will be created; refer to Figure-127.

Figure-126. Slice apart tool

Figure-127. Sliced part of base object created

Slice to compound

The **Slice** tool also known as **Slice to compound** tool is used to split shapes by intersection with other shapes. The procedure to use this tool is discussed next.

- First select the base object to be sliced and then select the other object to slice with from the **Model Tree** or from the 3D view area; refer to Figure-128.

Figure-128. Selection of objects to slice

- Click on the **Slice to compound** tool from **Split objects** drop-down in the **Toolbar** of **Part** workbench; refer to Figure-129. The sliced object will be created; refer to Figure-130.

Figure-129. Slice to compound tool

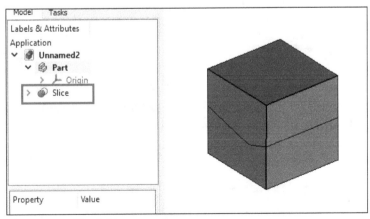

Figure-130. Sliced object created

- If you want to break the sliced object into individual pieces then select the object from the **Model Tree** and explode the object using **Explode compound** tool which we have discussed earlier. The sliced object will be exploded.

- If you want to view the exploded slice object individually then select the object from the **Model Tree** or from the 3D view area and move apart manually using the **Property editor**. The exploded view of sliced object will be displayed; refer to Figure-131.

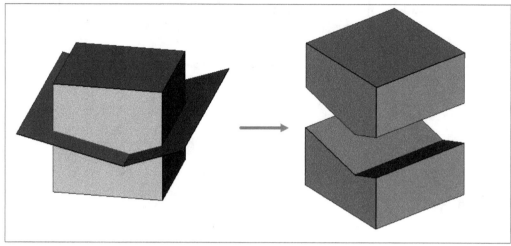

Figure-131. Exploded view of sliced object

Boolean XOR

The **Boolean XOR** tool is used to remove geometry shared by objects and leaves a void space between the involved objects. The procedure to use this tool is discussed next.

- Select two or more objects from the **Model Tree** or from the 3D view area; refer to Figure-132.

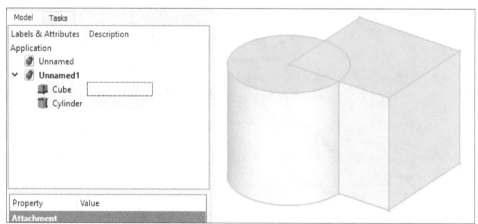

Figure-132. Selecting the objects

- Click on the **Boolean XOR** tool from **Split objects** drop-down in the **Toolbar** of **Part** workbench; refer to Figure-133. The overlapping geometry will be removed from the selected objects; refer to Figure-134.

Figure-133. Boolean XOR tool

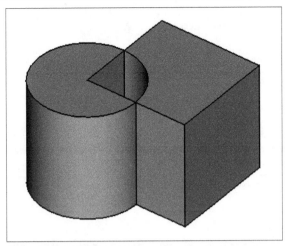

Figure-134. Overlapping geometry removed

Check Geometry

The **Check Geometry** tool runs a verification and reports if geometry is a valid solid. The procedure to use this tool is discussed next.

* Select the geometry from the **Model Tree** which you want to check for valid solid; refer to Figure-135.

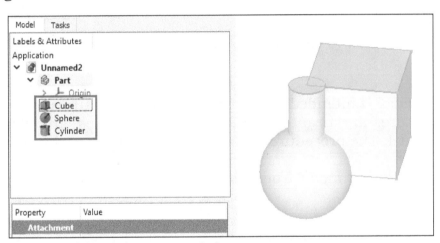

Figure-135. Selecting the geometry to check

* Click on the **Check geometry** tool from **Toolbar** in the **Part** workbench; refer to Figure-136. The **Check Geometry Results** and **Shape Content** dialogs will be displayed in the **Tasks** panel of **Combo View** showing the result of the scan; refer to Figure-137.

Figure-136. Check geometry tool

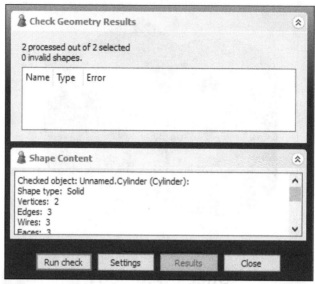

Figure-137. Check Geometry Results and Shape Content dialogs

- If the result of the scan showing errors, click on a specific error message in the dialog. The corresponding geometric object will be highlighted in the 3D view area.

Note that till the FreeCAD version 0.21, it has no automatic repair methods for solids, so you need to look at the steps used in FreeCAD and try to fix the error yourself.

- Click on **Close** button to close the dialog.

Defeaturing

The **Defeaturing** tool is intended for removal of selected features from the model. In this context, features are meant as holes, protrusions, gaps, chamfers, fillets, etc. found on the model. The procedure to use this tool is discussed next.

- Select the faces on the model from the 3D view area which you want to defeature; refer to Figure-138.

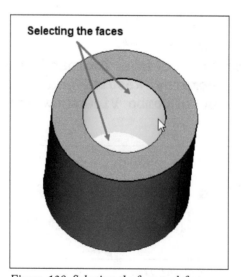

Figure-138. Selecting the faces to defeature

- Click on the **Defeaturing** tool from **Toolbar** in the **Part** workbench; refer to Figure-139. The selected feature will be removed from the model; refer to Figure-140.

Figure-139. Defeaturing tool

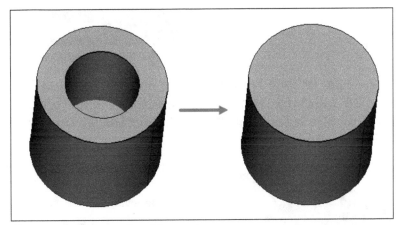

Figure-140. Defeatured model created

Measure Tools

The **Measure** tools allow linear and angular measurement between points, edges, and faces. These tools are available in the **Toolbar** of **Part** workbench; refer to Figure-141. The procedures to use these tools are discussed next.

Figure-141. Measure tools

Measure Linear

The **Measure Linear** tool allows you to make linear measurements. It measures the distance between two selected topology elements (vertex, edge, face) and displays the measurement in the 3D view. The shortest distance between the two elements is shown in red and delta measurements (distances parallel to standard X, Y, Z axes) are shown in green. The procedure to use this tool is discussed next.

- Click on the **Measure Linear** tool from **Toolbar** in the **Part** workbench; refer to Figure-142. The **Selections** and **Control** dialog will be displayed in the **Tasks** panel of **Combo View** asking you to select the elements; refer to Figure-143.

Figure-142. Measure linear tool

Figure-143. Selections and Control dialog

- Select the vertex, edge, or face from both the shapes in the 3D view area between which you want to measure the linear distance. The right tick mark will be displayed in front of both **Selection 01** and **Selection 02** buttons of the **Selections** dialog; refer to Figure-144.

Figure-144. Selecting the edges to measure distance

- In **Control** dialog, click on the **Reset selection** button to reset all the selections.
- Click on the **Toggle direct dimensions** button to measure the shortest distance between the two elements; refer to Figure-145.

Figure-145. Shortest distance between elements

- Click on the **Toggle orthogonal dimensions** button to create delta measurement between the elements; refer to Figure-146.

- Click on the **Clear all dimensions** button to clear all the measurements created between the elements.
- Click on **Close** button to close the dialog.

Figure-146. Delta measurement between elements

Measure Angular

The **Measure Angular** tool measures the angle between two straight edges, two planar faces, or one straight edge and a planar face and displays the measurement in the 3D view. The procedure to create angular measurement is same as discussed for linear measurement.

If you want the quicker access to the options used in **Control** dialog, use **Clear All** button, **Toggle All** button, **Toggle 3d** button, and **Toggle Delta** button from **Toolbar** in the **Part** workbench.

Measure Refresh

The **Measure Refresh** tool updates the 3D view to display all measurements created with **Measure Linear** and **Measure Angular** tool. The procedure to use this tool is discussed next.

- After creating the measurement of the model with the measurement tools, the measurements of the model will be displayed as shown in Figure-147.

Figure-147. Measurements of the model displayed

- Now, modify the model as desired and click on the **Measure Refresh** tool from **Toolbar** in the **Part** workbench; refer to Figure-148. The measurements of the model will be updated; refer to Figure-149.

Figure-148. Measure Refresh tool

Figure-149. Measurements of the model updated

Other tools

There are some other tools available in the **Part** menu of the **Ribbon**; refer to Figure-150. The procedures to use these tools are discussed next.

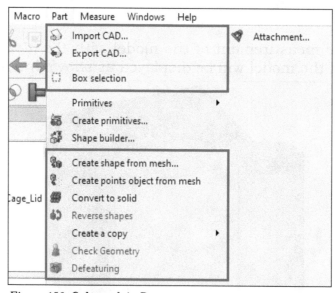

Figure-150. Other tools in Part menu

Import CAD

The **Import CAD** tool opens a dialog that allows you to import the following CAD file formats: **IGES**, **STEP**, and **BREP** into the current document. The procedure to use this tool is discussed next.

- Click on the **Import CAD** tool from the **Part** menu in the **Ribbon**. The **Open** dialog box will be displayed; refer to Figure-151.

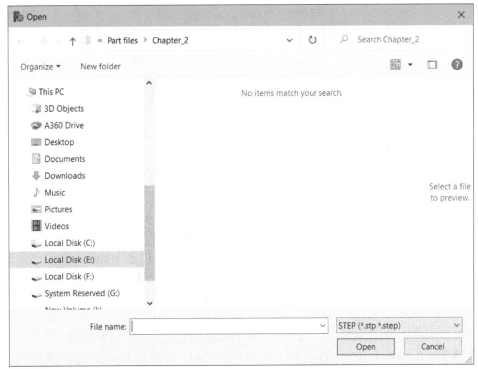

Figure-151. Open dialog box

- Select desired file of the supported format which you want to import from the directory and click on **Open** button from the dialog box. The file will be imported; refer to Figure-152.

Figure-152. File imported

Export CAD

The **Export CAD** tool opens a dialog that allows you to export objects in the document to the following CAD file formats: **IGES**, **STEP**, and **BREP**. The procedure to use this tool is discussed next.

- Click on the **Export CAD** tool from the **Part** menu in the **Ribbon**. The **Save As** dialog box will be displayed; refer to Figure-153.

Figure-153. Save as dialog box

- Specify desired directory in the dialog box in which you want to save the file.
- Specify desired name of the file to be save in the **File name** edit box.
- Specify the supported format to save the file from **Save as type** drop-down.
- Click on the **Save** button from the dialog box to save the file.

Box Selection

The **Box Selection** tool allows you to select the faces of the objects on the screen that are touched by the rectangular selection. The procedure to use this tool is discussed next.

- Click on the **Box Selection** tool from the **Part** menu in the **Ribbon**. You are asked to select the objects.
- Click **RMB** and drag the pointer in the 3D view area creating a rectangle that touches the objects which you want to select; refer to Figure-154. The objects will be selected.

Figure-154. Object selected

Create shape from mesh

The **Create shape from mesh** tool creates a shape from a mesh object. Mesh objects have limited editing capabilities in FreeCAD, converting them to shapes will allow their use with many more boolean and modification tools. The procedure to use this tool is discussed next.

- Select desired meshed object from the **Model Tree** or from the 3D view area which you want to convert to shape; refer to Figure-155.

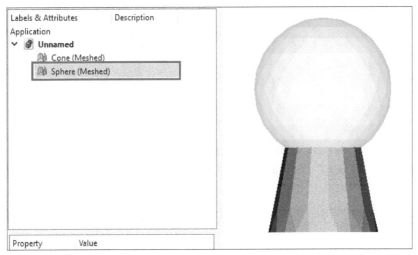

Figure-155. Meshed object selected

- Click on the **Create shape from mesh** tool from the **Part** menu in the **Ribbon**. The **Sewing Tolerance** dialog box will be displayed; refer to Figure-156.

Figure-156. Sewing Tolerance dialog box

- Specify desired tolerance value for the sewing shape in the **Enter tolerance** for sewing shape edit box and click on **OK** button from the dialog box. The selected meshed object will be created as a shape; refer to Figure-157.

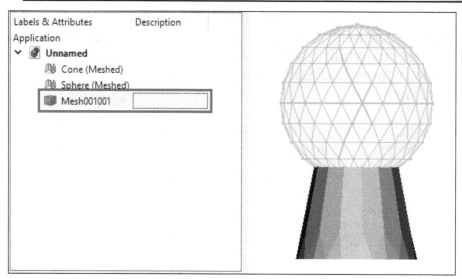

Figure-157. Shape created from meshed object

Create points object from mesh

The **Create points object from mesh** tool creates a shape object made of points from a mesh object imported or produced from the **Mesh** workbench. The resulting shape is a collection of vertices or points, which can be used as reference to further create lines, sketches, and faces with other tools. The procedure to use this tool is discussed next.

- Select desired meshed object from the **Model Tree** or from the 3D view area which you want to create points shape object; refer to Figure-158.

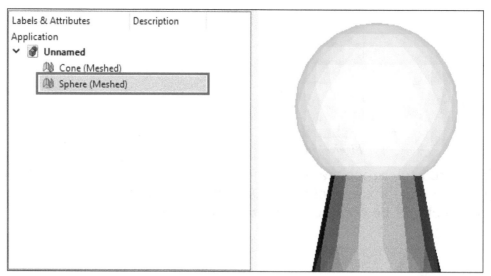

Figure-158. Meshed object selected

- Click on the **Create points object from mesh** tool from the **Part** menu in the **Ribbon**. The selected meshed object will be created as a shape object made of points; refer to Figure-159.

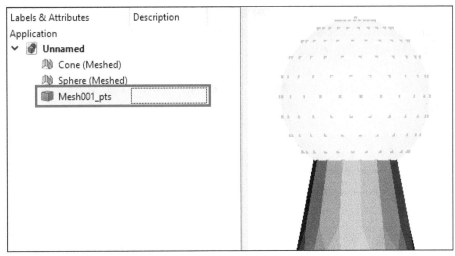

Figure-159. Points shape object created

Convert to Solid

The **Convert to solid** tool converts a shape object to a solid object. The procedure to use this tool is discussed next.

- Select the shape object from the **Model Tree** which you have created from the mesh object; refer to Figure-160.

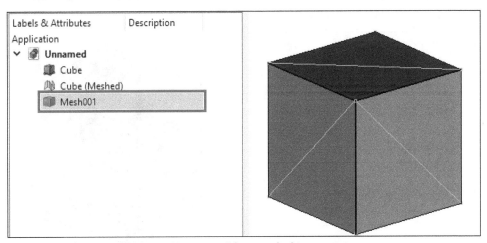

Figure-160. Selecting the shape object created from mesh object

- Click on the **Convert to solid** tool from the **Part** menu in the **Ribbon**. The selected shape object will be converted into a solid object; refer to Figure-161.

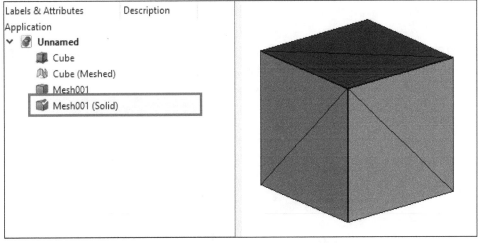

Figure-161. Shape object converted to solid object

Reverse shapes

The **Reverse shapes** tool flips the normals of all faces of the selected object. The procedure to use this tool is discussed next.

• Select desired shape from the **Model Tree** or from the 3D view area which you want to reverse; refer to Figure-162.

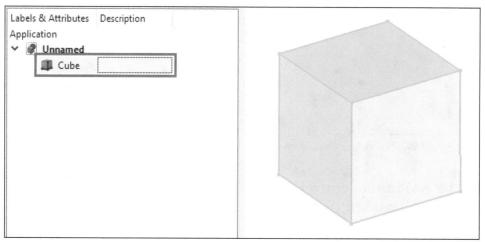

Figure-162. Selecting the shape to reverse

• Click on the **Reverse shapes** tool from the **Part** menu in the **Ribbon**. The selected shape will be reversed; refer to Figure-163.

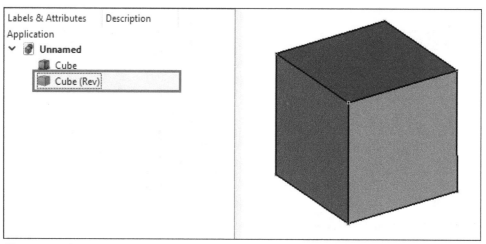

Figure-163. Reverse shape created

Now, if you want to verify that the shape is reversed or not then follow the steps given below:

• Hide the shape selected to reverse from the **Model Tree** by pressing the **Space Bar** from the keyboard; refer to Figure-164.

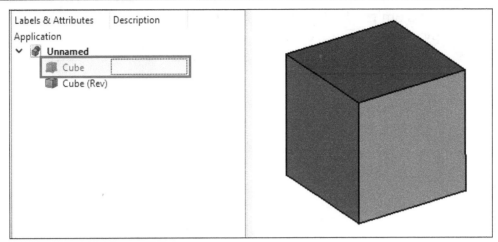

Figure-164. Hiding the shape selected to reverse

- Select the reversed shape from the **Model Tree** as shown in Figure-165.

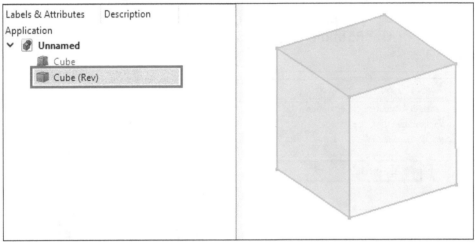

Figure-165. Reversed shape selected

- Select **One side** option from **Lighting** drop-down in the **Object Style** section of **View** tab in the **Property editor** dialog; refer to Figure-166.
- Click **RMB** in the 3D view area, the reversed shape will be turned black, it means that you are now looking at the back of the faces; refer to Figure-167.

Figure-166. Selecting Lighting option

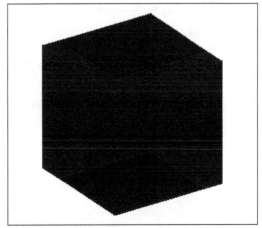

Figure-167. Reversed shape turned black

Create simple copy

The **Create simple copy** tool creates a simple copy of a selected object. The procedure to use this tool is discussed next.

- Select the object from the **Model Tree** or from the 3D view area of which you want to create a copy; refer to Figure-168.

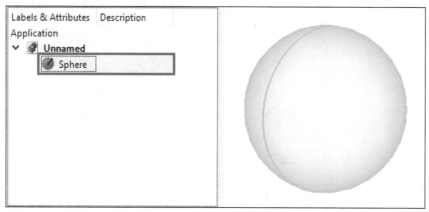

Figure-168. Selecting the object to copy

- Click on the **Create simple copy** tool from **Create a copy** cascading menu in the **Part** menu of the **Ribbon**. A copy of the selected object will be created; refer to Figure-169.

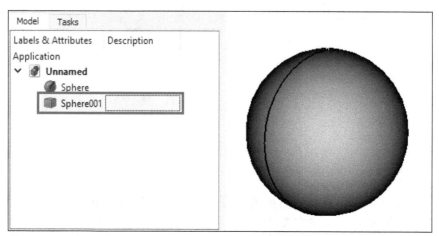

Figure-169. Copy of the selected object created

- You can reposition the copied object by selecting the object from **Model Tree** and specifying desired values in the **x, y,** or **z** coordinates from **Data** tab in the **Property editor** dialog; refer to Figure-170. The copied object will be repositioned; refer to Figure-171.

Figure-170. Specifying values to reposition the object

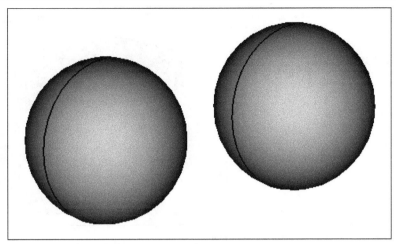

Figure-171. Copied object repositioned

Create transformed copy

The **Create transformed copy** tool creates transformed copy of a selected object that has been displaced from its original position. The procedure to use this tool is same as discussed in previous tool.

Create shape element copy

The **Create shape element copy** tool creates a copy from a sub-element (vertex, edge, face) of a selected object. The procedure to use this tool is same as discussed earlier.

Refine shape

The **Refine shape** tool creates a copy of selected object with a refined shape, that is, with certain edges and faces cleaned up. The procedure to use this tool is discussed next.

- Select the objects and perform boolean operations like, **Part Fuse**, as discussed earlier; refer to Figure-172. The objects will be fused but still some lines from the previous objects will remain visible; refer to Figure-173.

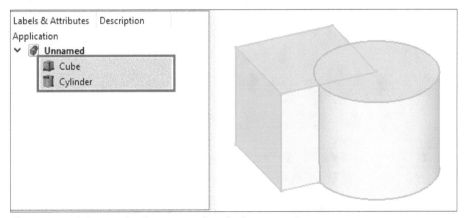

Figure-172. Selecting the objects to perform boolean operation

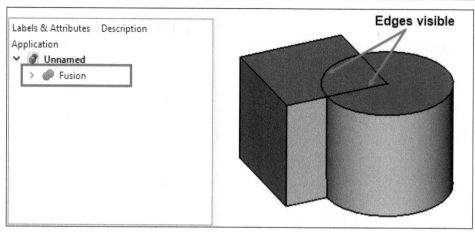

Figure-173. Objects fused but edges remain visible

- Select the fused feature from the **Model Tree** and click on the **Refine shape** tool from **Create a copy** cascading menu in the **Part** menu of the **Ribbon**. The objects will be refined; refer to Figure-174.

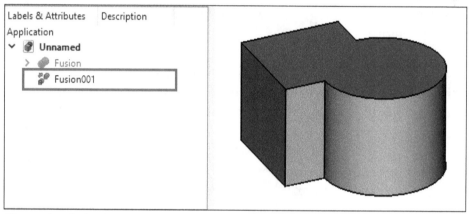

Figure-174. Refined object

- You can reposition the fused features with the **Property editor** dialog as discussed earlier.

Chapter 5

Solid Modeling Practice

Topics Covered

The major topics covered in this chapter are:

- *Practical 1*
- *Practical 2*
- *Practical 3*
- *Practices*

PRACTICAL 1

Create the model (isometric view) as shown in Figure-1. The views of the model with dimensions are given in Figure-2.

Figure-1. Practical Model 1

Figure-2. Views for practical 1

Before we start working on the practical, it is important to understand two terms; First angle projection and Third angle projection. These are the standards of placing views in the engineering drawing. The views placed in the above figure are using third angle projection. In first angle projection, the top view of model is placed below the front view and right side view is placed at left of the front view. You will learn more about projection in chapter related to drafting.

Starting Part Design Environment and Creating Sketch

- Double-click on the FreeCAD icon from desktop if you have not started FreeCAD yet.
- Start a new document by clicking on the **New** button from **Toolbar** or press **CTRL+N** shortcut key.
- Select the **Part Design** option from **Switch between workbenches** drop-down in the **Toolbar**. The tools to create solid model will be displayed.
- Click on the **Create sketch** tool from the **Toolbar** or **Part Design** menu. The **Select feature** dialog will be displayed in **Combo View** and you will be asked to select a plane for creating sketch.
- Select the XY plane from the list in the dialog and click on the **OK** button. The sketching environment will be displayed.
- Create a circle of diameter **200** with its center at origin; refer to Figure-3. Note that as we create a circle, we have also applied the diameter dimension to it. In FreeCAD, it is better to apply dimension as soon as you create the entity if the entity is independent.

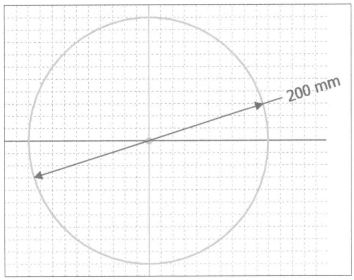

Figure-3. Circle to be created

- Create a circle of diameter **50** at the center of earlier created circle and then draw a rectangle as shown in Figure-4.

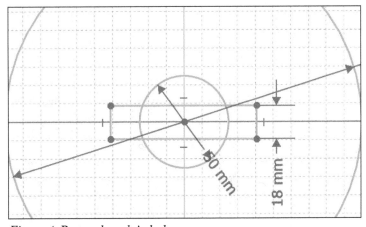

Figure-4. Rectangle and circle drawn

- Trim the inner portion of circle and rectangle, and then apply the dimensions as shown in Figure-5.

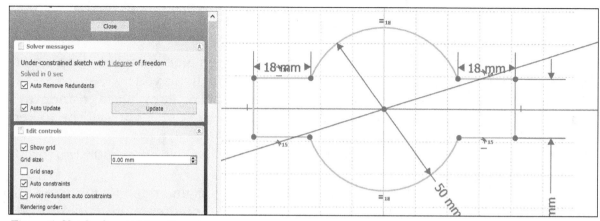

Figure-5. Sketch after applying dimensions

- Note that there is 1 degree of freedom left in the sketch. If you drag the vertical line of rectangle then you will find that it is free to move up and down. To constraint this movement, select the points of rectangle as shown in Figure-6 and click on the **Constrain vertical** tool from the **Toolbar**. The sketch will become fully constraint; refer to Figure-7.

Figure-6. Applying vertical constrain

Figure-7. After applying constraints

- Click on the **Close** button to exit the sketching environment.

Creating Pad and Thick Solid Features

- Select the newly created sketch from the **Model Tree** if not selected by default and then click on the **Pad** tool from **Toolbar** or **Part Design** menu. Preview of pad feature will be displayed.
- Specify the length of pad feature as **50** in the **Length** edit box of dialog; refer to Figure-8 and click on the **OK** button. The feature will be created.

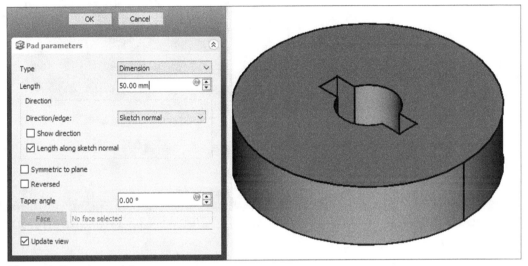

Figure-8. Creating pad feature

- Select the top flat face of the model from drawing area and click on the **Thickness** tool from the **Toolbar** or **Part Design** menu; refer to Figure-9. Preview of thickness feature will be displayed. (Do not worry if everything goes blank!!)

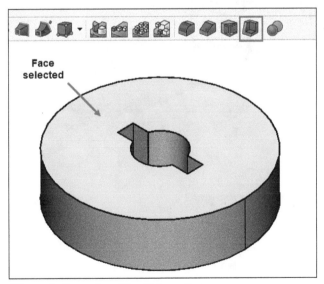

Figure-9. Face selected for thickness feature

- Specify the parameters shown in Figure-10 and click on the **OK** button to create the final model.

Figure-10. Preview of thickness feature

PRACTICAL 2

Create the model (isometric view) as shown in Figure-11. The dimensions of the model are given in Figure-12.

Figure-11. Model for Practical 2

Figure-12. Practical 2 drawing views

Creating first pad feature

This model is a combination of pad and loft features. We will create the model in multiple steps with first feature acting as base for all the other features.

- Start FreeCAD if not started yet and create a new document.
- Switch to the **Part Design** workbench using **Part Design** option from **Switch between workbenches** drop-down in the **Toolbar**.
- Click on the **Create sketch** tool from the **Toolbar** or **Part Design** menu. The **Select feature** dialog box will be displayed and you will be asked to specify a plane for creating sketch.
- Select the **XY** plane from dialog or drawing area and click on the **OK** button. The sketching environment will be displayed.
- Create the sketch as shown in Figure-13.

Figure-13. Sketch created on XY plane

- Click on the **Close** button from dialog after creating sketch to exit sketching mode.
- Click on the **Pad** tool from **Toolbar** after selecting newly created sketch. The **Pad parameters** dialog will be displayed with preview of feature.
- Select the **Symmetric to plane** check box from dialog and specify length of extrude feature as **120** mm; refer to Figure-14.

Figure-14. Pad parameters specified

- Click on the **OK** button from the dialog to create the feature.

Creating Loft feature

- Select the flat side face of model as shown in Figure-15 and click on the **Create sketch** tool from **Toolbar**. The sketching environment will be displayed.

Figure-15. Face selected for sketching

- Click on the **External geometry** tool from the **Sketcher geometries** cascading menu or **Toolbar**. You will be asked to select edges to be projected in the sketch.
- Select the boundary edges of flat face selected earlier; refer to Figure-16. Note that the projected edges are not part of sketch, they are for reference only.
- Click on the **Create rectangle** tool from **Toolbar** or **Sketcher geometries** cascading menu from the **Sketch** menu and create a rectangle overlapping the projected edges; refer to Figure-17.

Figure-16. Edges of face projected in sketch

Figure-17. Rectangle to be created

- Click on the **Close** button to exit sketching environment. Click in the empty area to exit any entity selected.
- Click on the **Create a datum plane** tool from the **Part Design** menu or **Toolbar**. The **Plane parameters** dialog will be displayed in **Combo View**.
- Select flat side face of model as reference and specify **Z** offset distance as **180**; refer to Figure-18.

Figure-18. Plane to be created

- Click on the **OK** button from the dialog to create plane.
- Select the newly created plane and click on the **Create sketch** tool from **Part Design** menu or **Toolbar**. The sketching environment will be displayed.
- Create a circle of diameter **90** mm at the origin; refer to Figure-19.

Figure-19. Circle to create

- Click on the **Close** button from the dialog to exit sketching mode.
- Click on the **Additive loft** tool from **Part Design** menu or **Toolbar**. The **Select feature** dialog will be displayed in **Combo View**.
- Select the sketch earlier created on flat face of model from the list; refer to Figure-20 and click on the **OK** button from the dialog. The **Loft parameters** dialog will be displayed and you will be asked to add section sketches for loft feature.

Figure-20. Sketch selected for base

- Click on the **Add Section** button from the dialog and select the circle created earlier. Preview of loft feature will be displayed; refer to Figure-21.

Figure-21. Preview of loft

- Click on the **OK** button from the dialog to create the feature.

Creating Second Pad Feature

- Select the datum plane earlier created and press **SPACEBAR** to hide it.
- Select the flat face of recently created loft feature and click on the **Create sketch** tool from **Part Design** menu. The sketching environment will be activated.
- Create a circle of diameter **70** at the origin; refer to Figure-22.

Figure-22. Circle created on loft face

- Click on the **Close** button from the dialog to exit sketching environment.
- Click on the **Pad** tool from the **Toolbar** or **Part Design** menu. Preview of pad feature will be displayed using the recently created sketch.
- Specify the length of feature as **220** in the **Length** edit box; refer to Figure-23 and click on the **OK** button to create the feature.

Figure-23. Pad feature after applying length value

Creating Pocket Feature

- Make sure nothing is selected from the model and then click on the **Create sketch** tool from **Part Design** menu. The **Select feature** dialog will be displayed with list of planes available.
- Select the **XZ Plane** from the dialog and click on the **OK** button. The sketching environment will be displayed; refer to Figure-24.

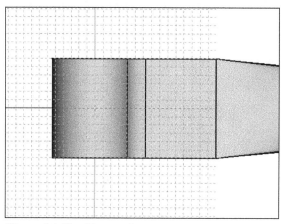

Figure-24. Sketching environment

- Create a rectangle as shown in Figure-25. You may need to switch draw style to **Wireframe** for creating rectangle.

Figure-25. Rectangle to create

- After creating sketch, click on the **Close** button.
- Make sure the sketch is selected and click on the **Pocket** tool from **Part Design** menu. Preview of the pocket feature will be displayed; refer to Figure-26.

Figure-26. Preview of pocket feature

- Click on the **OK** button from the dialog to create the feature.

Applying Fillets and Chamfers

- Select the edges of model and click on the **Fillet** tool from the **Part Design** menu. Preview of the fillet will be displayed; refer to Figure-27.

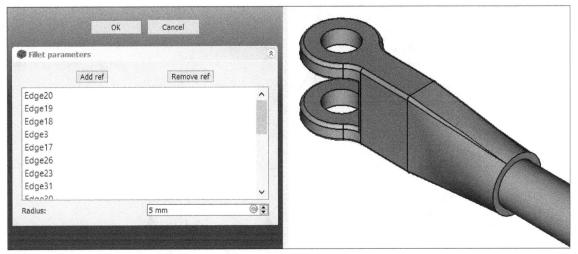

Figure-27. Preview of model after fillet

- Click on the **OK** button from the dialog to apply fillets.
- Select the round edge of pad feature created earlier and click on the **Chamfer** tool from **Part Design** menu.
- Specify the parameters as shown in Figure-28 and click on the **OK** button to apply chamfer.

Figure-28. Preview of chamfer applied

- Save the file at desired location.

PRACTICE 1 TO 4

Create 3D models for drawings shown in Figure-29, Figure-30, Figure-31, and Figure-32.

Figure-29. Practice 1

Figure-30. Practice 2

Figure-31. Practice 3

Figure-32. Practice 4

PRACTICAL 3

Create the model (isometric view) as shown in Figure-33. The dimensions and view are given in Figure-34.

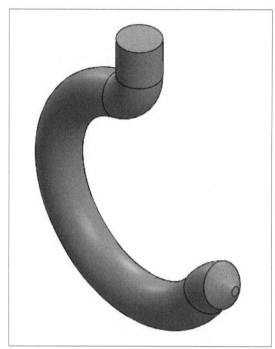

Figure-33. Model for Practical 3

Figure-34. Practical 3 drawing view

Creating Sketches for Hook

- Start FreeCAD application using desktop icon if not started yet.
- Create a new document and switch to **Part Design** workbench.
- Click on the **Create sketch** tool from **Toolbar**. The **Select feature** dialog will be displayed in **Combo View** and you will be asked to select sketching plane.
- Select the **XY Plane** from the list band click on the **OK** button. The sketching environment will be displayed.
- Create the sketch as shown in Figure-35.

Figure-35. Sketch to be created

- Click on **Close** button from the dialog to exit sketching environment.
- Click in the empty area to make sure nothing is selected.
- Select straight line of sketch as shown in Figure-36 and click on the **Create a datum plane** tool from **Toolbar** or **Part Design** menu. Preview of plane will be displayed normal to selected line segment; refer to Figure-37.

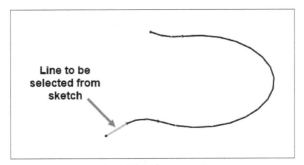

Figure-36. Line selected for plane creation

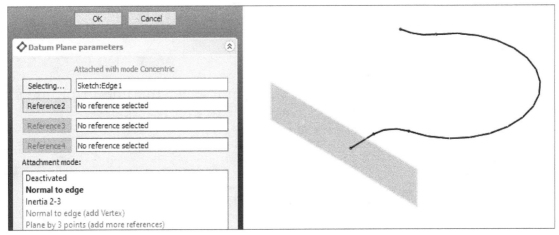

Figure-37. Preview of plane

- Click on the **OK** button from the dialog to create plane.
- Select the newly created plane from drawing area and click on the **Create sketch** tool from **Part Design** menu or **Toolbar**. The sketching environment will become active.

- Create a circle of diameter **25** mm at the origin; refer to Figure-38.

Figure-38. Second sketch to be created

- Click on the **Close** button from the dialog to exit sketching environment. Click in the empty area of drawing to make sure nothing is selected.

Creating Sweep Feature

- Select the newly created circle sketch and click on the **Additive pipe** tool from the **Part Design** menu. The **Pipe parameters** dialog will be displayed with earlier selected sketch defined as section profile; refer to Figure-39.

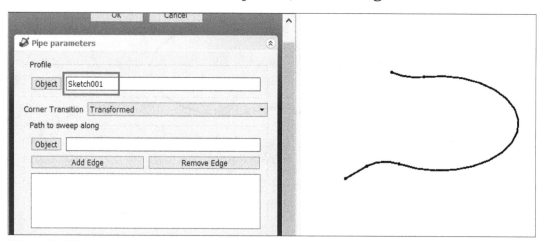

Figure-39. Pipe parameters dialog

- Click on the **Add Edge** button from dialog and select the line in profile sketch. Preview of pipe feature will be displayed; refer to Figure-40.

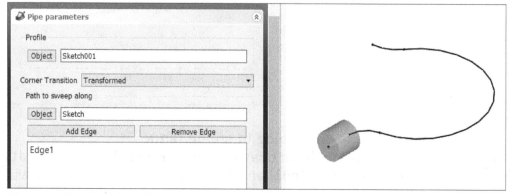

Figure-40. Preview of pipe feature

- Click on the **Add Edge** button again and select next curve in the sketch. Repeat the steps until all the curves of sketch are added as path; refer to Figure-41.
- Click on the **OK** button from dialog to create the feature.

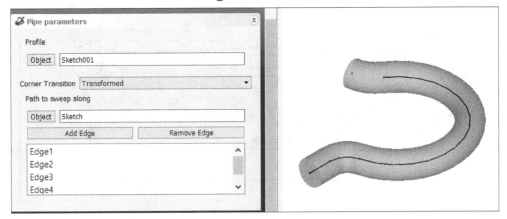

Figure-41. Preview of feature

- Apply a fillet of radius **10** mm at the end edge of model; refer to Figure-42.

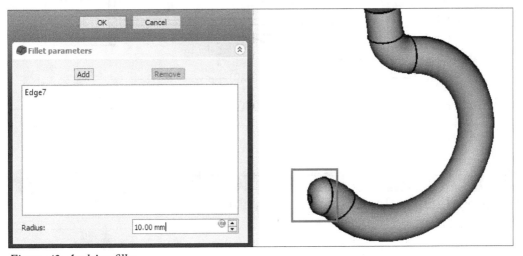

Figure-42. Applying fillet

PRACTICE 5

Create the model as shown in Figure-43. The dimensions are given in Figure-44.

Figure-43. Practice 5

Figure-44. Dimensions of the Practice 5 model

PRACTICE 6

Create the model using the drawings shown in Figure-45.

Figure-45. Rope Pulley

PRACTICE 7

Create the model as shown in Figure-46. Dimensions are given in Figure-47. Assume the missing dimensions.

Figure-46. Practice 7 model

Figure-47. Dimensions of the Practice 7 model

PRACTICE 8

Create the model by using the dimensions given in Figure-48.

Figure-48. Practice 8

PRACTICE 9

Create the model by using the dimensions given in Figure-49.

Figure-49. Practice 9

PRACTICE 10

Create the model by using the dimensions given in Figure-50.

Figure-50. Practice 10

PRACTICE 11

Create a ring nut with value of **D** as **5, 6, 8**, and **10** using equation and design table. Dimensions are given in Figure-51.

Figure-51. Ring Nut

FREE STUDENT NOTES

Chapter 6

Drafting

Topics Covered

The major topics covered in this chapter are:

- *Introduction to Drafting*
- *Drawing Objects Tools*
- *Annotation Objects Tools*
- *Modifying Objects Tools*
- *Draft Tray Toolbar*
- *Draft Snap Toolbar*

INTRODUCTION

The tools in **Draft Workbench** are used to draw simple 2D objects and modify them afterwards. It also provides tools to define a working plane, a grid, and a snapping system to precisely control the position of your geometry. Example of the functionality of the **Draft Workbench** includes many elements like lines, curves, arcs, and polygons; refer to Figure-1.

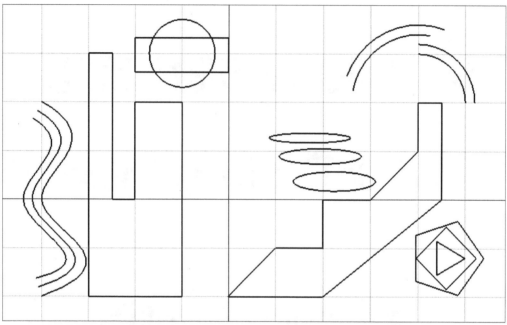

Figure-1. Draft workbench example

The created 2D objects can be used for general drafting in a way similar to Inkscape or AutoCAD. These 2D shapes can also be used as base features of 3D objects created with other workbenches, for example, the **Part** and **Arch** workbenches. Conversion of draft objects to sketches is also possible which means that the shapes can also be used with the **PartDesign** workbench for the creation of solid bodies.

STARTING DRAFT WORKBENCH

FreeCAD is primarily a 3D modeling application and thus its 2D tools aren't as advanced as in other drawing programs. If your primary goal is the production of complex 2D drawings and DXF files and you don't need 3D modeling, you may wish to consider a dedicated software program for technical drafting such as LibreCAD, QCad, or others.

• To start a new draft file, click on the **New** button from **File** menu and select **Draft** workbench from **Switch between workbenches** drop-down in the **Toolbar**; refer to Figure-2. The tools to perform drafting will be displayed in the **Toolbar**; refer to Figure-3.

Figure-2. Draft workbench

Figure-3. Draft workbench tools

Drawing objects tools

The **Drawing objects** tools are used for creating 2D objects. These tools are available in the **Toolbar** of **Draft** workbench; refer to Figure-4. The procedures to use these tools are discussed next.

Figure-4. Drawing objects tools

Creating Line

The **Line** tool creates a straight line defined by two points. The procedure to use this tool is discussed next.

• Click on the **Line** tool from **Toolbar** in the **Draft** workbench; refer to Figure-5. The **Line** dialog will be displayed in the **Tasks** panel of **Combo View** along with the plus 🔲 sign in place of original cursor; refer to Figure-6. You will be asked to specify the first point.

Figure-5. Line tool

Figure-6. Line dialog

- Click in the 3D view area to specify the first point or enter desired values for x, y, and z coordinates in the **Local X**, **Local Y**, and **Local Z** edit boxes of the dialog, respectively.
- After specifying coordinates for the first point in dialog, click on the **Enter point** button from the dialog. You will be asked to specify the second point.
- Move the cursor away and click at desired location to specify the second point or enter desired values for the coordinates in their respective edit boxes.
- After specifying coordinates for the second point in dialog, click on the **Enter point** button again. The line will be created; refer to Figure-7.

Figure-7. Line created

- Specify desired length of the line in the **Length** edit box.
- Specify desired value for the angle of line in the **Angle** edit box. Select the check box next to the **Angle** to constrain the pointer to the specified angle.
- On selecting the **Relative (R)** check box, the coordinates of second point are relative to the first one. If not selected, they are absolute taken from the origin.
- Select **Global (G)** check box to specify coordinates relative to global coordinate system.
- On selecting the **Continue (T)** check box, the **Line** tool will restart after you give the second point allowing you to draw another line segment without clicking on the tool button again.
- Click on **Undo (CTRL+Z)** button to undo the last point.

Creating Polyline

The **Wire** or **Polyline** tool creates a sequence of several line segments. This tool allows you to enter more than two points. The procedure to use this tool is discussed next.

- Click on the **Polyline** tool from **Toolbar** in the **Draft** workbench; refer to Figure-8. The **Polyline** dialog will be displayed in the **Tasks** panel of **Combo View** along with the plus ⊞ sign in place of original cursor; refer to Figure-9. You will be asked to specify first point of the line.

Figure-8. Polyline tool

Figure-9. Polyline dialog

- Click in the 3D view area to specify the first point or enter desired values for x, y, and z coordinates in the **Local X**, **Local Y**, and **Local Z** edit boxes of the dialog, respectively.
- After specifying coordinates for the first point in dialog, click on the **Enter point** button from the dialog. You will be asked to specify the additional points.
- Move the cursor away and click at desired locations to specify the additional points for the polyline or enter desired values for the coordinates in their respective edit boxes.
- After specifying coordinates for additional points, click on the **Enter point** button again. The polyline will be created; refer to Figure-10.

Figure-10. Polyline created

- On selecting the **Filled** check box, a closed wire or a polyline will create a face.
- Click on the **Finish (A)** button to finish the polyline, leaving it open.
- Click on the **Close (O)** button to close the wire, that is, a line segment will be added from the last point to the first one to form a face.
- Click on the **Wipe (W)** button to remove the line segments already placed but keep editing the wire from the last point.
- Click on the **Set WP (U)** button to adjust the current working plane in the orientation of the last point.
- Other parameters of the **Polyline** tool have been discussed in the **Line** tool.
- Click on **Close** button to close the dialog.

Creating Fillet

The **Fillet** tool creates a fillet, a rounded corner, or a chamfer, a straight edge, between two draft lines. The procedure to use this tool is discussed next.

• Select two lines that meet in a single point from the **Model Tree** or from the 3D view area between which you want to create a fillet or a chamfer; refer to Figure-11.

Figure-11. Selecting the lines to create fillet or chamfer

• Click on the **Fillet** tool from **Toolbar** in the **Draft** workbench; refer to Figure-12. The **Fillet** dialog will be displayed in the **Tasks** panel of **Combo View**; refer to Figure-13.

Figure-12. Fillet tool

Figure-13. Fillet dialog

• Specify desired radius for the fillet in the **Fillet radius** edit box of the dialog.
• Select **Delete original objects** check box to delete the original object after creating fillet.
• Select **Create chamfer** check box to create the chamfer feature between the selected lines.
• After specifying desired parameters, press **Enter** key from the keyboard. The fillet feature will be created; refer to Figure-14.

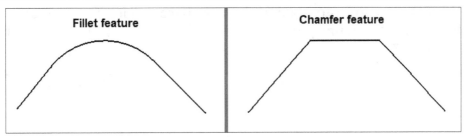

Figure-14. Fillet and chamfer features created between the lines

Creating Arc

The **Arc** tool creates a circular arc in the current work plane by entering four points, the center, a point that defines the radius, the first end point and the second end point, or by picking tangents, or any combination of these. The procedure to use this tool is discussed next.

- Click on the **Arc** tool from **Arc tools** drop-down in the **Toolbar** of **Draft** workbench; refer to Figure-15. The **Arc** dialog will be displayed in the **Tasks** panel of **Combo View** along with the plus ⊞ sign in place of original cursor; refer to Figure-16. You will be asked to specify the first point.

Figure-15. Arc tool

Figure-16. Arc dialog

- Click in the 3D view area to specify the first point or enter desired values for x, y, and z coordinates in the **Local X**, **Local Y**, and **Local Z** edit boxes of the dialog, respectively.
- After specifying coordinates in the dialog, click on the **Enter point** button. You will be asked to specify the radius of arc.
- Move the cursor away and click at desired location to specify the second point or enter desired radius value for the arc in the **Radius** edit box. You will be asked to specify the start angle.
- Click in the 3D view area to specify the start angle for the arc or enter desired angle value in the **Start angle** edit box of the dialog. You will be asked to specify the aperture angle.

- Click in the 3D view area to specify the aperture angle of an arc or enter desired angle value in the **Aperture angle** edit box of the dialog. The arc will be created; refer to Figure-17.

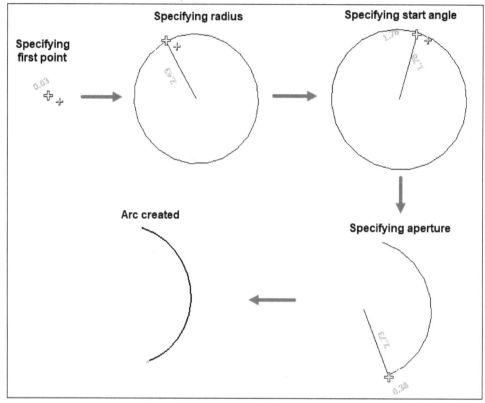

Figure-17. Arc created

- Other parameters in the **Arc** dialog have been discussed earlier.

Creating Arc by 3 Points

The **Arc by 3 points** tool creates a circular arc in the current working plane using three points that define its circumference. The center and radius are calculated from these points. The procedure to use this tool is discussed next.

- Click on the **Arc by 3 points** tool from **Arc tools** drop-down in the **Toolbar** of **Draft** workbench; refer to Figure-18. The **Arc by 3 points** dialog will be displayed in the **Tasks** panel of **Combo View** along with the plus ⊕ sign in place of original cursor; refer to Figure-19. You will be asked to specify the first point.

Figure-18. Arc by 3 points tool

Figure-19. Arc by 3 points dialog

- Click in the 3D view area to specify the first point or enter desired values for x, y, and z coordinates in the **Local X**, **Local Y**, and **Local Z** edit boxes of the dialog, respectively.
- After specifying coordinates for the first point, click on the **Enter point** button. You will be asked to specify the additional two points.
- Move the cursor away and click at desired locations to specify the second and third points or enter desired values for the coordinates in their respective edit boxes.
- On specifying the third point, the arc by three points will be created; refer to Figure-20.

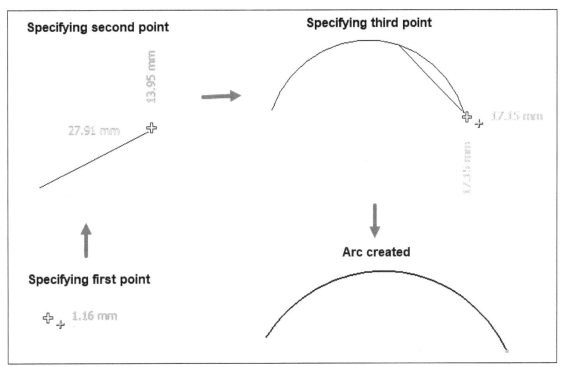

Figure-20. Three points arc created

- Other parameters in the dialog have been discussed earlier.

Creating Circle

The **Circle** tool creates a circle in the current work plane by entering two points, the center and the radius, or by picking tangents, or any combination of these. The procedure to use this tool is discussed next.

- Click on the **Circle** tool from **Toolbar** in the **Draft** workbench; refer to Figure-21. The **Circle** dialog will be displayed in the **Tasks** panel of **Combo View** along with the plus ⊞ sign in place of original cursor; refer to Figure-22. You will be asked to specify the first point.

Figure-21. Circle tool

Figure-22. Circle dialog

- Click in the 3D view area to specify the first point or enter desired values for x, y, and z coordinates in the **Local X**, **Local Y**, and **Local Z** edit boxes of the dialog, respectively.
- After specifying coordinates for the first point in dialog, click on the **Enter point** button. You will be asked to specify the radius of the circle.
- Move the cursor away and click at desired location to specify second point for the circle or enter desired radius value in the **Radius** edit box of the dialog. The circle will be created; refer to Figure-23.

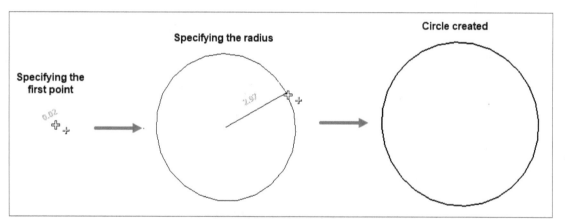

Figure-23. Circle created

- Other parameters in the **Circle** dialog have been discussed earlier.

Creating Ellipse

The **Ellipse** tool creates an ellipse in the current work plane by entering two points, defining the corners of a rectangular box in which the ellipse will fit. This tool can also be used to create elliptical arcs by specifying the start and end angles. The procedure to use this tool is discussed next.

- Click on the **Ellipse** tool from **Toolbar** in the **Draft** workbench; refer to Figure-24. The **Ellipse** dialog will be displayed in the **Tasks** panel of **Combo View** along with the plus ⊞ sign in place of original cursor; refer to Figure-25. You will be asked to specify the first point.

Figure-24. Ellipse tool

Figure-25. Ellipse dialog box

- Click in the 3D view area to specify the first point or enter desired values for the x, y, and z coordinates in the **Local ΔX**, **Local ΔY**, **Local ΔZ** and edit boxes of the dialog, respectively.
- After specifying coordinates for the first point in dialog, click on the **Enter point** button from the dialog. You will be asked to specify the second point.
- Move the cursor away and click at desired location to specify the second point or enter desired values for the coordinates in their respective edit boxes.
- After specifying coordinates for the second point in dialog, click on the **Enter point** button again. The ellipse will be created; refer to Figure-26.

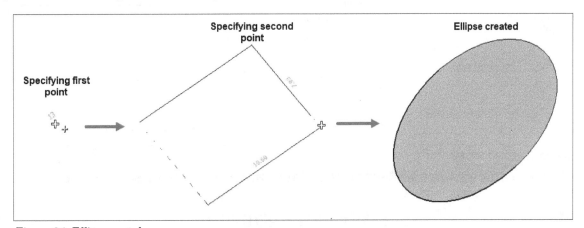

Figure-26. Ellipse created

- Other parameters in the **Ellipse** dialog have been discussed earlier.

Creating Rectangle

The **Rectangle** tool creates a rectangle by picking two points. The procedure to use this tool is discussed next.

- Click on the **Rectangle** tool from **Toolbar** in the **Draft** workbench; refer to Figure-27. The **Rectangle** dialog will be displayed in the **Tasks** panel of **Combo View** along with the plus ⊞ sign in place of original cursor; refer to Figure-28. You will be asked to specify the first point.

Figure-27. Rectangle tool

Figure-28. Rectangle dialog

- Click in the 3D view area to specify the first point or enter desired values for the x, y, and z coordinates in the **Local ΔX**, **Local ΔY**, and **Local ΔZ** edit boxes of the dialog, respectively.
- After specifying coordinates for the first point in dialog, click on the **Enter point** button from the dialog. You will be asked to specify the second point.
- Move the cursor away and click at desired location to specify the second point or enter desired values for the coordinates in their respective edit boxes of the dialog.
- After specifying coordinates for the second point in dialog, click on the **Enter point** button again. The rectangle will be created; refer to Figure-29.

Figure-29. Rectangle created

- Other parameters in the **Rectangle** dialog have been discussed earlier.

Creating Polygon

The **Polygon** tool creates a regular polygon inscribed in a circumference by picking two points, the center, and the radius. The procedure to use this tool is discussed next.

- Click on the **Polygon** tool from **Toolbar** in the **Draft** workbench; refer to Figure-30. The **Polygon** dialog will be displayed in the **Tasks** panel of **Combo View** along with the plus ⊕ sign in place of original cursor; refer to Figure-31. You will be asked to specify the first point.

Figure-30. Polygon tool

Figure-31. Polygon dialog

- Click in the 3D view area to specify the first point or enter desired values for the x, y, and z coordinates in the **Local ΔX**, **Local ΔY**, and **Local ΔZ** edit boxes of the dialog, respectively.
- After specifying coordinates in the dialog, click on the **Enter point** button from the dialog. You will be asked to specify the sides and radius.
- Specify desired number of sides for the polygon in the **Sides** edit box of the dialog.
- Move the cursor away and click at desired location to specify the radius of the polygon or enter desired value of radius in the **Radius** edit box of the dialog. The regular polygon will be created; refer to Figure-32.

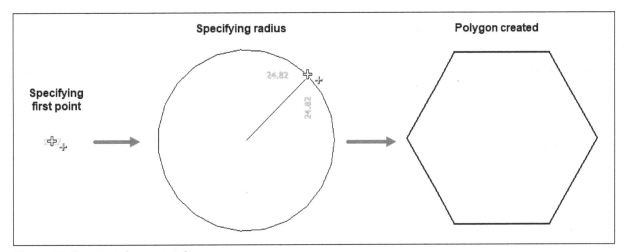

Figure-32. Regular polygon created

- Other parameters in the **Polygon** dialog have been discussed earlier.

Creating B-Spline

The **B-Spline** tool creates a B-spline curve from several points. The procedure to use this tool is discussed next.

- Click on the **B-spline** tool from **Toolbar** in the **Draft** workbench; refer to Figure-33. The **Bspline** dialog will be displayed in the **Tasks** panel of **Combo View** along with the plus ⊞ sign in place of original cursor; refer to Figure-34. You will be asked to specify the first point.

Figure-33. B-Spline tool

Figure-34. Bspline dialog

- Click in the 3D view area to specify the first point or enter desired values for the x, y, and z coordinates in the **Local ΔX**, **Local ΔY**, and **Local ΔZ** edit boxes of the dialog, respectively.
- After specifying coordinates for the first point in the dialog, click on the **Enter point** button from the dialog. You will be asked to specify the additional points.
- Move the cursor away and click at desired locations to specify the additional points or enter desired values for the coordinates in their respective edit boxes.
- After specifying coordinates for the additional points in the dialog, click on the **Enter point** button again and press **Esc** key. The B-spline will be created; refer to Figure-35.

Figure-35. B-spline created

- Other parameters in the **B-spline** dialog have been discussed earlier.

Creating Cubic bezier curve

The **Cubic bezier curve** tool creates a bezier curve of third degree by dragging two points. The procedure to use this tool is discussed next.

- Click on the **Cubic bezier curve** tool from **Bezier tools** drop-down in the **Toolbar** of **Draft** workbench; refer to Figure-36. The **CubicBezCurve** dialog will be displayed in the **Tasks** panel of **Combo View** along with the plus ⊞ sign in place of original cursor; refer to Figure-37. You will be asked to specify the first point.

Figure-36. Cubic bezier curve tool

Figure-37. CubicBezCurve dialog box

Note that it is not possible to create the curve by entering coordinates value in the edit boxes of dialog.

- To specify the starting point of curve, click & hold the **LMB**, move the cursor in desired direction to define slope of curve. Release the **LMB**, click & hold again to specify the next point and slope. Repeat the procedure to specify consecutive points. Press **ESC** button to exit the tool. The cubic bezier curve will be created; refer to Figure-38.

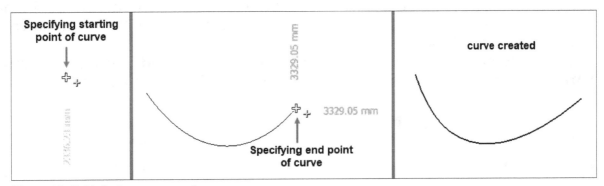

Figure-38. Cubic bezier curve created

- Other parameters in the dialog have been discussed earlier. Press **ESC** key to exit the tool.

Creating Bezier Curve

The **Bezier Curve** tool creates a bezier curve or a piecewise bezier curve from several points. This tool uses control points to define the direction of the curve. The procedure to use this tool is discussed next.

- Click on the **Bezier curve** tool from **Bezier tools** drop-down in the **Toolbar** of **Draft** workbench; refer to Figure-39. The **BezCurve** dialog will be displayed in the **Tasks** panel of **Combo View** along with the plus ⊞ sign in place of original cursor; refer to Figure-40. You will be asked to specify the first point.

Figure-39. Bezier curve tool

Figure-40. BezCurve dialog box

- Click in the 3D view area to specify the first point or enter desired values for the x, y, and z coordinates in the **Local ΔX**, **Local ΔY**, and **Local ΔZ** edit boxes of the dialog, respectively.
- After specifying coordinates for the first point in the dialog, click on the **Enter point** button from the dialog.
- Move the cursor away and click at desired locations to specify the additional points to create the bezier curve or enter desired values for the coordinates in their respective edit boxes of the dialog.
- After specifying coordinates for additional points in the dialog, click on the **Enter point** button again. The bezier curve will be created; refer to Figure-41.

Figure-41. Bezier curve created

- Other parameters in the **BezCurve** dialog have been discussed earlier. Press **Esc** key from keyboard to exit the tool.

Creating Point

The **Point** tool creates a simple point in the current work plane, handy to serve as reference for placing lines, wires, or other objects later. The procedure to use this tool is discussed next.

- Click on the **Point** tool from **Toolbar** in the **Draft** workbench; refer to Figure-42. The **Point** dialog will be displayed in the **Tasks** panel of **Combo View** along with the plus ⊕ sign in place of original cursor; refer to Figure-43. You will be asked to specify the location of point.

Figure-42. Point tool

Figure-43. Point dialog

- Click in the 3D view area to specify the location of point or enter desired values for x, y, and z coordinates in the **Local ΔX**, **Local ΔY**, and **Local ΔZ** edit boxes of the dialog, respectively.
- After specifying coordinates in the dialog, click on the **Enter point** button from the dialog. The point will be created; refer to Figure-44.

Figure-44. Point created

- Other parameters in the **Point** dialog have been discussed earlier. Click on **Close** button to close the dialog.

Creating Facebinder

The **Facebinder** tool creates a surface object from the selected faces of a solid object. It is parametric tool, meaning that if you modify the original object, the facebinder updates accordingly. It can be used to create an extrusion from a collection of faces to other objects. The procedure to use this tool is discussed next.

- Select desired faces from the solid object which you want to create the facebinder; refer to Figure-45.

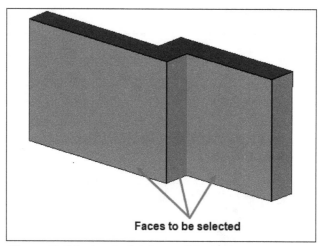

Figure-45. Selecting the faces to create facebinder

- Click on the **Facebinder** tool from **Toolbar** in the **Draft** workbench; refer to Figure-46. The facebinder object will be created; refer to Figure-47.

Figure-46. Facebinder tool

Figure-47. Facebinder object created

- If you want to see the facebinder object, select the original solid object from the **Model Tree** and hide it using **Space** key from the keyboard. The facebinder object will be displayed in the 3D view area; refer to Figure-48.

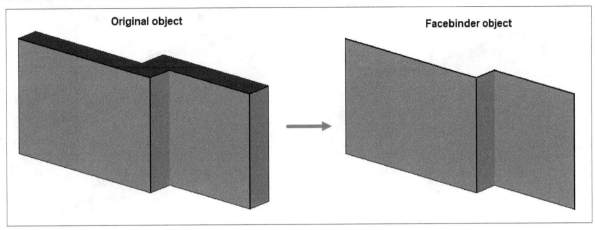
Figure-48. Facebinder object displayed

- You can edit the facebinder object using **Faces** dialog.
- To add the face to the facebinder object, press and hold the **CTRL** key from the keyboard and select the face or multiple faces from the solid object in the 3D view area; refer to Figure-49.

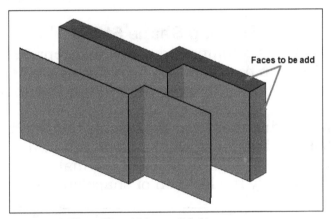
Figure-49. Faces to be add from solid object

- Holding the **CTRL** key, double-click on the facebinder object from the **Model Tree**. The **Faces** dialog will be displayed in the **Tasks** panel of **Combo View**; refer to Figure-50.

Figure-50. Faces dialog

- Click on **Add** button from the dialog, the faces added in the dialog will be added in the facebinder object; refer to Figure-51.

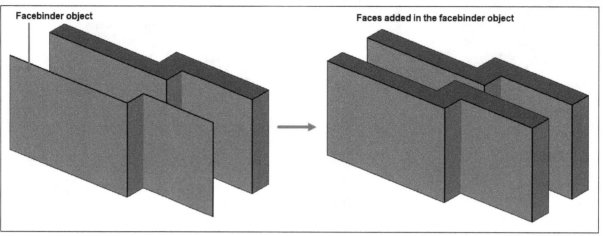

Figure-51. Faces added in the facebinder object

- If you want to remove the face from the facebinder object then select the face from the **Facebinder elements** list view area of the dialog and click on **Remove** button. The face will be removed.
- Click on **OK** button from the dialog to complete the procedure.

Creating Shape String

The **Shape String** tool inserts a compound shape that represents a text string. Text height, tracking, and font can be specified. The resulting shape can be used with the **Extrude** tool to create 3D letters. The procedure to use this tool is discussed next.

- Click on the **Shape String** tool from **Toolbar** in the **Draft** workbench; refer to Figure-52. The **ShapeString** dialog will be displayed in the **Tasks** panel of **Combo View** along with the plus ⊞ sign in place of original cursor; refer to Figure-53. You will be asked to specify the location of shape string.

Figure-52. Shape String tool

Figure-53. ShapeString dialog

- Click in the 3D view area to specify the location of shape string or enter desired values for x, y, and z coordinates in the **X**, **Y**, and **Z** edit boxes of the dialog, respectively.
- If you want to reset the coordinates, click on the **Reset Point** button from the dialog.
- Specify desired text you want to create in the **String** edit box.
- Specify the height of the text in the **Height** edit box.
- Click on the button ⬚ available next to the **Font file** edit box. The **Select a file** dialog box will be displayed; refer to Figure-54.

Figure-54. Select a file dialog box

- Select desired font file from the dialog box and click on **Open** button. The selected font file will be displayed in the **Font file** edit box of the dialog; refer to Figure-55. The font file is available in the installation folder of FreeCAD software.

Figure-55. Font file selected

- Click on **OK** button from the dialog. The shape string will be created; refer to Figure-56.

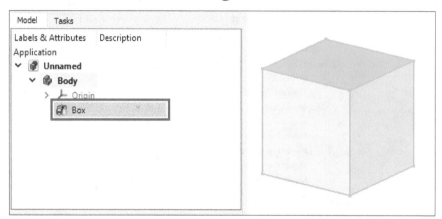

Figure-56. Shape string created

Creating Hatch

The **Hatch** tool is used to create hatches on the planar faces of a selected object. The procedure to use this tool is discussed next.

• Select desired object with planar faces from the **Model Tree** or from the 3D view area to create hatches on it; refer to Figure-57.

Figure-57. Object with planar faces selected

• Click on the **Hatch** tool from **Toolbar** in the **Draft** workbench; refer to Figure-58. The **Form** dialog will be displayed in the **Tasks** panel of **Combo View**; refer to Figure-59.

Figure-58. Hatch tool

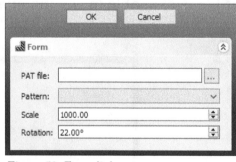

Figure-59. Form dialog

• Click on the ⋯ button to specify a PAT file for **PAT file** edit box. The **Select a file** dialog box will be displayed; refer to Figure-60.

Figure-60. Select a file dialog box

- Select a PAT file which is installed with FreeCAD and click on the **Open** button from the dialog box. (On windows, PAT file is usually located in: **C:\Program Files\ FreeCAD 0.21\data\Mod\TechDraw\PAT**).
- Select desired type of pattern from **Pattern** drop-down.
- Specify desired scale value for the pattern in the **Scale** edit box.
- Specify desired rotation value for the pattern in the **Rotation** edit box.
- After specifying desired parameters, click on the **OK** button from the **Form** dialog. The hatching will be created on the object; refer to Figure-61.

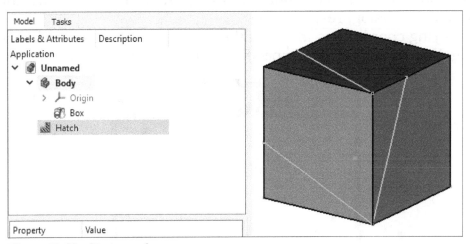

Figure-61. Hatching created

Annotation objects tools

The **Annotation objects** tools are used for annotating the objects. These tools are available in the **Toolbar** of **Draft** workbench; refer to Figure-62. The procedures to use these tools are discussed next.

Figure-62. Annotation objects tools

Creating Text

The **Text** tool inserts a multi-line text box at a given point. The procedure to use this tool is discussed next.

- Click on the **Text** tool from **Toolbar** in the **Draft** workbench; refer to Figure-63. The **Text** dialog will be displayed in the **Tasks** panel of **Combo View** along with the plus ⊕ sign in place of original cursor; refer to Figure-64. You will be asked to specify the location of text.

Figure-63. Text tool

Figure-64. Text dialog

- Click in the 3D view area to specify desired location to place text or enter desired values for the x, y, and z coordinates in the **Local ΔX**, **Local ΔY**, and **Local ΔZ** edit boxes of the dialog, respectively.
- After specifying coordinates in the dialog, click on the **Enter point** button from the dialog. You will be asked to specify text in the **Text** dialog; refer to Figure-65.

Figure-65. Text dialog asking to specify text

- Specify desired text in the edit box of the dialog and click on **Create text** button or press **ENTER** key twice from the keyboard. The text will be created; refer to Figure-66.

CADCAMCAE Works

Figure-66. Text created

- Other parameters in the dialog have been discussed earlier.

Creating Dimension

The **Dimension** tool creates an object that measures and displays the distance between two points; a third point specifies the position of the dimension line. The procedure to use this tool is discussed next.

- Click on the **Dimension** tool from **Toolbar** in the **Draft** workbench; refer to Figure-67. The **Dimension** dialog will be displayed in the **Tasks** panel of **Combo View** along with the plus ⊞ sign in place of original cursor; refer to Figure-68. You will be asked to specify the first point or select the edge.

Figure-67. Dimension tool

Figure-68. Dimension dialog

- Click in the 3D view area to specify first point from where you want to measure the distance or enter desired values for the x, y, and z coordinates in the **Local ΔX**, **Local ΔY**, and **Local ΔZ** edit boxes of the dialog, respectively.
- After specifying coordinates for the first point in the dialog, click on the **Enter point** button from the dialog. You will be asked to specify the second point.
- Move the cursor away and click at desired location to specify second point up to which you want to measure the distance or enter desired values for the coordinates in their respective edit boxes of the dialog. The first two points define the measured distance.

- After specifying coordinates for the second point in the dialog, click on the **Enter point** button again from the dialog. You will be asked to specify the third point.
- Move the cursor away and click at desired location to specify the third point or enter desired values for the coordinates in their respective edit boxes of the dialog. The third point defines the position of the measurement line.
- After specifying coordinates for the third point in the dialog, click on the **Enter point** button again from the dialog. The dimension will be created; refer to Figure-69.

Figure-69. Dimension created

- If you want to create the dimension by selecting the edge then on selecting the tool, the **Dimension** dialog will be displayed as discussed earlier.
- Click on the **Select edge** button from the **Dimension** dialog. You will be asked to select the edge to be measured.
- Select the edge of the model whose distance is to be measured by this dimension; refer to Figure-70. You will be asked to specify the point to place the measurement line.

Figure-70. Selecting the edge to be measured

- Move the cursor away and click at desired location to specify the point to position the measurement line or enter desired values for the x, y, and z coordinates in the **Local X**, **Local Y**, and **Local Z** edit boxes of the dialog, respectively.
- After specifying coordinates for the point in the dialog, click on the **Enter point** button from the dialog. The dimension will be created; refer to Figure-71.

Figure-71. Dimension created

- Other parameters in the **Dimension** dialog have been discussed earlier.
- Click on the **Close** button to close the dialog.

Creating Label

The **Label** tool inserts a multi-line text box with a 2-segment leader line and an arrow. If an object or a sub-element (face, edge, or vertex) is selected when starting the command, the Label can be made to display a certain attribute of the selected element including position, length, area, volume, or material. The procedure to use this tool is discussed next.

- Click on the **Label** tool from **Toolbar** in the **Draft** workbench; refer to Figure-72. The **Label** and **Label type** dialogs will be displayed in the **Tasks** panel of **Combo View** along with the plus ⊕ sign in place of original cursor; refer to Figure-73. You will be asked to specify the first point.

Figure-72. Label tool

Figure-73. Label and Label type dialogs

- Select desired label type from the drop-down available in the **Label type** dialog. Select **Custom** option to display the contents of custom text. Select **Name** option to display the internal name of target object. Select **Label** option to display the label of target object. Select **Position** option to display the coordinates of the base point of the target object. Select **Length** option to display the length of the target sub-element. Select **Area** option to display the area of the target sub-element. Select **Volume** option to display the volume of target object. Select **Tag** option to display the tag attribute of the target object. Select **Material** option to display the label of the material of the target object.

- Click in the 3D view area to specify the first point of label or enter desired values for the x, y, and z coordinates in the **Local ΔX**, **Local ΔY**, and **Local ΔZ** edit boxes of the dialog, respectively.

- After specifying coordinates for the first point in the dialog, click on the **Enter point** button from the dialog. You will be asked to specify the second point of label or the start point of a horizontal or vertical leader.

- Move the cursor away and click at desired location to specify the second point or enter desired value for the coordinates in their respective edit boxes.

- After specifying coordinates for the second point, click on the **Enter point** button from the dialog. You will be asked to specify the third point of label or the base point of the text.

- Move the cursor away and click at desired location to specify the third point or enter desired values for the coordinates in their respective edit boxes.

- After specifying coordinates for the third point in the dialog, click on the **Enter point** button again. The label will be created; refer to Figure-74.

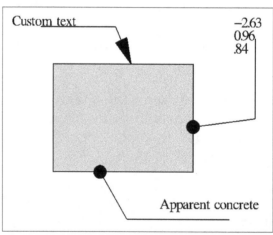

Figure-74. Labels created

Annotation Style Editor

The **Annotation style editor** tool allows you to define styles that affect the visual properties of annotation objects like dimension lines, text, and so on. The procedure to use this tool is discussed next.

- Click on the **Annotation style editor** tool from **Toolbar** in the **Draft** workbench; refer to Figure-75. The **Annotation Styles Editor** dialog box will be displayed; refer to Figure-76.

Figure-75. Annotation style editor tool

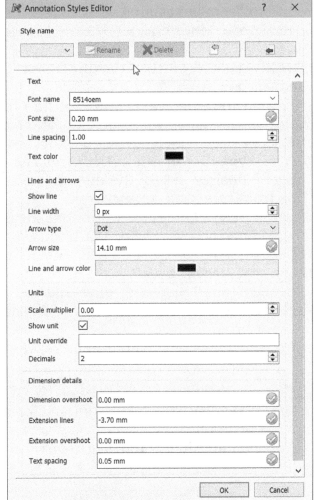

Figure-76. Annotation Styles Editor dialog box

- Select desired annotation style from the drop-down in the **Style name** area of the dialog box or select **Add new** option from the drop-down to define the new style.
- Click on the **Rename** button from **Style name** area to rename the selected style.
- Click on the **Delete** button from **Style name** area to delete the selected style.
- Click on the ⬜ button from **Style name** area to import all styles from **.json** file.
- Click on the ⬜ button from **Style name** area to export all styles to **.json** file.
- Specify desired font name to use for texts and dimensions in the **Font name** edit box from **Text** area.
- Specify desired size of the font in the **Font size** edit box from **Text** area in the dialog box.
- Specify desired spacing between lines in the **Line spacing** edit box from the **Text** area.
- Specify desired value in the **Scale multiplier** edit box from **Units** area to change the size of texts and markers.
- Select the **Show unit** check box from **Units** area to show the unit next to the dimension value.
- Specify a valid length unit like mm, m, in, ft in the **Unit override** edit box from **Units** area to force displaying the dimension value in this unit.

- Specify desired number of decimals to show for dimension values in the **Decimals** edit box from **Units** area.
- Select **Show lines** check box from **Line and arrows** area to display the dimension line.
- Specify desired width of the dimension lines in the **Line width** edit box from **Line and arrows** area.
- Click on the **Line/text color** button from **Line and arrows** area to specify the color of dimension lines, arrows, and texts.
- Select desired type of arrows or markers to use at the end of dimension lines from **Arrow type** drop-down in the **Line and arrows** area.
- Specify desired size of the dimension arrows or markers in the **Arrow size** edit box from **Line and arrows** area.
- Specify desired distance value that the dimension line is additionally extended in the **Dimension overshoot** edit box from **Line and arrows** area.
- Specify desired length of the extension lines in the **Extension lines** edit box from **Line and arrows** area.
- Specify desired distance value that the extension lines are additionally extended beyond the dimension line in the **Extension overshoot** edit box from **Line and arrows** area.
- After specifying desired parameters, click on the **OK** button to close the dialog box.

Modifying objects tools

The **Modifying objects** tools are used for modifying existing objects. They work on selected objects but if no object is selected, you will be asked to select one. These tools are available in the **Toolbar** of **Draft** workbench; refer to Figure-77. These tools are discussed next.

Figure-77. Modifying object tools

Moving object

The **Move** tool moves or copies the selected objects from one point to another. The procedure to use this tool is discussed next.

- Select the object from the **Model Tree** which you want to move; refer to Figure-78.

Figure-78. Selecting the object to move

- Click on the **Move** tool from **Toolbar** in the **Draft** workbench; refer to Figure-79. The **Move** dialog will be displayed in the **Tasks** panel of **Combo View** along with the plus ⊞ sign in place of original cursor; refer to Figure-80. You will be asked to specify the base point to move or copy an object.

Figure-79. Move tool

Figure-80. Move dialog

- Click in the 3D view area to specify the base point of moving the object or enter desired values for x, y, and z coordinates in the **Local ΔX**, **Local ΔY**, and **Local ΔZ** edit boxes of the dialog, respectively.
- After specifying coordinates in the dialog, click on the **Enter point** button from the dialog. You will be asked to specify the new position for the base point.
- Move the cursor away and click at desired location to specify the new position of the base point or enter desired values for the coordinates in their respective edit boxes of the dialog.
- After specifying coordinates in the dialog, click on the **Enter point** button again. The object will be moved to the new position; refer to Figure-81.

Figure-81. Object moved

- Specify desired length of the current segment in the **Length** edit box.
- Specify desired angle of the current segment in the **Angle** edit box and select the **Angle** check box to lock the current angle.
- Select **Copy** check box from the dialog to keep the original object in its place and make a copy at the second point.
- Select the **Modify subelements** check box to modify the sub-elements instead of entire objects.
- Other parameters in the **Move** dialog have been discussed earlier.

Rotating object

The **Rotate** tool rotates or copies the selected objects by a given angle around a reference point. The procedure to use this tool is discussed next.

- Select the object from the **Model Tree** which you want to rotate; refer to Figure-82.

Figure-82. Selecting the object to rotate

- Click on the **Rotate** tool from **Toolbar** in the **Draft** workbench; refer to Figure-83. The **Rotate** dialog will be displayed in the **Tasks** panel of **Combo View** along with the plus sign in place of original cursor; refer to Figure-84. You will be asked to specify the base point through which the axis of rotation will pass.
- Click in the 3D view area to specify the base point of rotating the object or enter desired values for x, y, and z coordinates in the **Local ΔX**, **Local ΔY**, and **Local ΔZ** edit boxes of the dialog, respectively.

Figure-83. Rotate tool

Figure-84. Rotate dialog

- After specifying coordinates in the dialog, click on the **Enter point** button from the dialog. You will be asked to specify the base angle.
- Move the cursor away and click at desired location to specify the base angle for the object that will rotate around the base point or enter desired value in the **Base angle** edit box of the dialog. You will be asked to specify the rotation angle.
- Rotate the cursor away and click at desired location to specify the rotation angle for the object or specify desired value in the **Rotation** edit box of the dialog. The object will be rotated; refer to Figure-85.

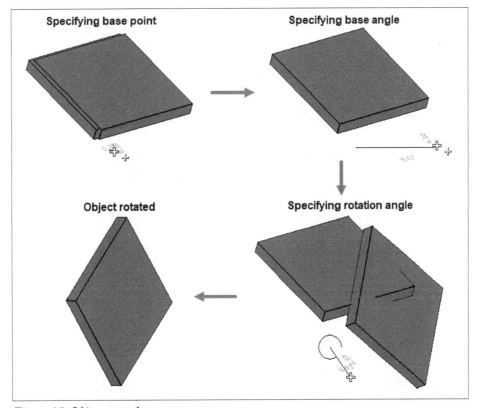

Figure-85. Object rotated

- Other parameters in the **Rotate** dialog have been discussed earlier.

Scaling object

The **Scale** tool scales or copies selected objects around a base point. The procedure to use this tool is discussed next.

- Select the object from the **Model Tree** or from the 3D view area which you want to scale; refer to Figure-86.

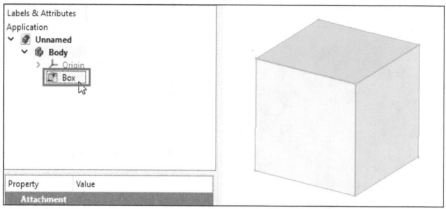

Figure-86. Selecting the object to scale

- Click on the **Scale** tool from **Toolbar** in the **Draft** workbench; refer to Figure-87. The **Scale** dialog will be displayed in the **Tasks** panel of **Combo View** along with the plus sign in place of original cursor; refer to Figure-88. You will be asked to specify the base point.

Figure-87. Scale tool

Figure-88. Scale dialog

- Click in the 3D view area to specify the base point for scaling the object or enter desired values for the x, y, and z coordinates in the **Local X**, **Local Y**, and **Local Z** edit boxes of the dialog, respectively.
- Other parameters in the **Scale** dialog have been discussed earlier.
- After specifying coordinates in the dialog, click on **Enter point** button from the dialog. The **Scale** dialog will be modified; refer to Figure-89.

Figure-89. Modified Scale dialog

- Specify desired values for scaling the object along x, y, and z directions in the **X factor**, **Y factor**, and **Z factor** edit boxes of the dialog, respectively. The preview of the scaled object will be displayed; refer to Figure-90.

Figure-90. Preview of scaled object

- Select **Uniform scaling** check box to lock the same value for X, Y, and Z factors.
- Select **Working plane orientation** check box to lock the scaling of an object along the current working plane.
- Select **Copy** check box to create the scaled copy of the original object.
- Select **Modify subelements** check box to modify the selected subelements instead of the whole object.
- Select **Create a clone** check box to create the clone of original object.
- After specifying desired parameters, click on **OK** button from the dialog. The object will be scaled; refer to Figure-91.

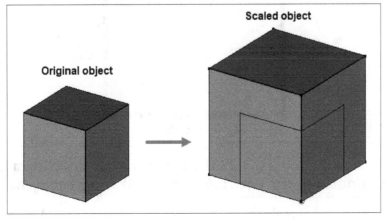

Figure-91. Original object scaled

Mirroring object

The **Mirror** tool produces a mirrored copy of selected object using the mirror operation. The copy just like a **Draft Clone**, is linked to the original object. This means that if the original object changes its shape and properties, the mirrored shape changes as well. The procedure to use this tool is discussed next.

• Select the object from the **Model Tree** or from the 3D view area which you want to be mirrored; refer to Figure-92.

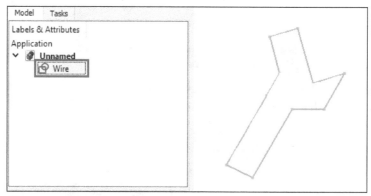

Figure-92. Selecting the object to be mirrored

• Click on the **Mirror** tool from **Toolbar** in the **Draft** workbench; refer to Figure-93. The **Mirror** dialog will be displayed in the **Tasks** panel of **Combo View** along with the plus ⊕ sign in place of original cursor; refer to Figure-94. You will be asked to specify the first point.

Figure-93. Mirror tool

Figure-94. Mirror dialog

• Click in the 3D view area to specify the first point or enter desired values for the x, y, and z coordinates in the **Local ΔX**, **Local ΔY**, and **Local ΔZ** edit boxes of the dialog, respectively.

- After specifying coordinates for the first point in dialog, click on the **Enter point** button from the dialog. You will be asked to specify the second point.
- Move the cursor away and click at desired location to specify the second point or enter desired values for the coordinates in their respective edit boxes.
- After specifying coordinates for the second point in dialog, click on the **Enter point** button again. The object will be mirrored; refer to Figure-95.

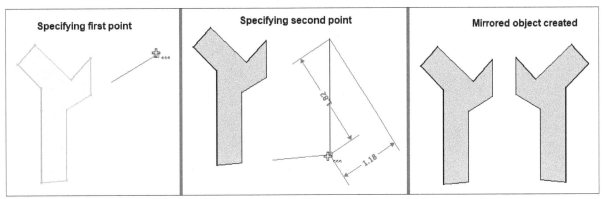

Figure-95. Mirrored object created

- Other parameters in the **Mirror** dialog have been discussed earlier.

Offsetting object

The **Offset** tool moves the selected object by a given distance (offset) perpendicular to itself. The procedure to use this tool is discussed next.

- Select the object from the **Model Tree** which you want to offset; refer to Figure-96.

Figure-96. Selecting the object to offset

- Click on the **Offset** tool from **Toolbar** in the **Draft** workbench; refer to Figure-97. The **Offset** dialog will be displayed in the **Tasks** panel of **Combo View** along with the plus ⊕ sign in place of original cursor; refer to Figure-98. You will be asked to specify the offset distance.

Figure-97. Offset tool

Figure-98. Offset dialog

- Move the cursor away and click in the 3D view area to specify the offset distance or enter desired value in the **Distance** edit box of the dialog and press **ENTER** key from keyboard. The offset of the object will be created; refer to Figure-99.

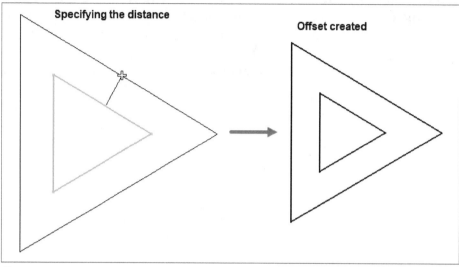

Figure-99. Offset created

Trimming/Extending object

The **Trimex** tool trims or extends lines and wires so that they end at an intersection with another curve or edge. This tool also extrudes faces created from closed wires. The procedure to use this tool is discussed next.

- Select the line from the 3D view area which you want to extend; refer to Figure-100.

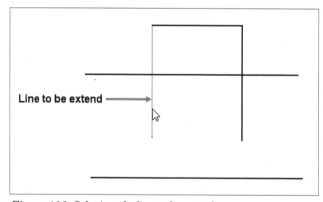

Figure-100. Selecting the line to be extend

- Click on the **Trimex** tool from **Toolbar** in the **Draft** workbench; refer to Figure-101. The **Trimex** dialog will be displayed in the **Tasks** panel of **Combo View** and the cursor will be attached to the selected line; refer to Figure-102. You will be asked to specify the distance.

Figure-101. Trimex tool

Figure-102. Trimex dialog

- Move the cursor away to the distance up to which you want to extend the line and click **LMB** or enter desired value in the **Distance** edit box of the dialog and press **Enter** key from keyboard. The line will be extended; refer to Figure-103.

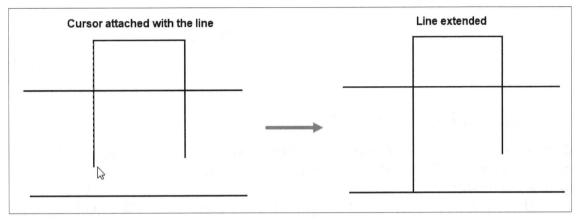

Figure-103. Line extended

- If you want to trim the line then select the line from the 3D view area which you want to trim; refer to Figure-104.

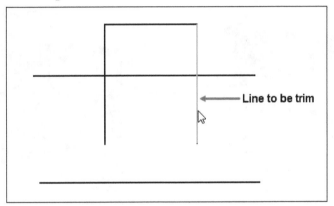

Figure-104. Selecting the line to be trim

- After selecting the line, click on the **Trimex** tool. The **Trimex** dialog will be displayed and the cursor will be attached to the selected line as discussed earlier. You will be asked to specify the distance.
- Move the cursor away to the distance upto which you want to trim the line and click **LMB** or enter desired value in the **Distance** edit box of the dialog and press **Enter** key from keyboard. The line will be trimmed; refer to Figure-105.

Figure-105. Line trimmed

- If you want to extrude the face created from closed wire then select the face from the **Model Tree** or from the 3D view area; refer to Figure-106.

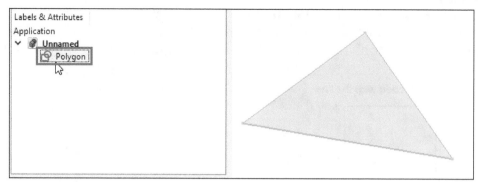

Figure-106. Selecting the face to extrude

- After selecting the face, click on the **Trimex** tool. The Trim dialog will be displayed and the cursor will be attached to the extrusion of face. You will be asked to specify the distance.
- Move the cursor away to the distance up to which you want to extrude the face and click **LMB** or enter desired value in the **Distance** edit box of the dialog and press **Enter** key from keyboard. The face will be extruded; refer to Figure-107.

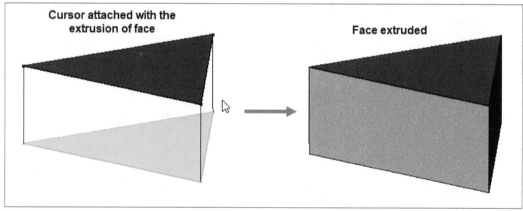

Figure-107. Face extruded

Stretching object

The **Stretch** tool stretches an object by moving some of its selected vertices. The equivalent action is editing the object and moving the points manually to a new position. The procedure to use this tool is discussed next.

- Select the object from the **Model Tree** or from the 3D view area which you want to stretch; refer to Figure-108.

Figure-108. Selecting the object to be stretched

- Click on the **Stretch** tool from **Toolbar** in the **Draft** workbench; refer to Figure-109. The **Stretch** dialog will be displayed in the **Tasks** panel of **Combo View** along with the plus sign in place of original cursor; refer to Figure-110. You will be asked to specify the first point.

Figure-109. Stretch tool

Figure-110. Stretch dialog

- Click in the 3D view area to specify the first point or enter desired values for the x, y, and z coordinates in the **Local ΔX**, **Local ΔY**, and **Local ΔZ** edit boxes of the dialog, respectively.
- After specifying coordinates for the first point in dialog, click on the **Enter point** button from the dialog. You will be asked to specify the second point.
- Move the cursor away and click at desired location to specify the second point or enter desired values for the coordinates in their respective edit boxes.
- After specifying coordinates for the second point in dialog, click on the **Enter point** button again. You will be asked to specify the third point. The first and second point define a selection rectangle. The vertices of the original object enclosed by this rectangle will become highlighted; refer to Figure-111.

Figure-111. Selecting the vertices to be stretch

- Click in the 3D view area to specify the third point or enter desired values for the coordinates in their respective edit boxes.
- After specifying coordinates for the third point in dialog, click on the **Enter point** button again. You will be asked to specify the fourth point.

- Move the cursor away and click at desired location to specify the fourth point or enter desired values for the coordinates in their respective edit boxes. The third and fourth point define a line whose distance and direction will be used to stretch the figure attached to the highlighted points.
- After specifying coordinates for the fourth point in dialog. Click on the **Enter point** button again. The object will be stretched; refer to Figure-112.

Figure-112. Object stretched

- Other parameters in the **Stretch** dialog have been discussed earlier.

Creating Clone

The **Clone** tool produces linked copies of selected shape. This means that if the original object changes its shape and properties, all clones change as well. The procedure to use this tool is discussed next.

- Select the object from the **Model Tree** or from the 3D view area for which you want to create clone; refer to Figure-113.

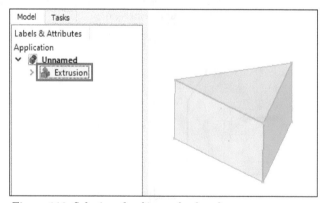

Figure-113. Selecting the object to be cloned

- Click on the **Clone** tool from **Toolbar** in the **Draft** workbench; refer to Figure-114. The cloned object will be created; refer to Figure-115.

Figure-114. Clone tool

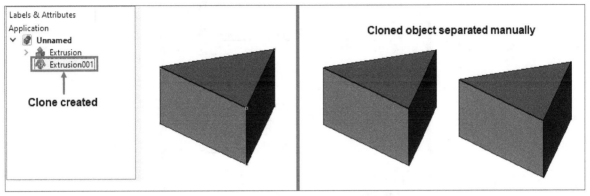

Figure-115. Cloned object created

- You can change the properties of clone created from the **Property editor** dialog as discussed earlier.

Creating Array

The **Array** tool creates an orthogonal (3-axes), polar, or circular array of selected object. The procedure to use this tool is discussed next.

- Select the object from **Model Tree** or from the 3D view area of which you want to create an array; refer to Figure-116.

Figure-116. Selecting the object to create array

- Click on the **Array** tool from **Array tools** drop-down in the **Toolbar** of **Draft** workbench; refer to Figure-117. The **Orthogonal array** dialog will be displayed in the **Tasks** panel of **Combo View**; refer to Figure-118.

Figure-117. Array tool

Figure-118. Orthogonal array dialog

- Specify desired number of elements for the x, y, and z directions in the **X**, **Y**, and **Z** edit boxes, respectively from **Number of elements** area of the dialog.
- Enter desired values in **X**, **Y**, and **Z** edit boxes of **X intervals**, **Y intervals**, and **Z intervals** area to specify the displacement for the elements in the X direction, Y direction, and Z direction, respectively.
- For a rectangular array along X direction, the **Y** and **Z** values must be **0**; along Y direction, the **X** and **Z** values must be **0**; along Z direction, the **X** and **Y** values must be **0**.
- Click on the **Reset X**, **Reset Y**, or **Reset Z** button to reset the displacement in the given direction to the default values.
- Select the **Fuse** check box to fuse the overlapping elements in the array.
- Select the **Link array** check box to create the link array instead of a regular array.

- After specifying desired parameters in the dialog, click on the **OK** button. The orthogonal array will be created; refer to Figure-119.

Figure-119. Orthogonal array created

Creating Polar Array

The **Polar array** tool creates an array of selected object by placing copies along a circumference. The procedure to use this tool is discussed next.

- Select the object from the **Model Tree** or from the 3D view area of which you want to create a polar array; refer to Figure-120.

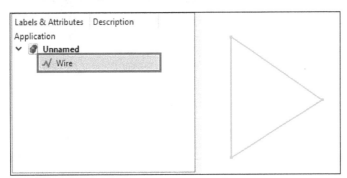

Figure-120. Selecting the object to create polar array

- Click on the **Polar array** tool from **Array tools** drop-down in the **Toolbar** of **Draft** workbench; refer to Figure-121. The **Polar array** dialog will be displayed in the **Tasks** panel of **Combo View** along with the plus ⊕ sign in place of original cursor; refer to Figure-122.

Figure-121. Polar array tool

Figure-122. Polar array dialog

- Enter desired angle value in the **Polar angle** edit box to specify the total angle of the array. The angle is positive in the counter-clockwise direction.
- Enter desired number of elements in the array, including a copy of the original object in the **Number of elements** edit box of the dialog.
- Click in the 3D view area to specify the point through which the rotation axis of the array will pass or enter desired coordinates for the center of rotation in the **X**, **Y**, and **Z** edit boxes from **Center of rotation** area of the dialog. The polar array will be created; refer to Figure-123.

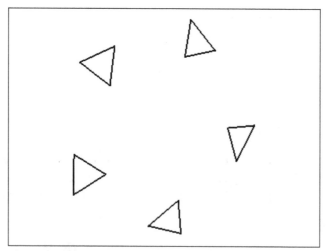

Figure-123. Polar array created

- Click on the **Reset point** button to reset the coordinates of the center of rotation to the origin.
- Other parameters in the dialog have been discussed earlier.

Creating Circular Array

The **Circular array** tool creates an array of selected object by placing copies along concentric circumferences. The procedure to use this tool is discussed next.

- Select the object from the **Model Tree** or from the 3D view area of which you want to create a circular pattern; refer to Figure-124.

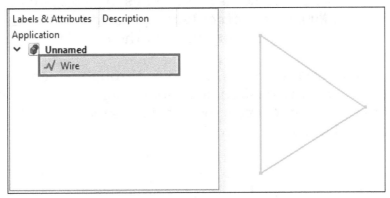

Figure-124. Selecting the object to create circular array

- Click on the **Circular array** tool from **Array tools** drop-down in the **Toolbar** of **Draft** workbench; refer to Figure-125. The **Circular array** dialog will be displayed in the **Tasks** panel of **Combo View** along with the plus ⊕ sign in place of original cursor; refer to Figure-126.

Figure-125. Circular array tool

Figure-126. Circular array dialog

- Enter desired value in **Radial distance** edit box to specify the distance from one layer of objects to the next layer of objects.

- Enter desired value in **Tangential distance** edit box to specify distance between the elements on the same circular layer. It must be larger than **0**.
- Specify desired number of circular layers or rings to create, including a copy of the original object in the **Number of circular layers** edit box. It must be at least **2**.
- Specify desired number of symmetry lines in the circular array in the **Symmetry** edit box.
- Click in the 3D view area to specify the point through which the rotation axis of the array will pass or enter desired coordinates for the center of rotation in the **X**, **Y**, and **Z** edit boxes from **Center of rotation** area of the dialog. The circular array will be created; refer to Figure-127.

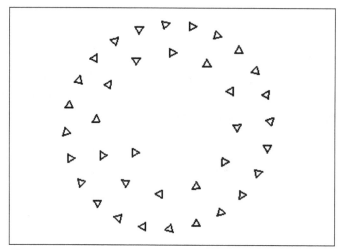

Figure-127. Circular array created

- Click on the **Reset point** button to reset the coordinates of the center of rotation to the origin.
- Other parameters in the dialog have been discussed earlier.

Creating Path array

The **Path array** tool places copies of selected shape along selected path which can be a wire, a B-Spline, and similar edges. The procedure to use this tool is discussed next.

- First, select the object from the **Model Tree** or from the 3D view area that you want to array and then select the object to be used as path along which the object will be distributed while holding the **CTRL** key; refer to Figure-128.

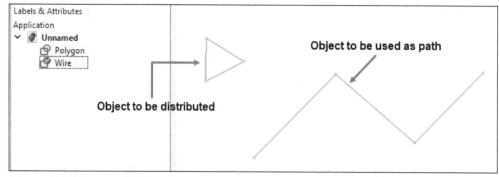

Figure-128. Selecting the object for path array

- Click on the **Path array** tool from **Array tools** drop-down in the **Toolbar** of **Draft** workbench; refer to Figure-129. The array objects will be created along path; refer to Figure-130.

Figure-129. Path array tool

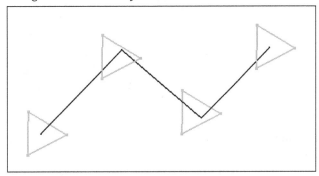

Figure-130. Array object created along path

- You can change the properties of array created from the **Property editor** dialog as discussed earlier.

Creating Path Link Array

The **Path Link array** tool creates a link array from selected object by placing copies along a path. The procedure to use this tool is same as discussed for **Path array** tool.

Creating Point array

The **Point array** tool places copies of selected object on selected points. The procedure to use this tool is discussed next.

- First, select the object from the **Model Tree** or from the 3D view area that you want to distribute and then select the point along which the object will be distributed; refer to Figure-131.

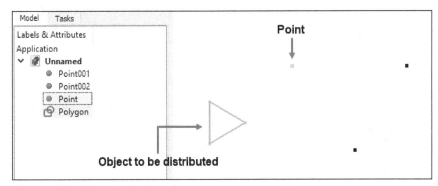

Figure-131. Selecting the object for point array

- Click on the **Point array** tool from **Array tools** drop-down in the **Toolbar** of **Draft** workbench; refer to Figure-132. The array object will be created along point; refer to Figure-133.

Figure-132. Point array tool

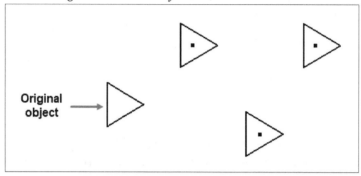

Figure-133. Array object created along point

- You can change the properties of array created from the **Property editor** dialog as discussed earlier.

Creating Point Link Array

The **Point Link array** tool creates a link array from a selected object by placing copies at the points from a point compound. The procedure to use this tool is same as discussed for **Point array** tool.

Path Twisted Array

The **Path twisted array** tool is used to create copies of selected object along selected path with defined twist angle. The procedure to use this tool is discussed next.

- Create a sketch and path in the 3D view area refer to Figure-134. The path can be a polyline, B-spline, Bezier curve or even edges of other object.

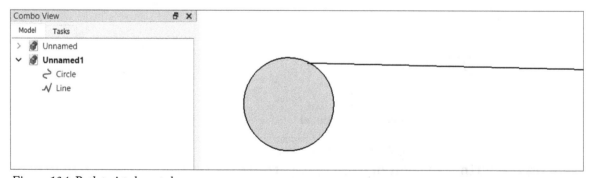

Figure-134. Path twisted created

- Select the object and then select the path from drawing area or **Model** tab of **Combo View**; refer to Figure-135.

Figure-135. Path twisted sketch select

- Click on the **Path twisted array** from the **Array** tools drop-down in the **Toolbar** of **Draft** workbench; refer to Figure-136. The array object will be created; refer to Figure-137.

Figure-136. Path array tool

Figure-137. Path twisted sketch created

- You can change the properties of array in the **Data** tab of **Property** dialog as discussed earlier.

Path Twisted Link Array

The **Path twisted link array** tool works similar to the **Path twisted array** tool but creates a link array instead of copy array. The procedure to use this tool is discussed next.

- Create a sketch and path in the 3D view area; refer to Figure-138.

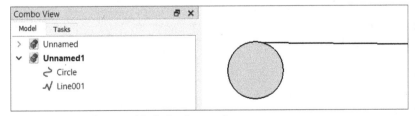

Figure-138. Path twisted link sketch created

- Select the object and then select the path; refer to Figure-139.

Figure-139. Path twisted link sketch selected

- Click on the **Path twisted link array** tool from the **Array** drop-down in the **Toolbar** of **Draft** workbench; refer to Figure-140. The array will be created; refer to Figure-141.

Figure-140. Path twisted link tool

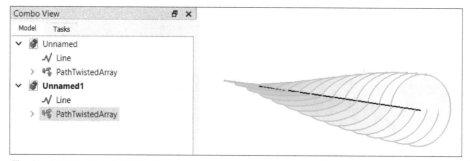

Figure-141. Path twisted link array

Editing Object

The **Edit** tool allows you to graphically edit certain properties of the selected object such as, the vertices of a wire, the length and width of a rectangle, the radius of the circle, and so on. The procedure to use this tool is discussed next.

- Select the object you want to edit from the **Model Tree** or from the 3D view area; refer to Figure-142.

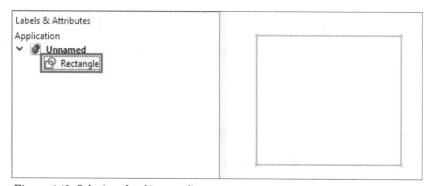

Figure-142. Selecting the object to edit

- Click on the **Edit** tool from **Toolbar** in the **Draft** workbench; refer to Figure-143. The **Edit** dialog will be displayed in the **Tasks** panel of **Combo View** and the snapping points will be displayed on the object; refer to Figure-144.

Figure-143. Edit tool

Figure-144. Edit dialog with the snapping points on object

- Click on the snapping point from where you want to edit the object. The **Line** dialog will be displayed in the **Tasks** panel of **Combo View** along with the plus sign in place of original cursor asking you to specify another point; refer to Figure-145.

Figure-145. Line dialog to edit the object

- Move the cursor away and click **LMB** at desired location to specify the another point of object or enter desired values for the x, y, and z coordinates in the **Local ΔX**, **Local ΔY**, and **Local ΔZ** edit boxes of the dialog, respectively.
- After specifying coordinates for another point in the dialog, click on **Enter point** button from the dialog. The object will be edited and the **Edit** dialog will be displayed again; refer to Figure-146.

Figure-146. Object edited

- Click on **Close** button to finish the editing operation as well as to close the dialog.

Highlighting the Subelement

The **Subelement highlight** tool temporarily highlights selected objects, or the base objects of selected objects to be used in conjunction with the subelement mode of the **Move** tool, **Rotate** tool, or **Scale** tool. The procedure to use this tool is discussed next.

- Select the object from the **Model Tree** or from the 3D view area whose subelement you want to highlight; refer to Figure-147.

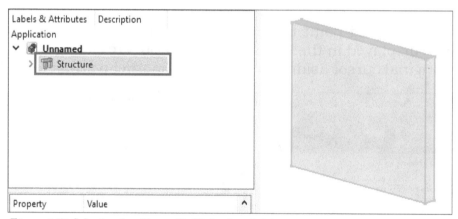
Figure-147. Selecting the object to highlight the subelement

- Click on the **Subelement highlight** tool from **Toolbar** in the **Draft** workbench; refer to Figure-148. The subelement of the selected object will be highlighted; refer to Figure-149.

Figure-148. Subelement highlight tool

Subelement highlighted

Figure-149. Subelement highlighted

- Now, if you want to move, rotate, or scale the subelement then select the point of highlighted subelement; refer to Figure-150 and click on the **Move** tool.

Figure-150. Highlighted point selected

- Move the cursor away and click at desired location to specify the new position of the subelement or enter desired values for the coordinates in their respective edit boxes of the dialog and click **Enter point** button. The subelement will be moved and copied to the new position; refer to Figure-151.

Figure-151. Subelement moved and copied

- Click on the **Subelement highlight** tool again to exit the tool.

Joining objects

The **Join** tool attempts to join all wires currently in the selection into a single wire. The procedure to use this tool is discussed next.

- Select two or more lines or curves from the **Model Tree** or from the 3D view area to make a single wire; refer to Figure-152.

Figure-152. Selecting the lines to create single wire

- Click on the **Join** tool from **Toolbar** in the **Draft** workbench; refer to Figure-153. The single wire will be created; refer to Figure-154.

Figure-153. Join tool

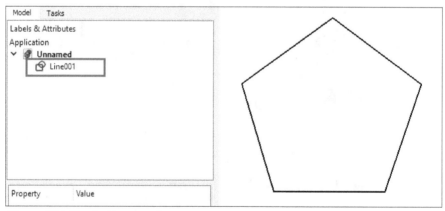

Figure-154. Single wire created

Splitting object

The **Split** tool attempts to split an existing wire at a specified edge or point. The procedure to use this tool is discussed next.

- Click on the **Split** tool from **Toolbar** in the **Draft** workbench; refer to Figure-155.

Figure-155. Split tool

- Click on the wire at the location from where you want to split the wire. The wire will be split from that location; refer to Figure-156.

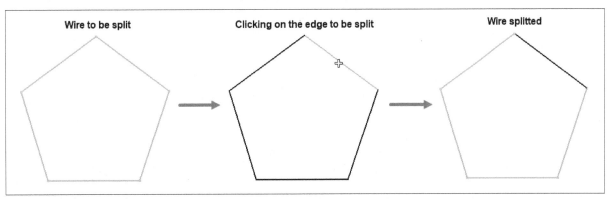

Figure-156. Wire split

Upgrading object

The **Upgrade** tool performs things such as creating faces and fusing different elements. The procedure to use this tool is discussed next.

- Select the object from the **Model Tree** or from the 3D view area which you want to upgrade; refer to Figure-157.

Figure-157. Selecting the object to upgrade

- Click on the **Upgrade** tool from **Toolbar** in the **Draft** workbench; refer to Figure-158. The opened wire will be upgraded to a closed wire; refer to Figure-159.

Figure-158. Upgrade tool

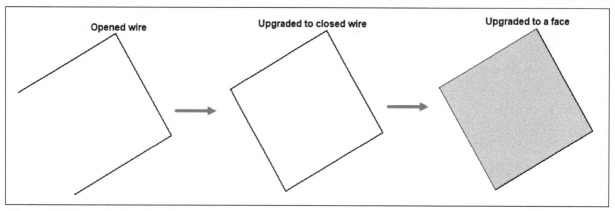

Figure-159. Upgradation of an object

- If you want to upgrade the closed wire then click on the **Upgrade** tool again. The closed wire will be upgraded to a face; refer to Figure-159.

Downgrading object

The **Downgrade** tool performs things such as breaking faces and de-constructing wires into their individual objects. The procedure to use this tool is discussed next.

- Select the object from the **Model Tree** or from the 3D view area which you want to downgrade; refer to Figure-160.

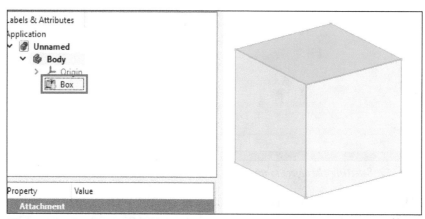

Figure-160. Selecting the object to downgrade

- Click on the **Downgrade** tool from **Toolbar** in the **Draft** workbench; refer to Figure-161. The object will be downgraded into the split faces; refer to Figure-162.

Figure-161. Downgrade tool

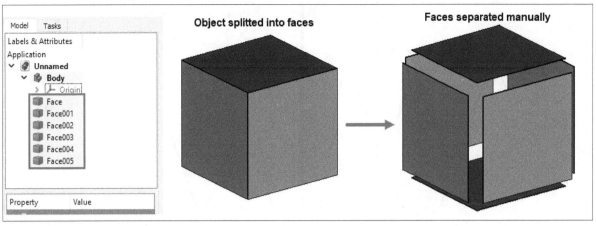

Figure-162. Object downgraded into the split faces

Converting Wire to B-Spline

The **Wire to B-Spline** tool converts wire to B-Spline and vice-versa. The procedure to use this tool is discussed next.

- Select the wire from the **Model Tree** or from the 3D view area which you want to convert into B-Spline; refer to Figure-163.

Figure-163. Selecting the wire to create B Spline

- Click on the **Wire to BSpline** tool from **Toolbar** in the **Draft** workbench; refer to Figure-164. The wire will be converted into B-Spline; refer to Figure-165.

Figure-164. Wire to BSpline tool

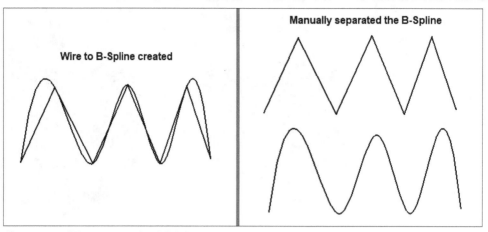

Figure-165. Wire converted to BSpline

Converting Draft to Sketch

The **Draft to Sketch** tool converts objects of **Draft** workbench into the objects of **Sketcher** workbench with constraints and vice-versa. The procedure to use this tool is discussed next.

- Select the object you want to convert from the **Model Tree** or from the 3D view area; refer to Figure-166.

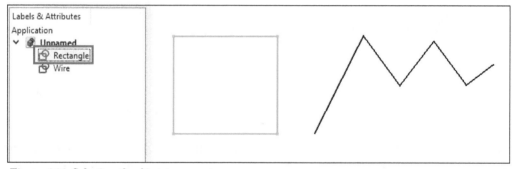

Figure-166. Selecting the object to convert

- Click on the **Draft to Sketch** tool from **Toolbar** in the **Draft** workbench; refer to Figure-167. The objects of **Draft** workbench will be converted into sketches of **Sketcher** workbench with constraints; refer to Figure-168.

Figure-167. Draft to Sketch tool

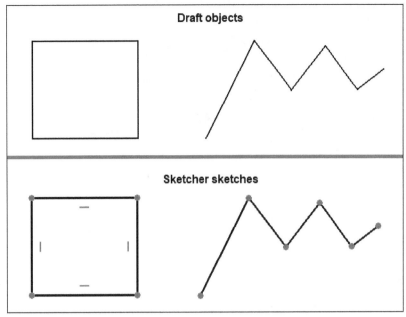

Figure-168. Objects converted into sketches

Set Slope

The **Set Slope** tool slopes/inclines selected lines or wires by increasing, or decreasing, the Z coordinate of all points after the first one. The procedure to use this tool is discussed next.

- Create and select one or more lines or wires from the **Model Tree** or from the 3D view area; refer to Figure-169.

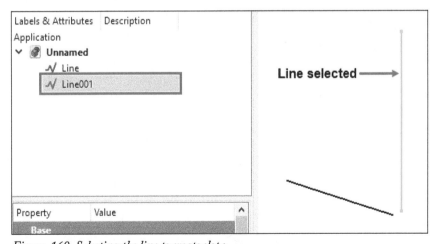

Figure-169. Selecting the line to create slope

- Click on the **Set Slope** tool from **Toolbar** in the **Draft** workbench; refer to Figure-170. The **Slope** dialog will be displayed; refer to Figure-171.

Figure-170. Slope tool

Figure-171. Slope dialog

- Specify desired slope factor value for the selected lines in the **Slope** edit box of the dialog. Note that entering **0** value means each segment is horizontal, entering **1** value means each segment is **45** degree up, and entering **-1** value means each segment is **45** degree down.
- After entering desired value, click on the **OK** button from the dialog. The slope of the line will be created; refer to Figure-172 .

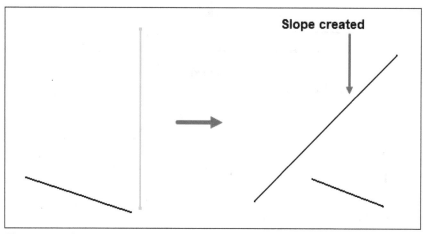

Figure-172. Slope of the line created

Creating Flip Dimension

The **Flip Dimension** tool rotates the dimension text of selected dimensions 180 degree around the dimension line. It can be used to correct dimensions whose text appears mirrored. The procedure to use this tool is discussed next.

- Create and select one or more dimensions of the object from the **Model Tree** or from the 3D view area; refer to Figure-173.

Figure-173. Selecting the dimensions

- Click on the **Flip Dimension** tool from **Toolbar** in the **Draft** workbench; refer to Figure-174. The dimension will be flipped; refer to Figure-175.

Figure-174. Flip Dimension tool

Figure-175. Dimension flipped

Creating Shape 2D View

The **Shape 2D View** tool produces a 2D projection from selected 3D solid object such as those created with the **Part**, **Part Design**, and **Arch** workbenches. The procedure to use this tool is discussed next.

- Rotate the view of 3D object to be projected from the 3D view area, so it reflects the direction of the desired projection. For example, a top view will project the object on the XY plane.
- Select the 3D object from the **Model Tree** or from the 3D view area; refer to Figure-176.

Figure-176. Selecting the 3D object to be project

- Click on the **Shape 2D View** tool from **Toolbar** in the **Draft** workbench; refer to Figure-177. The 2D projection will be created below the selected 3D object, lying on the XY plane; refer to Figure-178.

Figure-177. Shape 2D View tool

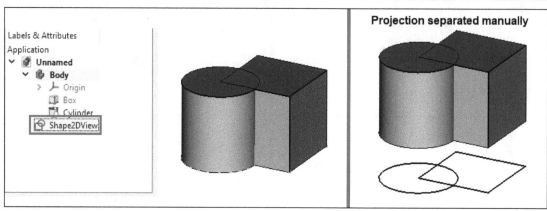

Figure-178. 2D projection created

- You can manually change the position of projection created from the **Property editor** dialog.

Draft Tray Toolbar

The **Draft tray toolbar** allows selecting the working plane together with some visual properties like the line color, shape color, text size, line width, and automatic group. This toolbar is available in the **Draft** workbench; refer to Figure-179. The procedures to use the tools of this toolbar are discussed next.

Figure-179. Draft tray toolbar

Creating Working Plane

The **Working Plane** tool sets a working plane on a standard view or selected face. The working plane in the 3D view area indicates where a draft shape will be built. The procedure to use this tool is discussed next.

- Click on the **Select Plane** tool from **Toolbar** in the **Draft** workbench; refer to Figure-180. The **Working plane setup** dialog will be displayed in the **Tasks** panel of **Combo View**; refer to Figure-181.

Figure-180. Select Plane tool

Figure-181. Working plane setup dialog

- Click on the **Top (XY)**, **Front (XZ)**, or **Slide (YZ)** button from **Select a face or working plane proxy or 3 vertices** area to set the working plane on XY, XZ, or YZ plane, respectively.
- Click on the **Align to view** button to set the working plane to the current 3D view, perpendicular to the camera axis and passing through the origin.
- Click on the **Automatic** button to unset any current working plane and automatically set a working plane when a tool is used.
- Specify desired offset value in the **Offset** edit box to set the working plane at specified perpendicular distance from the plane you selected.
- Select **Center plane on view** check box to put the origin of the working plane in the center of the current 3D view.

Or,

- Select desired vertex in the 3D view area to move the current working plane so that its origin matches the position of the selected vertex and click on the **Move working plane** button from **Select a single vertex to move the current working plane without changing its orientation** area of the dialog.
- Specify desired value in **Grid spacing** edit box to define the space between two consecutive lines in the grid.
- Specify desired value in **Main line every** edit box to draw a slightly thicker line in the grid at the set value.
- Specify desired value in **Grid extension** edit box to determine the number of grid lines in the X and Y direction of the grid.
- Specify desired maximum distance in **Snapping radius** edit box at which snap grid detects the intersections of grid lines.

- Click on the **Center view** button to use the origin of the current working plane as the center of the 3D view.
- Click on the **Previous** button to reset the working plane to its previous position.
- After specifying desired parameters, on clicking desired plane, the active plane will be displayed in the toolbar along with the grid view; refer to Figure-182 .

Figure-182. Active plane displayed along with grid view

Applying Style Settings

The **Set style** tool allows to set default visual properties like line color or line width to be used by all draft objects created after this tool is used and bulk apply these styles to selected objects. The procedure to use this tool is discussed next.

- Click on the **Set style** tool from **Toolbar** in the **Draft** workbench; refer to Figure-183. The **Style settings** dialog will be displayed in the **Tasks** panel of **Combo View**; refer to Figure-184.

Figure-183. Set style tool

Figure-184. Style settings dialog

- Click on the **Line color** button from **Lines and faces** area and select desired color for the line from the **Select Color** dialog box displayed.
- Specify desired width for the line in the **Line width** edit box.
- Select desired line style from **Draw style** drop-down.
- Select desired display mode for the faces from **Display mode** drop-down.
- Click on the **Shape color** button and select desired color for the faces from **Select Color** dialog box displayed.
- Specify desired transparency of faces in the **Transparency** edit box.
- Select desired font to use for texts and dimensions from **Text font** drop-down in the **Annotations** area of the dialog.
- Specify desired size for the texts in the **Text size** edit box.
- Click on the **Text color** button and select desired color for the texts from **Select Color** dialog box displayed.
- Select desired type of dimension arrows from **Arrow style** drop-down.
- Specify desired size of dimension arrows in the **Arrow size** edit box.
- Select the **Show unit** check box to show the unit suffix on dimension texts.
- Specify desired unit to use for dimensions in the **Unit override** edit box.
- After setting desired parameters in the dialog, select the object or multiple objects from the **Model Tree** or from the 3D view area on which you want to apply the style; refer to Figure-185.

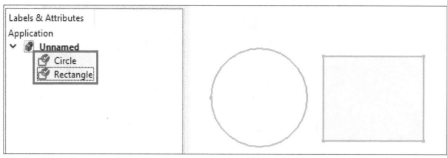

Figure-185. Selecting the objects to apply style

- Click on the **Apply to selected objects** button from the dialog. The style will be applied; refer to Figure-186.

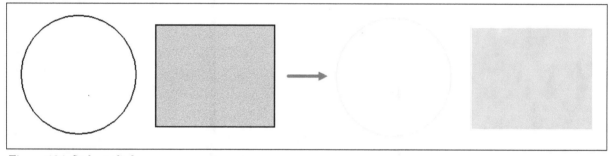

Figure-186. Style applied

Toggle Construction Mode

The **Toggle Construction Mode** toggles the draft construction mode on or off. The construction geometry is comprised of lines, points, and other shapes that serve as references or snapping elements which are helpful in building your main geometry. The procedure to use this tool is discussed next.

- Click on the **Toggle construction mode** button from **Toolbar** in the **Draft** workbench; refer to Figure-187.

Figure-187. Toggle construction mode button

- Create some objects in the 3D view area as desired. Now, it will be created in construction mode; refer to Figure-188.

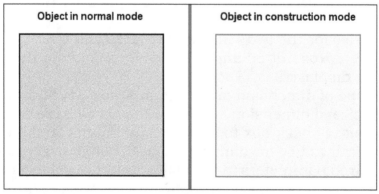

Figure-188. Objects in normal and construction mode

- Click on the **Toggle construction mode** button again to go back to normal mode.

Creating Auto Group

The **Auto Group** tool changes the active layer, or optionally, the active group or groups like arch object. New draft and arch objects will be automatically placed in this active layer or group. Assigning a group or layer can help in The procedure to use this tool is discussed next.

- First, create the layer in the active document with the help of **Layer** tool as discussed earlier.
- Click on the **Auto Group** tool from **Toolbar** in the **Draft** workbench; refer to Figure-189. The active layers created will be displayed in the drop-down; refer to Figure-190.

Figure-189. Auto Group tool

Figure-190. Active layers displayed

- Select **None** option to work without an active layer.
- Select desired layer from the existing layers to make it active.
- Select **Add new Layer** option to create a new layer.

Draft Snap Toolbar

The **Draft Snap toolbar** allows to select various snapping mode. All the buttons in this toolbar are toggle buttons. This toolbar is available in the **Draft** workbench; refer to Figure-191. The procedures to use the tools of this toolbar are discussed next.

Figure-191. Draft Snap toolbar

Toggle Snap

The **Toggle Snap** tool allows you to activate or deactivate globally the Draft Snap methods. The procedure to use this tool is discussed next.

- If you want to make the snapping methods available then click on the **Toggle Snap** button from **Toolbar** in the **Draft** workbench. The snapping will be displayed while creating the objects; refer to Figure-192.

Figure-192. Snapping displayed while creating objects

Endpoint 🖊

The **Endpoint** snap button activates snaps to the endpoints of line, arc, and spline segments.

Midpoint 🖊

The **Midpoint** snap button activates snaps to the middle point of line and arc segments.

Center ◎

The **Center** snap button activates snaps to the center point of arcs and circles.

Angle ◇

The **Angle** snap button activates snaps to the points of circles and arcs at specific increments of 30° and 45° on the arc; this includes 0°, 60°, 90°, 180°, 210°, 270°, and other integer multiples.

Intersection ✕

The **Intersection** snap button activates snaps to the intersection of two line or arc segments. Hover the mouse over the two desired objects to activate their intersection snaps.

Perpendicular ⊥

The **Perpendicular** snap button activates snaps to a line or edge or to an extension of it to produce a line that is perpendicular to that edge.

Extension ⚬⚬⚬

The **Extension** snap button activates snaps to a point on an imaginary line that extends beyond the endpoints of line segments. Hover the mouse over desired object to activate its extension snap.

Parallel ⫽

The **Parallel** snap button activates snaps on an imaginary line parallel to a line segment. Hover the mouse over desired object to activate its parallel snap.

Special ✎

The **Special** snap button activates snaps to the special location points defined by a particular object such as an Arch Wall.

Near

The **Near** snap button activates snaps to the closest point or edge on the nearest object.

Ortho

The **Ortho** snap button activates snaps to a point on an imaginary line that originates from the previous point and extends infinitely at specific increments of 45°; this includes 0°, 90°, 135°, 180°, 225°, 270°, and other integer multiples.

Grid

The **Grid** snap button activates snaps to the intersection of two grid lines, if the grid is visible.

Working Plane

The **Working Plane** button always places a snapped point on the current working plane even if you also use another snapping method and select a point outside of that working plane. In other words, it projects an external snapping point to the current working plane.

Dimensions

The **Dimensions** snap button shows temporary X and Y dimensions between the current point and the last snapped point on screen while snapping.

Toggle Grid

The **Toggle Grid** tool allows you to show and hide the grid defined in the **Draft Preferences** or with the **Draft Select Plane** tool. The procedure to use this tool is discussed next.

• If you want to set the appearance of the grid then click on the **Preferences** tool from the **Edit** menu; refer to Figure-193. The **Preferences** dialog box will be displayed; refer to Figure-194.

Figure-193. Preferences tool

Figure-194. Preferences dialog box

- Click on the **Draft** option from left area of the dialog box. The parameters related to **Draft** will be displayed.
- Click on the **Grid and snapping** tab from **Draft** option of the dialog box. The parameters related to **Grid and snapping** will be displayed.
- Specify desired values in the **Main lines every**, **Grid spacing**, **Grid size**, and **Grid color and transparency** edit boxes from **Grid** section of the tab; refer to Figure-195.

Figure-195. Specifying the grid parameters

- Click on **Apply** and then **OK** button from the dialog box to apply the changes.
- To make the grid visible, click on the **Toggle Grid** button from **Toolbar** in the **Draft** workbench. The grid will be displayed in the 3D view area; refer to Figure-196.

Figure-196. Grid displayed in the 3D view area

- Click on the **Toggle Grid** button again to turn off the draft grid.

Draft Utility Tools Toolbar

The **Draft utility tools toolbar** is available in the **Draft** workbench; refer to Figure-197. The procedures to use the tools of this toolbar are discussed next.

Figure-197. Draft utility tools toolbar

Manage Layer

The **Manage Layer** tool creates a new layer in the current document to which objects can be added to control object visibility and color. The procedure to use this tool is discussed next.

- Click on the **Manage Layer** tool from **Toolbar** in the **Draft** workbench; refer to Figure-198. A layer will be created in the layer container of the **Model Tree**; refer to Figure-199.

Figure-198. Layer tool

Figure-199. Layer created in the layer container

- Select the layer created, the **Property editor** dialog will be displayed with the properties of layer; refer to Figure-200.

Figure-200. Property editor dialog with layer parameters

- Change the properties of the layer like **Line Color**, **Line Width**, **Shape Color**, and so on as desired; refer to Figure-201.

Figure-201. Properties of the layer changed

- Now, create desired object and drag it into the layer. The properties of the object changed according to the properties of layer; refer to Figure-202.

Figure-202. Properties of the object changed

Add a new named group

The **Add a new named group** tool is used to create a named group and moves selected objects to that group. The procedure to use this tool is discussed next.

- Select one or more objects which you want to add to the group; refer to Figure-203.

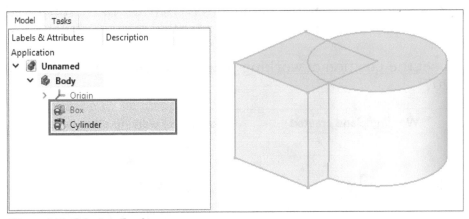

Figure-203. Selecting the objects

- Click on the **Add a new named group** tool from **Toolbar** in the **Draft** workbench; refer to Figure-204. The **Add group** dialog will be displayed in the **Tasks** panel of **Combo View**; refer to Figure-205.

Figure-204. Add a new named group tool

Figure-205. Add group dialog

- Specify desired name of the group in the **Group name** edit box and click on the **OK** button from the dialog. The selected objects will be moved to the created group; refer to Figure-206.

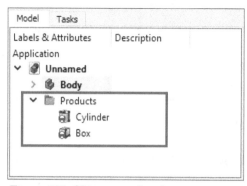

Figure-206. Objects moved to the group

Creating Working plane proxy

The **Working plane proxy** tool creates a working plane proxy to save the current working plane position. A working plane proxy can be used to quickly restore a working plane. The procedure to use this tool is discussed next.

- Create and set the position of working plane as desired in the 3D view area; refer to Figure-207.

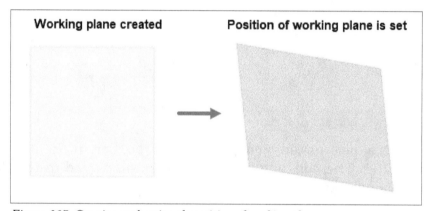

Figure-207. Creating and setting the position of working plane

- Click on the **Working plane proxy** tool from **Toolbar** in the **Draft** workbench; refer to Figure-208. The working plane proxy will be created; refer to Figure-209.

Figure-208. Working plane proxy tool

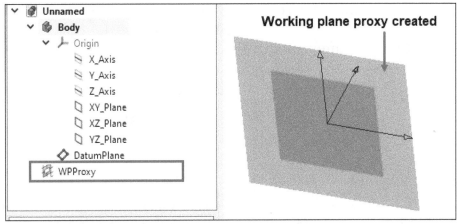

Figure-209. Working plane proxy created

• Now, change the position of working plane as desired and double-click on the working plane proxy created in the **Model Tree**. The position of the working plane will be restored; refer to Figure-210.

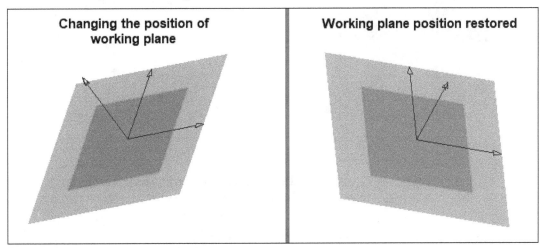

Figure-210. Working plane position restored

Toggle normal/wireframe display

The **Toggle normal/wireframe display** tool switches the display mode of selected objects between "**Flat Lines**" and "**Wireframe**". The procedure to use this tool is discussed next.

• Select desired object from the **Model Tree** or from the 3D view area of which you want to toggle the display mode; refer to Figure-211.

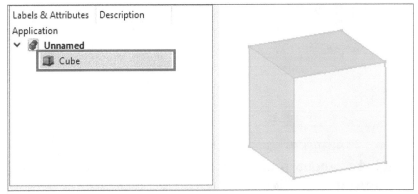

Figure-211. Selecting the object to toggle the display mode

- Click on the **Toggle normal/wireframe display** tool from **Toolbar** in the **Draft** workbench; refer to Figure-212. The display mode of the object will be changed from flat lines to wireframe; refer to Figure-213.

Figure-212. Toggle display mode tool

Figure-213. Display mode changed

- Select the object and click on **Toggle normal/wireframe display** tool again to change the display mode from wireframe to flat lines.

Moving to the group

The **Add to group** tool quickly adds selected objects to an existing group. The procedure to use this tool is discussed next.

- First, create a group with **Group** tool as discussed earlier in this book and select the object which you want to add to the group; refer to Figure-214.

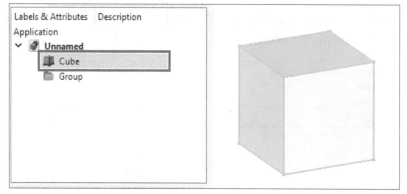

Figure-214. Selecting the object to add to the group

- Click on the **Add to group** tool from **Toolbar** in the **Draft** workbench; refer to Figure-215. A menu will be displayed with the two options near the cursor; refer to Figure-216.

Figure-215. Move to group tool

Figure-216. Two options displayed

- Click on the **+ Add new group** option from the menu displayed. The **Add group** dialog will be displayed as discussed earlier.
- Specify desired name of the group in the **Group name** edit box and click on the **OK** button. The object will be added to the group; refer to Figure-217.

Figure-217. Object added to the group

- If you want to ungroup the object then select the object and click on the **Ungroup** option from the menu displayed. The object will be moved out of the group.

Selecting group contents

The **Select group** tool selects the contents of a selected group or layer. The procedure to use this tool is discussed next.

- Create and select the group or one or more objects in the group from the **Model Tree**; refer to Figure-218.

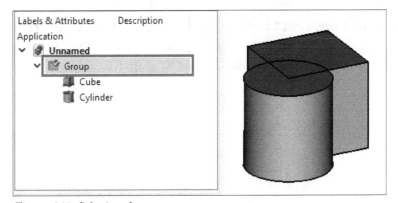

Figure-218. Selecting the group

- Click on the **Select group** tool from **Toolbar** in the **Draft** workbench; refer to Figure-219. All contents of the group will be selected; refer to Figure-220.

Figure-219. Select group tool

Figure-220. All contents of the group selected

Adding to the construction group

The **Add to Construction group** tool add the selected objects to the construction group. The procedure to use this tool is discussed next.

- Select the objects from the **Model Tree** or from the 3D view area which you want to add to the construction group; refer to Figure-221.

Figure-221. Selecting the objects to add to construction group

- Click on the **Add to Construction group** tool from **Toolbar** in the **Draft** workbench; refer to Figure-222. The construction group will be created and the objects will be moved to the construction group; refer to Figure-223.

Figure-222. Add to Construction group tool

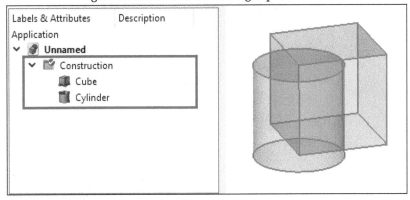

Figure-223. Objects added to the construction group

Healing the objects

The **Heal** tool heals the problematic objects found in old files of FreeCAD. It tries to recreate the old objects from scratch and transfer their properties to the new objects.

- Select one or more problematic objects from the **Model Tree** or from the 3D view area which you want to heal. If no objects are selected, the entire document will be processed.
- Click on the **Heal** tool from **Utilities** drop-down in the **Menu** bar of **Draft** workbench; refer to Figure-224. The healing process will be started or if no errors are found, the tool will do nothing.

Figure-224. Heal tool

PRACTICAL 1

Create the sketch as shown in Figure-225.

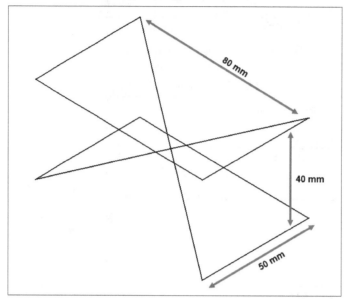

Figure-225. Practical 1

Steps to be performed:
- Start a new part file.
- Select a sketching plane and activate sketching mode.
- Create the sketch using **Rectangle** and **Line** tool.
- Apply the dimensions using dimensioning tools.
- Save the file.

Starting a New Sketch

- Start FreeCAD if not started yet.
- Select the **Draft** option from the **Switch between workbenches** drop-down in the **Toolbar**. The draft environment will be displayed.

Creating Sketch

- Click on the **Rectangle** tool from the **Toolbar** or from the **Drafting** menu to activate rectangle tool. You will be asked to specify start point of line.
- Create a rectangle of width **50** mm and length **80** mm; refer to Figure-226.
- Select the rectangle created, click on the **Clone** tool from the **Toolbar** to create the clone of rectangle and specify the distance of **40** mm between both the rectangles; refer to Figure-227.

Figure-226. Creating rectangle

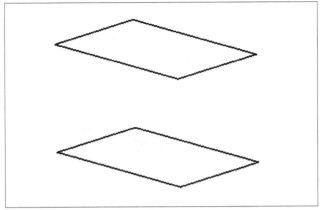

Figure-227. Distance specified between rectangles

- Select the rectangle one by one from the **Model Tree** and downgrade them into edges using **Downgrade** tool; refer to Figure-228.

Figure-228. Rectangles downgraded

Deleting the Edges

- Select the edges of the sketch and click on the **Delete** button from the keyboard to delete the selected edges; refer to Figure-229.

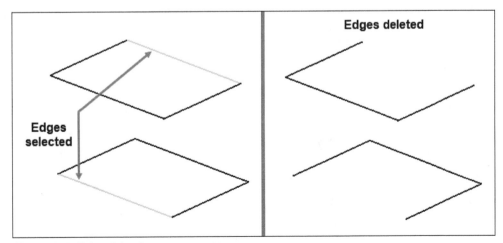

Figure-229. Edges deleted

Creating the lines

- Click on the **Line** tool from the **Toolbar** and create the lines joining open ends of the sketch. The sketch will be created; refer to Figure-230.

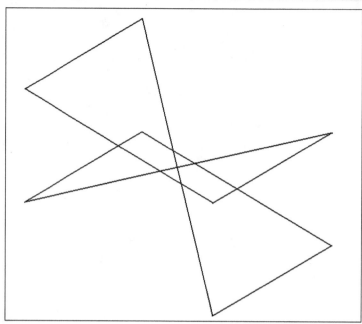

Figure-230. Sketch created

Chapter 7

Arch Modeling

Topics Covered

The major topics covered in this chapter are:

- *Introduction to Arch Workbench*
- *Arch Tools*
- *Axis Tools*
- *Panel Tools*
- *Pipe Tools*
- *Material Tools*

INTRODUCTION

The **Arch** workbench provides a modern building information modeling (BIM) workflow to FreeCAD with support for features like fully parametric architectural entities such as walls, beams, roofs, windows, stairs, pipes, and furniture. It supports industry foundation class (IFC) files and production of 2D floor plans in combination with the **TechDraw** workbench; refer to Figure-1.

Figure-1. Arch workbench

The BIM functionality of FreeCAD is now progressively split into this **Arch** workbench which holds basic architectural tools and the **BIM** workbench which is available from the **Addon Manager**. This **BIM** workbench adds a new interface layer on top of the **Arch** tools with the aim of making the BIM workflow more intuitive and user-friendly.

STARTING ARCH WORKBENCH

The **Arch** workbench imports all tools from the **Draft** workbench as it uses its 2D objects to build 3D parametric architectural objects. Nevertheless, Arch can also use solid shapes created with other workbenches like **Part** and **Part Design**.

- To start a new arch file, click on the **New** button from **File** menu and select **Arch** workbench from **Switch between workbenches** drop-down in the **Toolbar**; refer to Figure-2. The tools related to **Arch** workbench will be displayed in the **Toolbar**; refer to Figure-3.

Figure-2. Arch workbench

Figure-3. Arch workbench tools

ARCH TOOLS

The **Arch** tools are used for creating architectural objects. These tools are available in the **Toolbar** of **Arch** workbench; refer to Figure-4. These tools are discussed next.

Figure-4. Arch tools

Creating Wall

The **Wall** tool builds a wall object from scratch or on top of any other shape-based or mesh-based object. A wall can be built without any base object in which case it behaves as a cubic volume using length, width, and height properties. When built on top of an existing shape, a wall can be based on a linear 2D object, a flat face, a solid, or a mesh. The procedure to use this tool is discussed next.

• Click on the **Wall** tool from **Toolbar** in the **Arch** workbench; refer to Figure-5. The **First point of wall** and **Wall options** dialogs will be displayed in the **Tasks** panel of **Combo View** along with the plus ⊞ sign in place of original cursor; refer to Figure-6. You will be asked to specify the first point.

Figure-5. Wall tool

Figure-6. First point of wall and Wall options dialogs

- Specify desired width and height of the wall in the **Width** and **Height** edit boxes from **Wall options** dialog, respectively.
- Specify the alignment of a wall on its baseline by selecting desired option from **Alignment** drop-down in the **Wall options** dialog.
- After specifying parameters in **Wall options** dialog, click in the 3D view area to specify the first point or enter desired values for x, y, and z coordinates in the **Local X**, **Local Y**, and **Local Z** edit boxes from **First point of wall** dialog, respectively.
- After specifying coordinates for the first point in dialog, click on the **Enter point** button from **First point of wall** dialog. The **Next point** dialog will be displayed asking you to specify the second point.
- Specify desired angle in the **Angle** edit box at which the wall to be created and select the **Angle** check box to lock the current angle.
- Move the cursor away and click at desired location to specify the second point as well as length of the wall or enter desired values for the coordinates in their respective edit boxes.
- After specifying coordinates for the second point in dialog, click on the **Enter point** button again. The wall will be created; refer to Figure-7.

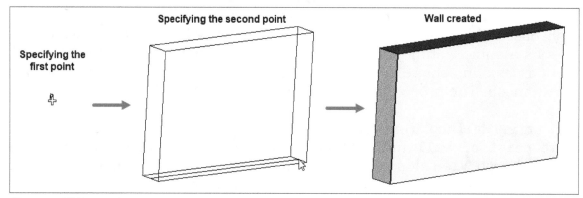

Figure-7. Wall created

- Click on **Close** button to close the dialog.
- If you want to create a wall using base object then select the base geometry objects (Draft object, sketch, etc.) from **Model Tree** or from the 3D view area to create a wall; refer to Figure-8.

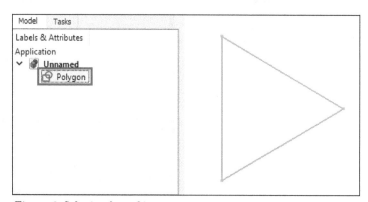

Figure-8. Selecting base object

- Click on the **Wall** tool as discussed earlier. The wall will be created; refer to Figure-9.

Figure-9. Wall created

- If you want to edit the properties of wall then select the wall created from the **Model Tree**. The **Property editor** dialog will be displayed in the **Model** panel of **Combo View** with the parameters related to wall; refer to Figure-10.

Figure-10. Property editor dialog with wall parameters

- Specify desired height and length of each block in the **Block Height** and **Block Length** edit boxes from **Blocks** section of the dialog, respectively.
- Specify the size of joints between each block in the **Joint** edit box.
- Select desired option from **Make Blocks** drop-down whether to make the wall generate blocks or not.
- Specify the horizontal offset of first line and second line of blocks in the **Offset First** and **Offset Second** edit boxes, respectively.
- Select desired option for the alignment of the wall on its baseline from **Align** drop-down in the **Wall** section of the dialog.
- Enter desired value in the **Face** edit box to specify index of the face from the base object to use. If the value is not set or 0, the whole object is used.
- Specify desired length, width, and height of the wall in the **Length**, **Width**, and **Height** edit boxes, respectively.
- Enter desired value in the **Offset** edit box to specify the distance between the wall and its baseline.
- Enter desired value in the **x**, **y**, and **z** edit boxes from **Normal** cascading menu to specify extrusion for the wall along x, y, and z direction, respectively.
- On specifying parameters in the **Property editor** dialog, the properties of wall will be modified; refer to Figure-11.

Figure-11. Wall edited

Creating Structure

The **Structure** tool allows you to build structural elements such as columns or beams, by specifying their width, length, and height, or by basing them on a 2D profile (face, wire, or sketch). The procedure to use this tool is discussed next.

- Click on the **Structure** tool in **Structure** drop-down from **Toolbar** in the **Arch** workbench; refer to Figure-12. Multiple dialogs will be displayed in the **Tasks** panel of **Combo View** along with the plus ⊞ sign and a structure of column in place of original cursor; refer to Figure-13. You will be asked to specify placement point for structure.

Figure-12. Structure tool

Figure-13. Multiple dialogs displayed

- Select desired drawing mode from **Drawing mode** area of the **Structure options** dialog. In **Beam** mode, you will be asked to specify two points in 3D view area or by entering coordinates. The new structural object will span between these two points. In **Column** mode, you will be asked to specify one point in 3D view area or by entering coordinates. The new structural object will be placed at that point.
- Select desired presets of structure from **Category** drop-down in the **Structure options** dialog.
- Select desired size of the preset from **Preset** drop-down in the **Structure options** dialog.
- On selecting the **Precast concrete** option from **Category** drop-down, the parameters related to the selected structure and preset will be displayed in **Precast elements** dialog and **Precast options** dialog in the **Tasks** panel of **Combo View**; refer to Figure-14 and Figure-15.

Figure-14. Precast elements dialog

Figure-15. Precast options dialog

- Specify desired parameters in **Precast elements** and **Precast options** dialogs.
- Specify desired length, width, and height of the structure in the **Length**, **Width**, and **Height** edit boxes from **Structure options** dialog, respectively.
- Click on the **Switch L/H** or **Switch L/W** button to switch between length and height values or length and width values, respectively.
- After specifying parameters in all the dialogs, click in the 3D view area to specify the point for creating the structure or enter desired values for the x, y, and z coordinates in the **Local ΔX**, **Local ΔY**, and **Local ΔZ** edit boxes of the **Base point of column** dialog, respectively.
- After specifying coordinates for the point in dialog, click on the **Enter point** button from **Base point of column** dialog. The structure object will be created; refer to Figure-16.

Figure-16. Structure object created

- Click on **Close** button to close the dialog.
- The procedure for creating a structure using base object is same as discussed for previous tool.
- If you want to edit the properties of structure object then select the structure object created from the **Model Tree**. The **Property editor** dialog will be displayed in the **Model** panel of **Combo View** with the parameters related to structure object; refer to Figure-17.

Property	Value	
Structure		
Face Maker	None	
Height	1025.19 mm	
Length	100.00 mm	
Nodes	[-0.00 0.00 0.00], ...	
Nodes Offset	0.00 mm	
> Normal	[0.00 0.00 0.00]	
Profile		
Tool		
Width	100.00 mm	
View Data		

Figure-17. Property editor dialog with structure parameters

- Select desired option from **Face Maker** drop-down in the **Structure** section of the dialog to specify the type of face generation algorithm to be used for building the profile.
- Specify desired height, length, and width of the structure in the **Height**, **Length**, and **Width** edit boxes, respectively.
- Specify desired structural nodes of this element in the **Vectors** dialog box displayed by clicking on the **Nodes** button.
- Enter desired value in the **Nodes Offset** edit box to specify the offset between the centerline and the nodes line.
- Enter desired values in **x**, **y**, and **z** edit boxes of **Normal** cascading menu to specify the extrusion of base face of this structure along x, y, and z directions, respectively.
- Click on the **Tool** button to select an optional extrusion path which can be any type of wire.

- On specifying the parameters in the **Property editor** dialog, the properties of structure object will be modified; refer to Figure-18.

Figure-18. Structure object edited

Creating Structural System

The **Structural System** tool is used to create a structural system object from a selected structure and axis. The procedure to use this tool is discussed next.

- Select a structure and an axis from which you want to create a structural system; refer to Figure-19.

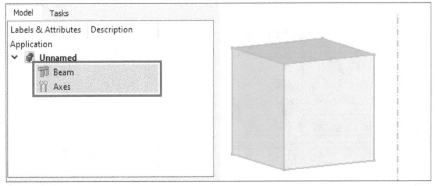

Figure-19. Structure and an axis selected

- Click on the **Structural System** tool from drop-down in the **Toolbar** of **Arch** workbench; refer to Figure-20. The structural system will be created; refer to Figure-21.

Figure-20. Structural System tool

Figure-21. Structural system created

Creating Multiple Structures

The **Multiple Structures** tool is used to create multiple structure objects from a selected base using each selected edge as an extrusion path. The procedure to use this tool is discussed next.

- First, select the base object and then select an edge as an extrusion path; refer to Figure-22.

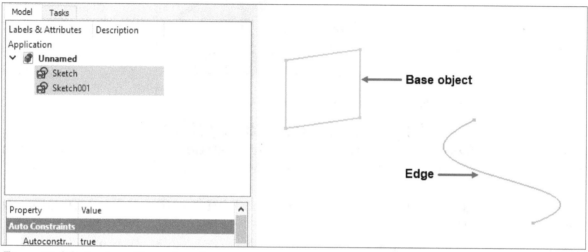

Figure-22. Selecting base object and an edge

- Click on the **Multiple Structures** tool from drop-down in the **Toolbar** of **Arch** workbench; refer to Figure-23. The structure object will be created; refer to Figure-24.

Figure-23. Multiple Structures tool

Figure-24. Structure object created

Creating Rebar

The **Rebar** tool allows you to place reinforcing bars inside structure objects. Rebar objects are based on 2D profiles such as draft objects and sketches that must be drawn on a face of the structural object. The procedure to use this tool is discussed next.

- First, create the structure object using **Structure** tool as discussed earlier; refer to Figure-25.

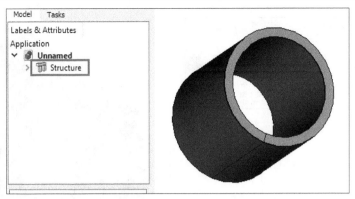

Figure-25. Structure object created

- Switch to **Sketcher** workbench from **Switch between workbenches** drop-down in the **Toolbar**. The tools related to **Sketcher** workbench will be displayed.
- Select desired face of the structural object on which you want to create the sketch; refer to Figure-26.

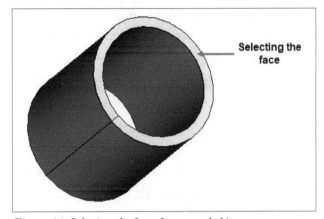

Figure-26. Selecting the face of structural object

- Click on the **New Sketch** tool from **Toolbar** in the **Sketcher** workbench and create desired sketch.
- After creating the sketch, switch to **Arch** workbench from **Switch between workbenches** drop-down.
- Select the newly created sketch from the **Model Tree** or from the 3D view area; refer to Figure-27.

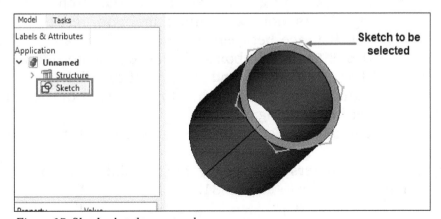

Figure-27. Sketch selected to create rebar

- Click on the **Rebar** tool from **Toolbar** in the **Arch** workbench; refer to Figure-28. The rebar object will be created; refer to Figure-29.

Figure-28. Rebar tool

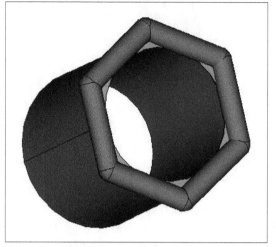

Figure-29. Rebar object created

- If you want to edit the properties of rebar object then select the rebar object created from the **Model Tree**. The **Property editor** dialog will be displayed in the **Model** panel of **Combo View** with the parameters related to rebar object; refer to Figure-30.

Figure-30. Property editor dialog with rebar parameters

- Specify desired amount of bars in the **Amount** edit box from **Rebar** section of the dialog.
- Specify desired diameter of the bars in the **Diameter** edit box.
- Enter desired value in **x**, **y**, and **z** edit boxes from **Direction** cascading menu to specify the spreading of bars along x, y, and z direction, respectively.
- Specify desired offset distance between the border of the structural object and the first bar in the **Offset Start** edit box and between the border of the structural object and the last bar in the **Offset End** edit box of the dialog.
- Specify desired rounding value to be applied to the corner of the bars, expressed in times the diameter in **Rounding** edit box.
- Enter desired value in **Spacing** edit box to specify the distance between the axis of each bar.
- On specifying the parameters in **Property editor** dialog, the properties of rebar object will be modified; refer to Figure-31.

Figure-31. Reber object edited

Creating Curtain Wall

The **Curtain Wall** tool creates a curtain wall by subdividing a base face into quadrangular faces, then creating vertical mullions on the vertical edges, horizontal mullions on the horizontal edges, and filling the spaces between mullions with panels. The procedure to use this tool is discussed next.

- Click on the **Curtain Wall** tool from **Toolbar** in the **Arch** workbench; refer to Figure-32. The **Point** dialog will be displayed in the **Tasks** panel of **Combo View** along with the plus ⊕ sign in place of original cursor; refer to Figure-33. You will be asked to specify the first point.

Figure-32. Curtain Wall tool

Figure-33. Point dialog

- Click in the 3D view area to specify the first point or enter desired values for the x, y, and z coordinates in the **Local ΔX**, **Local ΔY**, and **Local ΔZ** edit boxes of the **Point** dialog, respectively.
- After specifying coordinates for the first point, click on the **Enter point** button from the dialog. You will be asked to specify the second point.

- Move the cursor away and click at desired location to specify the second point or enter desired values for the coordinates in their respective edit boxes in the dialog.
- After specifying coordinates for the second point, click on the **Enter point** button again from the dialog. The curtain wall will be created; refer to Figure-34.

Figure-34. Curtain wall created

- If you want to create the wall from the selected object, sketch, etc. then create and select the object or sketch and click on the **Curtain Wall** tool. The curtain wall will be created.
- If you want to edit the properties of curtain wall then select the curtain wall created from the **Model Tree**. The **Property editor** dialog will be displayed in the **Model** panel of **Combo View** with the parameters related to curtain wall; refer to Figure-35.

Property	Value
Curtain Wall	
Center Profiles	true
Diagonal Mullion Number	0
Diagonal Mullion Profile	
Diagonal Mullion Size	50.00 mm
Height	3000.00 mm
Horizontal Mullion Alignment	false
Horizontal Mullion Height	50.00 mm
Horizontal Mullion Number	2
Horizontal Mullion Profile	
Horizontal Mullion Width	50.00 mm
Horizontal Sections	1
Host	
Panel Number	1
Panel Thickness	20.00 mm
Refine	false
Swap Horizontal Vertical	false
Vertical Direction	[0.00 0.00 1.00]
Vertical Mullion Alignment	false
Vertical Mullion Height	50.00 mm
Vertical Mullion Number	2
Vertical Mullion Profile	
Vertical Mullion Width	50.00 mm
Vertical Sections	1

Figure-35. Property editor dialog with curtain wall parameters

- Select desired option from **Center Profiles** drop-down from **Curtain Wall** area of the dialog to specify whether to center the profile over the edges or not.

- Specify desired number of diagonal mullions in the **Diagonal Mullion Number** edit box.
- Click on the **Diagonal Mullion Profile** button to specify the profile for diagonal mullions.
- Specify desired size of the diagonal mullions in the **Diagonal Mullion Size** edit box.
- Specify desired height of the curtain wall in the **Height** edit box.
- Select desired option from **Horizontal Mullion Alignment** drop-down to specify whether the profile of the horizontal mullions gets aligned with the surface or not.
- Specify desired number, height, and width of the horizontal mullions profile in the **Horizontal Mullion Number**, **Horizontal Mullion Height**, and **Horizontal Mullion Width** edit boxes of the dialog, respectively.
- Click on the **Horizontal Mullion Profile** button to specify the profile for horizontal mullions.
- Specify desired number of horizontal sections of this curtain wall in the **Horizontal Sections** edit box.
- Click on the **Host** button to specify the host object for this curtain wall.
- Specify desired number of panels in the **Panel Number** edit box.
- Specify desired thickness of the panels in the **Panel Thickness** edit box.
- Select desired option from **Refine** drop-down to specify whether to perform subtractions between components or not, so none overlap.
- Select desired option from **Swap Horizontal Vertical** drop-down to specify whether to swap horizontal and vertical lines or not.
- Enter desired values in the **x**, **y**, and **z** edit boxes from **Vertical Direction** cascading menu to specify the vertical direction reference to be used by this object to deduce vertical/horizontal directions.
- Select desired option from **Vertical Mullion Alignment** drop-down to specify whether the profile of the vertical mullions get aligned with the surface or not.
- Specify desired number, height, and width of the vertical mullions profile in the **Vertical Mullion Number**, **Vertical Mullion Height**, and **Vertical Mullion Width** edit boxes of the dialog, respectively.
- Click on the **Vertical Mullion Profile** button to specify the profile for vertical mullions.
- Specify desired number of vertical sections of this curtain wall in the **Vertical Sections** edit box.
- On specifying desired parameters in the **Property editor** dialog, the properties of the curtain wall will be modified; refer to Figure-36.

Figure-36. Curtain wall modified

Creating Building Part

The **Building Part** tool is used to create Floor/Storey/Levels as well as all kinds of situations where different Arch/BIM objects need to be grouped and that group might need to be handled as one object or replicated. The procedure to use this tool is discussed next.

- Select one or more objects from **Model Tree** or from the 3D view area which you want to include in a new building part; refer to Figure-37.

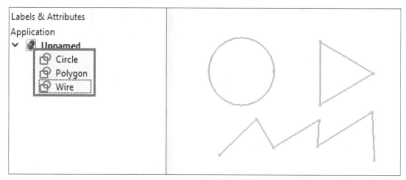

Figure-37. Selecting the objects to include

- Click on the **Building Part** tool from **Toolbar** in the **Arch** workbench; refer to Figure-38. The building part will be created in the **Model Tree** and all the selected objects will be included in the building part; refer to Figure-39.

Figure-38. Building Part tool

Figure-39. Objects included in building part

- If you want to add more objects to the building part then drag and drop the object to it or use **Add component** tool from the **Toolbar** which will be discussed later in this chapter.
- If you want to remove the objects then drag and drop the object out of it or use **Remove component** tool from the **Toolbar** which will be discussed later in this chapter.
- If you want to edit the properties of building part then select the building part created from **Model Tree**. The **Property editor** dialog will be displayed in the **Model** panel of **Combo View** with the parameters related to building part; refer to Figure-40.

Figure-40. Property editor dialog with building part parameters

- Specify computed floor area of the building part in the **Area** edit box from **Building Part** section of the dialog.
- Specify desired height of the building part and of its children objects in the **Height** edit box.
- Enter desired value in **Level Offset** edit box. This value is added to the **Placement. Base.z** attribute of the building part to indicate a vertical offset without actually moving the object.
- Select desired IFC type of this object from **Ifc Role** drop-down in the **Component** section of the dialog.
- Specify desired description and tag for this object in the **Description** and **Tag** edit boxes of the dialog, respectively.
- On specifying the parameters in **Property editor** dialog, the properties of building part object will be modified.

Creating Project

The **Project** tool creates a project including selected objects. It is a special object suitable to add better compatibility with IFC files. Every IFC file is required to contain an IFC project entity. The procedure to use this tool is discussed next.

- Click on the **Project** tool from **Toolbar** in the **Arch** workbench; refer to Figure-41. The project will be created and displayed in the **Model Tree**; refer to Figure-42.

Figure-41. Project tool

Figure-42. Project created

- Now, if you want to add object to the project then drag & drop the object to the project which you want to add in the **Model Tree**; refer to Figure-43.

Figure-43. Objects added to the project

Creating Site

The **Site** tool is a special object that combines properties of a standard FreeCAD group object and Arch objects. It is particularly suited for representing a whole project site or terrain. The procedure to use this tool is discussed next.

- Select one or more building objects from the **Model Tree** which you want to include in a site; refer to Figure-44.

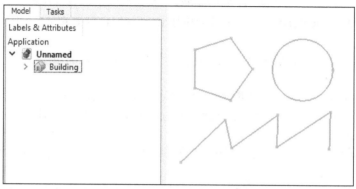

Figure-44. Selecting the building object

- Click on the **Site** tool from **Toolbar** in the **Arch** workbench; refer to Figure-45. The site will be created and the building object will be included in the site; refer to Figure-46.

Figure-45. Site tool

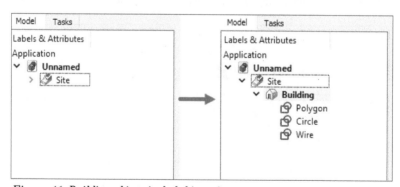

Figure-46. Building objects included in a site

- If you want to add more objects to the site then drag and drop the object to it or use **Add component** tool from the **Toolbar** which will be discussed later in this chapter.
- If you want to remove the objects then drag and drop the object out of it or use **Remove component** tool from the **Toolbar** which will be discussed later in this chapter.
- If you want to edit the properties of site object then select the site object created from the **Model Tree**. The **Property editor** dialog will be displayed in the **Model** panel of **Combo View** with the parameters related to site; refer to Figure-47.

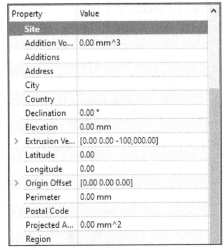

Figure-47. Property editor dialog with site parameters

- Enter desired value in **Addition Volume** and **Subtraction volume** edit boxes from **Site** section of the dialog to specify the volume of earth to be added to this terrain and to be removed from this terrain, respectively.
- Specify desired address, city, country, postal code, and region of this site in the **Address**, **City**, **Country**, **Postal code**, and **Region** edit boxes, respectively.
- Enter desired value in **Latitude** and **Longitude** edit boxes to specify the latitude and longitude of this site, respectively.
- Enter desired value in **x**, **y**, and **z** edit boxes from **Extrusion Vector** cascading menu to use the extrusion vector when performing boolean operations.
- Specify desired perimeter length of this terrain in the **Perimeter** edit box.
- Enter desired value in **Projected Area** edit box to specify area of the projection of this object onto the XY plane.
- Select desired option from **Remove Splitter** drop-down to specify whether to remove the splitters from resulting shape or not.
- Click on the **Terrain** button to specify base terrain of this site.
- Enter desired url in the **Url** edit box to show this site in a mapping website.
- On specifying the parameters in **Property editor** dialog, the properties of site object will be modified.

Creating Building

The **Building** tool is a special type of FreeCAD group object particularly suited for representing a whole building unit. They are mostly used to organize your model by containing floor objects. The procedure to use this tool is discussed next.

- Select one or more objects from the **Model Tree** or from the 3D view area which you want to include in a building; refer to Figure-48.

Figure-48. Selecting the objects to include in a building

- Click on the **Building** tool from **Toolbar** in the **Arch** workbench; refer to Figure-49. The building will be created in the **Model Tree** and all the selected objects will be included in the building; refer to Figure-50.

Figure-49. Building tool

Figure-50. Objects included in building

- If you want to add more objects to the building then drag and drop the object to it or use **Add component** tool from the **Toolbar** which will be discussed later in this chapter.
- If you want to remove the objects then drag and drop the object out of it or use **Remove component** tool from the **Toolbar** which will be discussed later in this chapter.
- The parameters in the **Property editor** dialog to edit the properties of building object are same as discussed for previous tool.

Creating Floor

The **Level** tool is a special type of FreeCAD group object that has a couple of additional properties particularly suited for building floors. The procedure to use this tool is discussed next.

- Select one or more objects from **Model Tree** or from the 3D view area which you want to include in a new floor; refer to Figure-51.

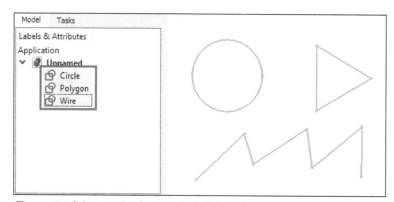

Figure-51. Selecting the objects to include in a floor

- Click on the **Level** tool from **Toolbar** in the **Arch** workbench; refer to Figure-52. The Floor will be created in the **Model Tree** and all the selected objects will be included in the floor; refer to Figure-53.

Figure-52. Floor tool

Figure-53. Objects included in a floor

- If you want to add more objects to the floor then drag and drop the object to it or use **Add component** tool from the **Toolbar** which will be discussed later in this chapter.
- If you want to remove the objects then drag and drop the object out of it or use **Remove component** tool from the **Toolbar** which will be discussed later in this chapter.
- The parameters in the **Property editor** dialog to edit the properties of floor object are same as discussed for previous tool.

Adding an External Reference

The **External reference** tool allows you to place an object in the current document that copies its shape and colors from a part-based object stored in another FreeCAD file. If that FreeCAD file changes, the reference object is marked to be reloaded. The procedure to use this tool is discussed next.

- Click on the **External reference** tool from **Toolbar** in the **Arch** workbench; refer to Figure-54. The **External reference** dialog will be displayed in the **Tasks** panel of **Combo View**; refer to Figure-55.

Figure-54. External reference tool

Figure-55. External reference dialog

- Click on the **Choose file** button from **External file** area of the dialog. The **Choose reference file** dialog box will be displayed; refer to Figure-56.

Figure-56. Choose reference file dialog box

- Select the existing FreeCAD file and click on **Open** button from the dialog box. The selected file will display in the **External file** area and the objects of selected reference file will be available in the **Part to use** drop down of **External reference** dialog; refer to Figure-57.

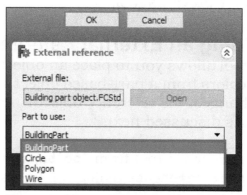

Figure-57. Objects of selected reference file

- Select one of the included part-based objects from the drop-down list and click on **OK** button from the dialog. The selected reference object will be added in the **Model Tree**; refer to Figure-58.

Figure-58. Reference object opened

- The reference object can be moved and rotated, the current position will be retained after reloading the object.
- If the original object gets moved in containing file, this movement will reflect in the reference object.
- If you want to open the containing file of reference object then select reference object and click **RMB** in the **Model Tree**. A shortcut menu will be displayed; refer to Figure-59.

Figure-59. Shortcut menu

- Click on the **Open reference** option from the menu. The file containing the reference object will be opened in the new tab; refer to Figure-60.

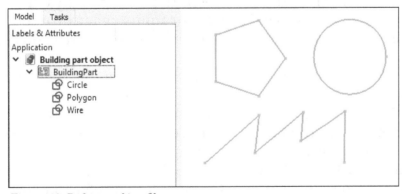

Figure-60. Reference object file

Creating Window

A **Window** is a base object for all kinds of embed-able objects such as windows and doors. The **Window** tool features several presets; this allows the user to create common types of windows and doors with certain edit-able parameters without the need for the user to create the base 2D objects and components manually. The procedure to use this tool is discussed next.

- Click on the **Window** tool from **Toolbar** in the **Arch** workbench; refer to Figure-61. The **Point** and **Window options** dialogs will be displayed in the **Tasks** panel of **Combo View** along with the plus ⊞ sign and structure of window in place of original cursor; refer to Figure-62.

Figure-61. Window tool

Figure-62. Point and Window
options dialogs

- Select **Auto include in host object** check box to insert the window into host object on creation from **Windows options** dialog.
- Specify the height of sill for the window in the **Sill height** edit box of the dialog.
- Select desired preset for the window or door from **Preset** drop-down in the dialog.
- Specify desired width and height of the window in the **Width** and **Height** edit boxes of the dialog.
- After specifying all the parameters in the dialog, you will be asked to specify the location of point.
- Click in the 3D view area to specify the point or enter desired values for the x, y, and z coordinates in the **Local ΔX**, **Local ΔY**, and **Local ΔZ** edit boxes of the **Point** dialog, respectively.
- After specifying coordinates for the point in dialog, click on the **Enter point** button from the **Point** dialog. The window will be created; refer to Figure-63.

Figure-63. Window created

- If you want to create, modify, or delete the components of a window then double click on the window created from the **Model Tree**. The **Window elements** and **Component** dialog will be displayed in the **Tasks** panel of **Combo View**; refer to Figure-64
- If you want to edit the properties of window then select the window created from **Model Tree**. The **Property editor** dialog will be displayed in the **Model** panel of **Combo View** with parameters related to window; refer to Figure-65.

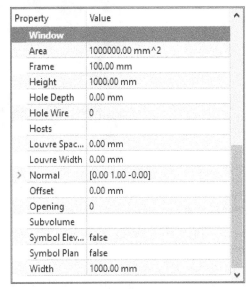

Figure-65. Property editor dialog with window parameters

Figure-64. Window elements and Component dialogs

- Specify desired height and width of the window in the **Height** and **Width** edit boxes from **Window** section of the dialog, respectively.
- Specify desired depth of the hole created by this window in its host object in the **Hole Depth** edit box.
- Enter desired value in **Host Wire** edit box to specify the number of wire from the base object that is used to create a hole in the host object of this window.
- Enter desired value in **Louvre Spacing** edit box to define the spacing between louvre elements.
- Enter desired value in **Louvre Width** edit box to define the size of the louvre elements.
- Enter desired value in **Opening** edit box to define opening mode of all the components and to define a hinge provided in them or in an earlier component in the list.
- Select desired value from **Symbol Elevation** drop-down to specify whether to show 2D opening symbol in elevation or not.
- Select desired value from **Symbol Plan** drop-down whether to show 2D opening symbol in plan or not.

- On specifying desired value in the **Property editor** dialog, the properties of window will be modified; refer to Figure-66.

Figure-66. Window edited

Creating Roof

The **Roof** tool is used to create a sloped roof using selected wire. The created roof object is parametric keeping its relationship with the base object. The procedure to use this tool is discussed next.

- Select the wire from the **Model Tree** or from the 3D view area on which you want to create the roof; refer to Figure-67.

Figure-67. Selecting the wire to create roof

- Click on the **Roof** tool from **Toolbar** in the **Arch** workbench; refer to Figure-68. The default roof will be created; refer to Figure-69.

Figure-68. Roof tool

Figure-69. Default roof created

- If you want to edit the roof then select the roof created from the **Model Tree** and double-click on it. The **Roof** dialog will be displayed in the **Tasks** panel of **Combo View**; refer to Figure-70.

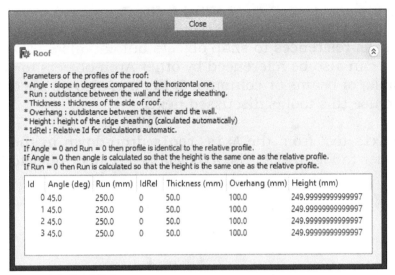

Figure-70. Roof dialog

- Specify desired values in the **Angle**, **Run**, **IdRel**, **Thickness**, **Overhang**, and **Height** edit boxes of the dialog. The roof will be edited.
- If you want to edit the properties of roof then select the roof created from the **Model Tree**. The **Property editor** dialog will be displayed in the **Model** panel of **Combo View** with the parameters related to roof object; refer to Figure-71.

Property	Value
Roof	
Angles	[45.00,45.00,45.00,45.00]
Border Leng...	0.00 mm
Face	0
Flip	false
Heights	[]
Id Rel	[0,0,0,0]
Overhang	[100.00,100.00,100.00,100.00]
Ridge Length	0.00 mm
Runs	[250.00,250.00,250.00,250.00]
Thickness	[50.00,50.00,50.00,50.00]

Figure-71. Property editor dialog with roof parameters

- Enter desired values in the **Angles** edit box to specify slope angle of the roof pane (an angle for each edge in the wire).
- Enter desired value in the **Face** edit box to specify face index of the base object to be used.
- Enter desired values in **Id Rel** edit box to specify relation Id of the slope angle of the roof.
- Enter desired values in **Overhang** edit box to specify overhang of the roof pane (an overhang for each edge in the wire).
- Enter desired values in **Runs** edit box to specify width of the roof pane (a run for each edge in the wire).
- Enter desired values in **Thickness** edit box to specify thickness of the roof pane (a thickness for each edge in the wire).
- On specifying parameters in the **Property editor** dialog, the properties of roof object will be modified.

Axis tools

The **Axis** tools allow you to place a series of axes in the current document.

Creating Axis

The **Axis** tool allows you to place a series of axes in the current document. The axes serve mainly as references to snap objects but can also be used together with **AxesSystems**, and can also be referenced by other Arch objects to create parametric arrays, for example, of beams or columns. **Grids** can also be used in place of axes. The procedure to use this tool is discussed next.

- Click on the **Axis** tool from the **Axis tools** drop-down in the **Toolbar** of **Arch** workbench; refer to Figure-72. The axis will be created; refer to Figure-73.

Figure-72. Axis tool

Figure-73. Axis created

- If you want to edit the parameters of axis then double-click on the axis created. The **Axes** dialog will be displayed in the **Tasks** panel of **Combo View**; refer to Figure-74.

Figure-74. Axes dialog

- Click on **Add** button from the dialog to add the number of axes and if you want to remove the axis then select the axis from the dialog which you want to remove and click on **Remove** button.

- Specify the distance and angle of axis in the **Distance** and **Angle** edit boxes of the dialog, respectively.
- Click on **Close** button to close the dialog.
- If you want to edit the properties of axis then select the axis created from the **Model Tree**. The **Property editor** dialog will be displayed in the **Model** panel of **Combo View** with the parameters related to axis; refer to Figure-75.

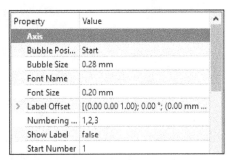

Figure-75. Property editor dialog with axis parameters

- Specify desired length of the axes in the **Length** edit box from **Axis** section of the dialog.
- Select desired option from **Bubble Position** drop-down to specify where the bubble is placed on the axis.
- Specify desired size of the axis bubbles in the **Bubble Size** edit box of the dialog.
- Specify desired font name to draw the bubble number and/or labels in the **Font Name** edit box.
- Specify desired size of the label text in the **Font Size** edit box of the dialog.
- Select desired option from **Numbering Style** drop-down to specify how the axes are numbered; 1, 2, 3, A, B, C, etc.
- Select desired option from **Show Label** drop down to specify whether to turn the display of the label texts on or off.
- On specifying parameters in the **Property editor** dialog, the properties of axis will be modified.

Creating Axis System

The **Axis System** tool allows you to combine several axis to the document. This is useful to define the intersection points between the different axes. Arch objects can then use this system to duplicate their shape on the different intersection points. The procedure to use this tool is discussed next.

- Select the axes from the **Model Tree** or from the 3D view area; refer to Figure-76.

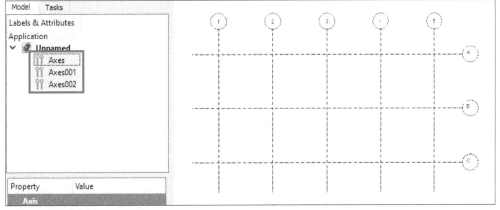

Figure-76. Selecting the axes

- Click on the **Axis System** tool from **Axis** tools drop-down in the **Toolbar** of **Arch** workbench; refer to Figure-77. The Axis system will be created and displayed in the **Model Tree**; refer to Figure-78.

Figure-77. Axis system tool

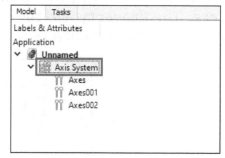

Figure-78. Axis system created

- If you want to edit the parameters of axis system then double-click on the axis system created from the **Model Tree**. The **Axes** dialog will be displayed in the **Tasks** panel of **Combo View**; refer to Figure-79.

Figure-79. Axes dialog

- If you want to add more axes in the axis system then select the axis from the **Model Tree** and click on **Add** button from the **Axes** dialog and if you want to remove the axis then select the axis from **Axis system components** area of the dialog and click on **Remove** button from the dialog.
- Click on **OK** button to close the dialog.

Creating Grid

The **Grid** tool allows you to place a grid-like object in the document. This object is meant to serve as a base to build Arch objects that need a regular but complex frame, such as windows, curtain walls, column grids, railings, etc. The procedure to use this tool is discussed next.

- Click on the **Grid** tool from **Axis tools** drop-down in the **Toolbar** of **Arch** workbench; refer to Figure-80. The grid will be created and displayed in the **Model Tree**; refer to Figure-81.

Figure-80. Grid tool

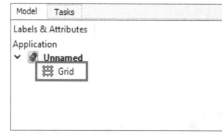

Figure-81. Grid created

- To edit the parameters of grid, double-click on the grid created from the **Model Tree**. The **Grid** dialog will be displayed in the **Tasks** panel of **Combo View**; refer to Figure-82.

- Specify desired width and height of the grid in the **Total width** and **Total height** edit boxes of the dialog, respectively.

- Click on the **Add row** button to add the number of rows in the grid and click on **Add col** button to add the number of columns in the grid. The rows and columns will be displayed in the **Rows/Columns** area of the dialog; refer to Figure-83.

Figure-82. Grid dialog

Figure-83. Grid dialog displaying rows and columns

- If you want to remove the row or column then select the row or column which you want to remove from the dialog and click on **Del row** or **Del col** buttons, respectively.

- Click on **OK** button from the dialog. The grid will be created; refer to Figure-84.

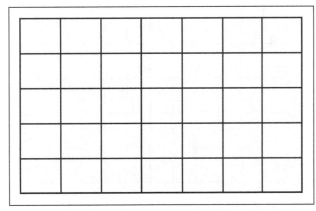

Figure-84. Grid created

- If you want to edit the properties of grid then select the grid created from the **Model Tree**. The **Property editor** dialog will be displayed in the **Model** panel of **Combo View** with the parameters related to grid; refer to Figure-85.

Figure-85. Property editor dialog with grid parameters

- Enter desired values in **Auto Height** and **Auto Width** edit boxes from **Grid** section of the dialog to create automatic row divisions and automatic column divisions, respectively.
- Specify desired size of columns and rows in the **Column Size** and **Row Size** edit boxes of the dialog, respectively.
- Specify desired number of columns and rows in the **Columns** and **Rows** edit boxes, respectively.
- Specify total height and total width of the grid in the **Height** and **Width** edit boxes, respectively.
- Enter desired value in **Hidden Faces** edit box to specify indices of faces to hide.
- Select desired option from **Points Output** drop-down to specify the type of 3D points produced by this grid object.
- Select desired option from **Reorient** drop-down to specify whether the grid must reorient its children along edge normals or not.
- On specifying parameters in the **Property editor** dialog, the properties of grid will be modified; refer to Figure-86.

Figure-86. Grid edited

Creating Section Plane

The **Section Plane** tool is used to create section plane for generating section views. You can define the placement of section plane according to the current working plane and it can be relocated and reoriented by moving and rotating it, until it describes the 2D view you want to obtain. The Section plane object will only consider a certain set of objects, not all the objects of the document. The procedure to use this tool is discussed next.

- Set desired working plane to define where you want to place the section plane as discussed earlier.
- Select the object from the **Model Tree** or from the 3D view area which you want to include in the section view; refer to Figure-87.

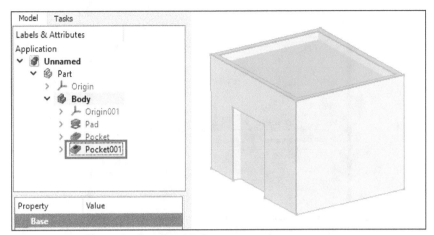

Figure-87. Selecting the object to include in section view

- Click on the **Section Plane** tool from **Toolbar** in the **Arch** workbench; refer to Figure-88. The section plane will be created; refer to Figure-89.

Figure-88. Section plane tool

Figure-89. Section plane created

- You can move or rotate the section plane in desired position using **Move** tool as discussed earlier.
- If you want to edit the section plane then double-click on the section plane created from the **Model Tree**. The **Section plane settings** dialog will be displayed in the **Tasks** panel of **Combo View**; refer to Figure-90.

Figure-90. Section plane settings dialog

- The object selected for section plane will be displayed in the **Objects seen by this section plane** list box of the dialog.
- If you want to add more objects to be section then select the object from the **Model Tree** and click on **Add** button from the dialog.
- If you want to remove the objects then select the object from the dialog and click on **Remove** button.
- To rotate the section plane along x, y, or z direction then click on the **Rotate X**, **Rotate Y**, or **Rotate Z** button from **Section plane placement** area of the dialog, respectively.
- Click on the **Resize** button to modify the size of section plane.
- Click on the **Center** button to place the section plane at the center of object.
- Click on **OK** button to close the dialog.
- You can edit the properties of section plane using **Property editor** as discussed earlier.
- To create a section view of the object, select the section plane from the **Model Tree** or from the 3D view area and click on the **Shape 2D View** tool from the **Toolbar**. The 2D section view will be created; refer to Figure-91.

Figure-91. Section view created

- If you want to edit the properties of section then select the section plane created from the **Model Tree**. The **Property editor** dialog will be displayed in the **Model** panel of **Combo View** with parameters related to section plane; refer to Figure-92.

Property	Value	^
Section Plane		
Arrow Size	50.00 mm	
Cut Distance	0.00 mm	
Cut Margin	1.00 mm	
Cut View	false	
Display Hei...	9.46 mm	
Display Len...	11.90 mm	
Line Color	■ [40, 49, 82]	
Line Width	1.00	
Transparency	85	v

Figure-92. Property editor dialog with section plane parameters

- Select desired option from **Only Solids** drop-down from **Section Plane** section of the dialog to specify whether the non-solid objects in the set will be disregarded or not.
- Specify desired size of the arrows of the section plane gizmo in the 3D view area in the **Arrow Size** edit box of the dialog.
- Specify desired height and length of the section plane gizmo in the 3D view area in the **Display Height** and **Display Length** edit boxes, respectively.
- Select desired option from **Cut View** drop-down to specify whether the whole 3D view will be cut at the location of this section plane or not.
- On specifying parameters in the **Property editor** dialog, the properties of section plane will be modified.

Creating Space

The **Space** tool allows you to define an empty volume either by basing it on a solid shape or by defining its boundaries or a mix of both. The space object always defines a solid volume. The procedure to use this tool is discussed next.

- Select the existing solid object or faces on the boundary object from **Model Tree** or from the 3D view area which you want to create a space object; refer to Figure-93.

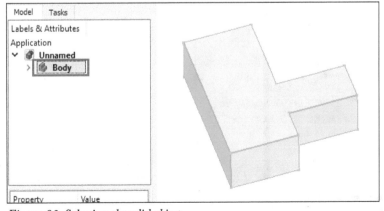

Figure-93. Selecting the solid object

- Click on the **Space** tool from **Toolbar** in the **Arch** workbench; refer to Figure-94. The space object will be created; refer to Figure-95.

Figure-94. Space tool

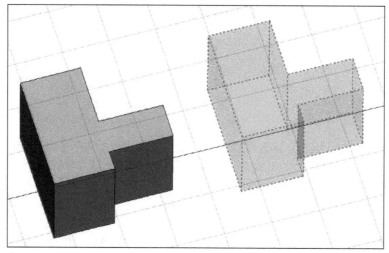

Figure-95. Space object created

- If you want to edit the space object then double-click on the space object created from the **Model Tree**. The **Component** dialog will be displayed in the **Tasks** panel of **Combo View**; refer to Figure-96.
- Specify desired parameters from the **Component** dialog as discussed earlier.
- Click on **OK** button to close the dialog.
- If you want to edit the parameters of space object then select the space object created from **Model Tree**. The **Property editor** dialog will be displayed in the **Model** panel of **Combo View** with parameters related to space object; refer to Figure-97.

Figure-96. Component dialog

Figure-97. Property editor dialog with space object parameters

- Select desired option from **Auto Power** drop-down in the **Space** section of the dialog to specify whether the equipment power will be automatically filled by the equipment included in this space or not.

- Select desired option from **Conditioning** drop-down to specify the type of air conditioning of this space.
- Enter desired value in **Equipment Power** edit box to specify the electric power needed by the equipment of this space in watts.
- Enter desired values in **Finish Ceiling**, **Finish Floor**, and **Finish Walls** edit boxes to specify finishing of ceiling, finishing of floor, and finishing of walls of this space, respectively.
- Specify desired thickness of the floor finish in the **Floor Thickness** edit box.
- Objects that are included inside this space such as furniture, will be displayed in the list box available by clicking on the **Group** button.
- Select desired option from **Internal** drop-down to specify whether this space is internal or external.
- Enter desired value in **Lighting Power** edit box to specify the electric power needed to light this space in watts.
- Specify desired number of people who typically occupy this space in the **Number Of People** edit box.
- Select desired type of this space from **Space Type** drop-down in the dialog.
- Specify desired number of decimals to use for calculated texts in the **Decimals** edit box.
- Specify desired size of the first line of text in the **First Line** edit box.
- Specify desired name, size, and color of the text in the **Font Name**, **Font Size**, and **Text Color** edit boxes of the dialog, respectively.
- Enter desired value in **Line Spacing** edit box to specify the space between lines of text.
- Select desired option from **Show Unit** drop-down to specify whether to show the unit suffix or not.
- Specify desired text to show in the **Text** edit box.
- Select desired option from **Text Align** drop-down to specify the justification of the text.
- Enter desired value in **x**, **y**, and **z** edit boxes from **Text Position** cascading menu to specify the position of text along x, y, and z directions, respectively.
- On specifying the parameters in **Property editor** dialog, the properties of space object will be modified.

Creating Stairs

The **Stairs** tool allows you to automatically build several types of stairs. Stairs can be built from scratch or from a straight line in which case, the stairs follow the line. If the line is not horizontal but has a vertical inclination, the stairs will also follow its slope. The procedure to use this tool is discussed next.

- Click on the **Stairs** tool from **Toolbar** in the **Arch** workbench; refer to Figure-98. The stairs will be created; refer to Figure-99.

Figure-98. Stairs tool

Figure-99. Stairs created

- If you want to edit the properties of stairs then select the stairs created from the **Model Tree**. The **Property editor** dialog will be displayed in the **Model** panel of **Combo View** with the parameters related to stairs; refer to Figure-100.

Figure-100. Property editor dialog with stairs parameters

- Select desired option from **Align** drop-down in the **Stairs** section of dialog to specify the alignment of these stairs on their baseline, if applicable.
- Specify desired height, length, and width of the stairs in the **Height**, **Length**, and **Width** edit boxes of the dialog, respectively.
- Specify desired size of nosing in the **Nosing** edit box from **Steps** section of the dialog.
- Specify number of steps (risers) in these stairs in the **Number of Steps** edit box.
- Specify desired height of the risers in the **Riser Height** edit box of the dialog.
- Specify depth of the treads in the **Tread Depth** edit box of the dialog.
- Specify thickness of the treads in the **Tread Thickness** edit box.
- Select desired type of landings from **Landings** drop-down in the **Structure** section of the dialog.
- Specify desired offset between the border of the stairs and the structure in the **Stringer Offset** edit box.
- Specify width of the stringers in the **Stringer Width** edit box of the dialog.
- Select desired type of structure of these stairs from **Structure** drop-down.
- Specify desired thickness of the structure from **Structure Thickness** edit box.
- Select desired type of winders from **Winders** drop-down in the dialog.
- On specifying the parameters in **Property editor** dialog. The properties of stairs will be modified; refer to Figure-101.

Figure-101. Stairs edited

Panel tools

The **Panel** tools allow you to build all kinds of panel-like elements.

Creating Panel

The **Panel** tool allows you to build all kinds of panel-like elements, typically for panel constructions but also for all kinds of objects that are based on a flat profile. The procedure to use this tool is discussed next.

- Click on the **Panel** tool from **Panel tools** drop-down in the **Toolbar** of **Arch** workbench; refer to Figure-102. The **Point** and **Panel options** dialog will be displayed in the **Tasks** panel of **Combo View** along with the plus ⊞ sign and a structure of panel in place of original cursor; refer to Figure-103. You will be asked to specify the location of panel.

Figure-102. Panel tool

Figure-103. Point and Panel op-tions dialogs

- Select desired preset of panel from **Preset** drop-down in the **Panel options** dialog.
- Specify desired length, width, and thickness of panel in the **Length**, **Width**, and **Thickness** edit boxes of the dialog, respectively.
- Click on **Rotate** button from the dialog to rotate the panel at desired location.
- After specifying parameters in the **Panel options** dialog, click in the 3D view area to specify the location of panel or enter desired values for x, y, and z coordinates in the **Local ΔX**, **Local ΔY**, and **Local ΔZ** edit boxes of the **Point** dialog, respectively.
- After specifying the coordinates in dialog, click on **Enter point** button. The panel will be created; refer to Figure-104.

Figure-104. Panel created

- If you want to edit the properties of panel then select the panel created from the **Model Tree**. The **Property editor** dialog will be displayed in the **Model** panel of **Combo View** with the parameters related to panel; refer to Figure-105.

Property	Value
Panel	
Area	250000.00 mm^2
Face Maker	None
Length	500.00 mm
> Normal	[0.00 0.00 0.00]
Offset	0.00 mm
Sheets	1
Thickness	10.00 mm
Wave Bottom	false
Wave Direct...	0.00 °
Wave Height	0.00 mm
Wave Length	0.00 mm
Wave Offset	0.00 mm
Wave Type	Curved
Width	500.00 mm

Figure-105. Property editor dialog with panel parameters

- Specify desired length, width, thickness, and area of the panel in the **Length**, **Width**, **Thickness**, and **Area** edit boxes of the dialog, respectively.
- Specify the number of sheets of material of which the panel is made in the **Sheets** edit box.
- Specify desired length, height, and orientation of the waves for corrugated panels in the **Wave Length**, **Wave Height**, and **Wave Direction** edit boxes of the dialog, respectively.
- Select desired type of the wave for corrugated panel from **Wave Type** drop-down in the dialog.
- Select desired option from **Wave Bottom** drop-down to specify whether the bottom wave of the panel is flat or not.

- On specifying the parameters in the **Property editor** dialog, the properties of panel will be modified; refer to Figure-106.

Figure-106. Panel edited

Creating Panel Cut

The **Panel Cut** tool creates a flat 2D view of a panel to be included in the 3D document of a panel sheet or directly exported to DXF. The procedure to use this tool is discussed next.

- Select one or more panel objects from **Model Tree** or from the 3D view area; refer to Figure-107.

Figure-107. Selecting the panel object

- Click on the **Panel Cut** tool from **Panel tools** drop-down in the **Toolbar** of **Arch** workbench; refer to Figure-108. The panel cut object will be created; refer to Figure-109.

Figure-108. Panel cut tool

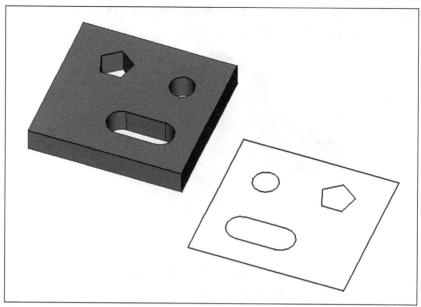

Figure-109. Panel cut object created

- If you want to edit the properties of panel cut then select the panel cut created from the **Model Tree**. The **Property editor** dialog will be displayed in the **Model** panel of **Combo View** with the parameters related to panel cut; refer to Figure-110.

Figure-110. Property editor dialog with panel cut parameters

- Select desired font file to specify font of the tag text from **Font File** edit box in the **Panel Cut** section of the dialog.
- Select desired option from **Make Face** drop-down to specify whether the panel cut will be a face or a wire.
- Enter desired values in **x**, **y**, and **z** edit boxes from **Tag Position** cascading menu to specify the position of tag text along x, y, and z directions, respectively.
- Enter desired value in **Tag Rotation** edit box to specify the rotation of tag text.
- Specify desired size of the tag text in the **Text Size** edit box.
- Specify desired text to display in the **Tag Text** edit box of the dialog.
- Enter desired value in **Margin** edit box from **Arch** section to specify the margin that can be displayed outside the panel cut shape.
- Select desired option from **Show Margin** drop-down to specify whether to turn the display of the margin on/off.
- On specifying parameters in the **Property editor** dialog, the properties of panel cut will be modified.

Creating Panel Sheet

The **Panel Sheet** tool allows you to build a 2D sheet, including any number of panel cut objects or any other 2D object such as those made by the **Draft** workbench and **Sketcher** workbench. The Panel Sheet is typically made to layout cuts to be made by a CNC machine. These sheets can then be exported to a DXF file. The procedure to use this tool is discussed next.

- Select one or more panel cut objects or any other 2D objects from the **Model Tree** or from the 3D view area from which you want to create a panel sheet; refer to Figure-111.

Figure-111. Selecting the panel cut object

- Click on the **Panel Sheet** tool from **Panel tools** drop-down in the **Toolbar** of **Arch** workbench; refer to Figure-112. A panel sheet of the object will be created; refer to Figure-113.

Figure-112. Panel Sheet tool

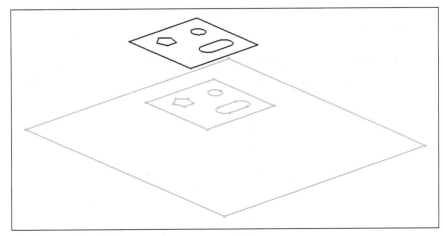

Figure-113. Panel sheet created

- If you want to edit the properties of panel sheet then select the panel sheet created from the **Model Tree**. The **Property editor** dialog will be displayed in the **Model** panel of **Combo View** with parameters related to panel sheet; refer to Figure-114.

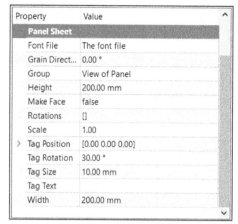

Figure-114. Property editor dialog with panel sheet parameters

- Select desired font file to specify font of the tag text from **Font File** edit box in the **Panel Sheet** section of the dialog.
- Enter desired value in **Grain direction** edit box which allows you to inform the main direction of the panel fiber.
- Specify desired height and width of the sheet in the **Height** and **Width** edit boxes of the dialog, respectively.
- Select desired option from **Make Face** drop-down to specify whether the panel cut will be a face or a wire.
- Enter desired value in **Scale** edit box to specify the scale applied to each panel view.
- Enter desired values in **x**, **y**, and **z** edit boxes from **Tag Position** cascading menu to specify the position of tag text along x, y, and z directions, respectively.
- Enter desired value in **Tag Rotation** edit box to specify the rotation of tag text.
- Specify desired size of the tag text in the **Text Size** edit box.
- Specify desired text to display in the **Tag Text** edit box of the dialog.
- On specifying parameters in the **Property editor** dialog, the properties of panel sheet will be modified.

Creating Nest

The **Nest** tool allows you to select a flat shape to be a container and a series of other flat shapes to be organized inside the space defined by the container shape. The procedure to use this tool is discussed next.

- Click on the **Nest** tool from **Toolbar** in the **Arch** workbench; refer to Figure-115. The **Nesting** dialog will be displayed in the **Tasks** panel of **Combo View**; refer to Figure-116.

Figure-115. Nest tool

Figure-116. Nesting dialog

- Select a flat, rectangular object from **Model Tree** or from the 3D view area which is to be container; refer to Figure-117.

Figure-117. Selecting the rectangular object

- Click on **Pick selected** button from **Container** area of the dialog. The selected object will be displayed in the **Container** selection box.
- Select the other flat objects by holding the **CTRL** key that you want to place inside the container from **Model Tree** or from the 3D view area. Note that these objects must all be flat and in the same plane as the container.
- After selecting the objects, click on **Add selected** button to add it in the **Shapes** list box of the dialog.
- Specify desired tolerance value in the **Tolerance** edit box from **Nesting parameters** area.
- Enter desired value in **Arcs subdivisions** edit box to specify the number of segments to divide non-linear edges for calculations.
- Enter desired values in **Rotations** edit box to specify the angles to rotate the shapes.
- After specifying the parameters, click on **Start** button from **Nesting operation** area to start the calculation process.

- If you want to stop the calculation process then click on the **Stop** button.
- At the end of calculation process, click on the **Preview** button to create a temporary preview of the result.
- Click on **OK** button from the dialog to apply the result. The nest of the objects will be created; refer to Figure-118.

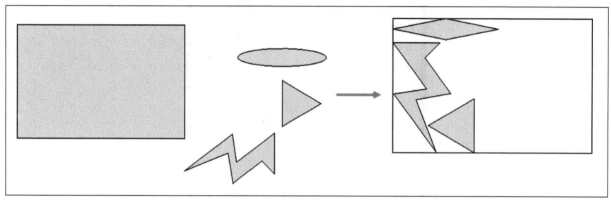

Figure-118. Nest created

Creating Equipment

The **Equipment** tool offers you a simple and convenient way to insert non-structural, standalone elements such as pieces of furniture, hydro-sanitary equipments, or electrical appliances to your projects. The procedure to use this tool is discussed next.

- Select the existing solid object or a mesh object from **Model Tree** or from the 3D view area; refer to Figure-119.

Figure-119. Selecting the existing solid object

- Click on the **Equipment** tool from **Toolbar** in the **Arch** workbench; refer to Figure-120. The non-structural element will be enclosed in an equipment object; refer to Figure-121.

Figure-120. Equipment tool

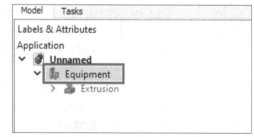

Figure-121. Equipment object created

- If you want to edit the properties of equipment then select the equipment created from **Model Tree**. The **Property editor** dialog will be displayed in the **Model** panel of **Combo View** with parameters related to equipment; refer to Figure-122.

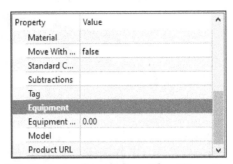

Figure-122. Property editor dialog with equipment parameters

- Enter desired value in **Equipment Power** edit box from **Equipment** section of the dialog to specify the electric power needed by the equipment of this model in watts.
- Specify desired description of the model for this equipment in the **Model** edit box.
- Enter desired URL of the product page where more information about this equipment can be found in the **Product URL** edit box of the dialog.
- On specifying parameters in the **Property editor** dialog, the properties of equipment will be modified.

Creating Frame

The **Frame** tool is used to build all kinds of frame objects based on a profile and a layout. The profile is extruded along the edges of the layout which can be any 2D object such as a sketch or a draft object. It is especially useful to create railings or frame walls. The procedure to use this tool is discussed next.

- First, create a layout object and a profile object from which you want to create a frame; refer to Figure-123.

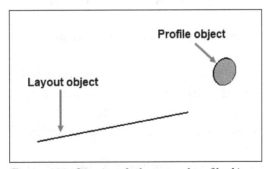

Figure-123. Creating the layout and profile object

- Select the created layout object and then while holding the **CTRL** key, select the profile object from the **Model Tree** or from the 3D view area.
- Click on the **Frame** tool from **Toolbar** in the **Arch** workbench; refer to Figure-124. The frame object will be created; refer to Figure-125.

Figure-124. Frame tool

Figure-125. Frame object created

- If you want to edit the properties of frame object then select the frame object created from the **Model Tree**. The **Property editor** dialog will be displayed in the **Model** panel of **Combo View** with the parameters related to frame object; refer to Figure-126.

Figure-126. Property editor dialog with frame parameters

- Select desired option from **Align** drop-down in the **Frame** section of the dialog to specify whether the profile must be rotated to have its normal axis aligned with each edge or not.
- Select desired option from the **Edges** drop-down to specify the type of edges to be considered.
- Select desired option from **Fuse** drop-down to specify whether the geometry is fused or not.
- Enter desired values in **x**, **y**, and **z** edit boxes from **Offset** cascading menu to specify the distance between layout object and the frame object.
- The **Profile** edit box specifies the profile used to build this frame.
- Enter desired value in **Rotation** edit box to specify the rotation of the profile around its extrusion axis.
- On specifying parameters in the **Property editor** dialog, the properties of the frame object will be modified.

Creating Fence

The **Fence** tool creates a fence object by repeating single fence post and section along a given path. Note that you need to create separate bodies for section and fence post using Part Design workbench and then convert them into two separate structures by using **Structure** tool in **Arch** workbench. For creating path, we have used Polyline tool of Draft workbench. The procedure to use this tool is discussed next.

- Create a single fence post, a single section, and a path to be followed by the fence using desired workbench; refer to Figure-127.

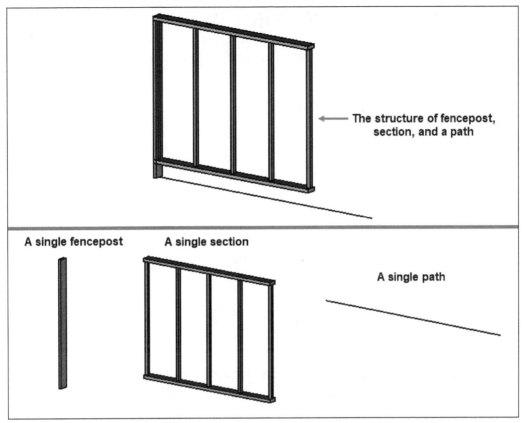

Figure-127. Structure to create fence

- Now, select the section, post, and path in exactly that order and click on the **Fence** tool from **Toolbar** in the **Arch** workbench; refer to Figure-128. The fence object will be created; refer to Figure-129.

Figure-128. Fence tool

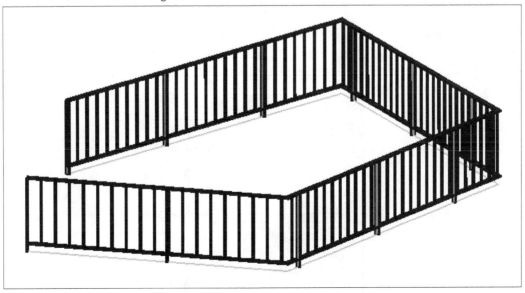

Figure-129. Fence object created

Creating Truss

The **Truss** tool creates a truss object, either from a selected linear object or from scratch, i.e. if no object is selected when launching the command. The procedure to use this tool is discussed next.

- Create and select the line from the **Model Tree** or from the 3D view area from which you want to create a truss; refer to Figure-130.

Figure-130. Selecting the line to create truss

- Click on the **Truss** tool from **Toolbar** in the **Arch** workbench; refer to Figure-131. The truss object will be created; refer to Figure-132.

Figure-131. Truss tool

Figure-132. Truss object created

Or,

- Click on the **Truss** tool from **Toolbar** in the **Arch** workbench as discussed earlier. The **Point** dialog will be displayed in the **Tasks** panel of **Combo View** along with the plus ⊞ sign in place of original cursor; refer to Figure-133. You will be asked to specify the first point.

Figure-133. Point dialog of truss tool

- Click in the 3D view area to specify the first point or enter desired values for the x, y, and z coordinates in the **Local ΔX**, **Local ΔY**, and **Local ΔZ** edit boxes of the dialog, respectively.
- After specifying coordinates for the first point, click on the **Enter point** button in the dialog. You will be asked to specify the second point.
- Move the cursor away and click at desired location to specify the second point or enter desired values for the coordinates in their respective edit boxes.
- After specifying coordinates for the second point, click on the **Enter point** button again from the dialog. The truss object will be created.
- If you want to edit the properties of truss object then select the truss object created from the **Model Tree**. The **Property editor** dialog will be displayed in the **Model** panel of **Combo View** with the parameters related to truss object; refer to Figure-134.

Property	Value
Truss	
Height End	200.00 mm
Height Start	100.00 mm
> Normal	[0.00 0.00 1.00]
Rod Direction	Forward
Rod End	false
Rod Mode	/|/|/|
Rod Sections	3
Rod Size	2.00 mm
Rod Type	Round
Slant Type	Simple
Strut End Offset	0.00 mm
Strut Height	10.00 mm
Strut Start Offset	0.00 mm
Strut Width	5.00 mm
Truss Angle	53.53 °

Figure-134. Property editor dialog with truss parameters

- Specify desired height of the truss at the end and start position in the **Height End** and **Height Start** edit boxes from **Truss** section of the dialog, respectively.
- Enter desired values in the **x, y,** and **z** edit boxes from **Normal** cascading menu to specify the normal direction of the truss.
- Select desired direction of the rods from **Rod Direction** drop-down.
- Select desired option from **Rod End** drop-down to specify whether the truss has a rod at its endpoint or not.
- Select desired rod mode to draw the rods from **Rod Mode** drop-down.
- Specify desired number of rod sections in the **Rod Sections** edit box.
- Specify desired diameter or side of the rods in the **Rod Size** edit box.
- Select desired type of the middle element of the truss from **Rod Type** drop-down.
- Select desired slant type of the truss from **Slant Type** drop-down.
- Enter desired values in **Strut End Offset** and **Strut Start Offset** edit boxes to specify end offset and start offset for the top strut, respectively.
- Specify desired height and width of the main top and bottom elements of the truss in the **Strut Height** and **Strut Width** edit boxes of the dialog, respectively.
- Specify desired angle of the truss in the **Truss Angle** edit box.
- After specifying desired parameters in the dialog, the properties of the truss object will be modified; refer to Figure-135.

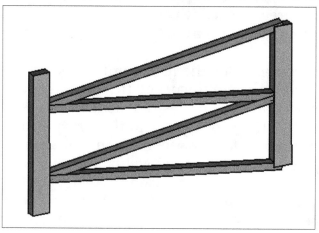

Figure-135. Truss object modified

Creating Profile

The **Profile** tool creates a parametric 2D profile object. This object can then be used as a base in different other tools that perform extrusions. The procedure to use this tool is discussed next.

- Click on the **Profile** tool from **Toolbar** in the **Arch** workbench; refer to Figure-136. The **Create profile** and **Profile settings** dialogs will be displayed in the **Tasks** panel of **Combo View**; refer to Figure-137.

Figure-136. Profile tool

Figure-137. Create profile and Profile settings dialogs

- Select desired category of the profile to be create from **Category** drop-down in the **Profile settings** dialog.
- Select desired preset of the profile from **Preset** drop-down in the **Profile settings** dialog.
- Click in the 3D view area to specify the location of profile or enter desired values for the x, y, and z coordinates in the **Local ΔX**, **Local ΔY**, and **Local ΔZ** edit boxes from the **Create profile** dialog, respectively.

- After specifying coordinates for the location of profile, click on the **Enter point** button from the dialog. The profile object will be created; refer to Figure-138.
- If you want to edit the properties of profile object then select the profile object created from the **Model Tree**. The **Property editor** dialog will be displayed in the **Model** panel of **Combo View** with the parameters related to profile object; refer to Figure-139.

Figure-138. Profile object created

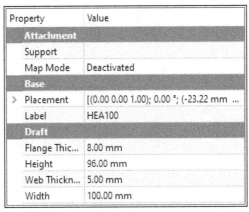

Figure-139. Property editor dialog with profile parameters

- Specify desired thickness of the flanges in the **Flange Thickness** edit box from **Draft** section of the dialog.
- Specify desired height and width of the beam in the **Height** and **Width** edit boxes of the dialog, respectively.
- Specify desired thickness of the web in the **Web Thickness** edit box of the dialog.
- After specifying desired parameters in the dialog, the properties of the profile object will be modified; refer to Figure-140.

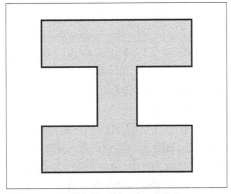

Figure-140. Profile object modified

Material tools

The **Material tools** allow you to add materials to the active document.

Creating Material

The **Material** tool allows you to add materials to the active document and attribute a material to an **Arch** object. Materials are stored into a **Materials** folder in the active document. The procedure to use this tool is discussed next.

- Select one or more objects from the **Model Tree** or from the 3D view area to which you want to attribute a new material; refer to Figure-141.

- Click on the **Material** tool from **Material tools** drop-down in the **Toolbar** of **Arch** workbench; refer to Figure-142. The **Arch material** dialog will be displayed in the **Tasks** panel of **Combo View**; refer to Figure-143.

Figure-141. Selecting the object to attribute material

Figure-142. Material tool

Figure-143. Arch material dialog

- Select one of the preset materials to be used from **Choose preset** drop-down in the dialog.
- Select desired option from **Copy existing** drop-down to copy the values from an existing material in the document.
- Specify desired name for the material in the **Name** edit box of the dialog.
- Click on the **Edit button** available next to the **Name** edit box. The **Material Editor** dialog box will be displayed which allows you to edit many additional properties and add your own custom ones; refer to Figure-144.
- Specify more detailed description of the material in the **Description** edit box of the **Arch material** dialog.
- Specify desired display color for the material which will be applied to all objects that use that material by clicking on the **Color** edit button.
- Specify desired display color for the material, which will be applied on **TechDraw** pages, when an object with this material is cut, and the **Display materials** property of the containing section plane is set to **True**.
- Specify desired transparency value for this material in the **Transparency** edit box.
- Specify a name and reference number of a specification system such as Masterformat or Omniclass in the **Standard code** edit box of the dialog.
- Enter desired url in the **URL** edit box where more information about the material can be found.
- Click on **OK** button from the dialog. The material will be applied to the object.

Figure-144. Material Editor dialog box

Creating Multi-Material

The **Multi-Material** tool defines a list of materials with their names and thickness values. This multi-materials list can then be added to an Arch object instead of a single Arch Material. The procedure to use this tool is discussed next.

- Select the object from the **Model Tree** or from the 3D view area which you want to attribute a new multi-material; refer to Figure-145.

Figure-145. Selecting the object to attribute multi material

- Click on the **Multi-Material** tool from **Material tools** drop-down in the **Toolbar** of **Arch** workbench; refer to Figure-146. The **Multimaterial definition** dialog will be displayed in the **Tasks** panel of **Combo View**; refer to Figure-147.

Figure-146. Multi Material tool

Figure-147. Multimaterial definition dialog

- Select desired option from **Copy existing** drop-down to copy the values from an existing material in the document.
- Specify desired name for the material in the **Name** edit box from **Edit definition** area of the dialog.
- The **Composition** list is the list of the different material layers that compose this multi-material.
- Click on the **Add** button to add a new layer to the list.
- Click on the **Up**, **Down**, **Del**, and **Invert** buttons to move a selected layer up, to move a selected layer down, to delete a selected layer, and to invert the selected layer, respectively.
- Double-click on the layer name to edit the name of layer.
- Select desired material available in the same document from the **Material** drop-down.
- Enter desired thickness value in the **Thickness** edit box of the dialog.
- Click on **OK** button to close the dialog. The Multi-material will be applied to the object.

Creating Schedule

The **Schedule** tool allows you to create and automatically populate a spreadsheet with contents gathered from the model. The procedure to use this tool is discussed next.

- First, you need to create or open a FreeCAD document which contains some objects; refer to Figure-148.

Figure-148. FreeCAD objects created

- Click on the **Schedule** tool from **Toolbar** in the **Arch** workbench; refer to Figure-149. The **Schedule definition** dialog box will be displayed; refer to Figure-150.

Figure-149. Schedule tool

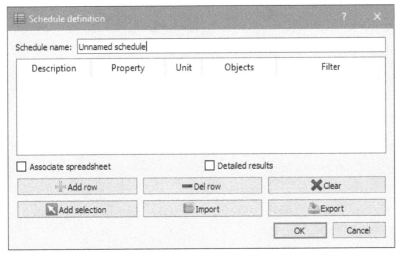

Figure-150. Schedule definition dialog box

- Specify desired name for the schedule in the **Schedule name** edit box.
- On selecting the **Associate spreadsheet** check box, an associated spreadsheet containing the results will be maintained together with this schedule object.
- On selecting the **Detailed results** check box, additional lines will be filled with each object considered.
- Click on **Add row** button to add a new row, click on **Del row** button to delete the selected row, and click on **Clear** button to delete all the existing rows.
- Click on the **Add selection** button to add objects currently selected in the document.
- Click on **Import** button to build the list created in another spreadsheet application and import that as a csv file here.
- Click on **Export** button to export the contents of Result spreadsheet to a csv file.
- Double-click each cell from that line in the dialog to specify desired values.
- Specify desired description for this operation in the **Description** edit box.
- Enter desired value in the **Property** edit box to retrieve for each object or to count the objects.
- Enter desired unit in the **Unit** edit box to express the resulting value.
- Specify the list of object names to be considered by this operation in the **Objects** edit box.
- Specify desired list of filters in the **Filter** edit box. Each filter is written in the form; filter:value.
- After specifying all the parameters, click on **OK** button from the dialog; refer to Figure-151. A new schedule object will be added to the document which contains a result spreadsheet; refer to Figure-152.

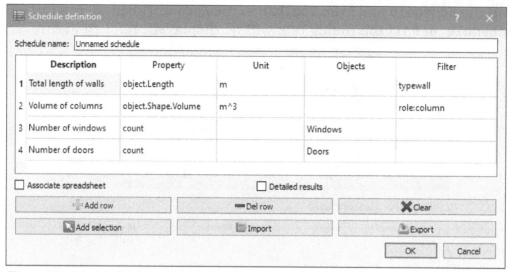

Figure-151. Schedule definition dialog box with specified parameters

Figure-152. Schedule object added

- If you want to edit the schedule then double-click on the **Schedule** created from the **Model Tree**.
- If you want to get the results in spreadsheet itself then double-click on the **Result** created from the **Model Tree**. A new result file in spreadsheet will be opened; refer to Figure-153.

Figure-153. Result in spreadsheet opened

Pipe tools

The **Pipe** tools are used to create pipes from selected objects and to create corner connection between pipes. These tools are discussed next.

Creating Pipe

The **Pipe** tool allows you to create pipes from scratch or from selected objects. You can use geometry tools in Draft workbench to create base wire for pipes. The procedure to use this tool is discussed next.

- Select a linear line, wire, or an open sketch from the **Model Tree** or from the 3D view area with which you want to create a pipe; refer to Figure-154.

Figure-154. Selecting the wire to create pipe

- Click on the **Pipe** tool from **Pipe tools** drop-down in the **Toolbar** of **Arch** workbench; refer to Figure-155. The pipe will be created; refer to Figure-156.

Figure-155. Pipe tool

Figure-156. Pipe created

- If you want to edit the properties of pipe then select the pipe created from **Model Tree**. The **Property editor** dialog will be displayed in the **Model** panel of **Combo View** with parameters related to pipe; refer to Figure-157.

Figure-157. Property editor dialog with pipe parameters

- Specify desired diameter and length of pipe in the **Diameter** and **Length** edit boxes of the dialog, respectively.
- Specify desired offset from the end point and offset from the start point in the **Offset End** and **Offset Start** edit boxes, respectively.
- The **Profile** edit box specifies the base profile of this pipe. If profile is not specified then the pipe will be cylindrical.
- Specify desired wall thickness of this pipe in the **Wall Thickness** edit box.
- On specifying parameters in the **Property editor** dialog, the properties of pipe will be modified; refer to Figure-158.

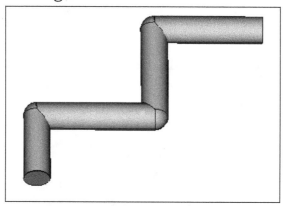

Figure-158. Pipe edited

Creating Pipe Connector

The **Pipe Connector** tool allows you to create corner or tee connection between two or three selected pipes. The procedure to use this tool is discussed next.

- Select two or three pipes joined at common vertex from the **Model Tree** or from the 3D view area.
- Click on the **Pipe Connector** tool from **Pipe tools** drop-down in the **Toolbar** of **Arch** workbench; refer to Figure-159. The pipe connector will be created; refer to Figure-160.

Figure-159. Pipe connector tool

Figure-160. Pipe connector created

- You can edit the curvature radius of connector in the **Radius** edit box from **Pipe Connector** section of **Property editor** dialog.

Cutting the object with Plane

The **Cut an object with a Plane** tool allows you to cut an Arch object according to a plane. The procedure to use this tool is discussed next.

- First, select the object which you want to cut from the **Model Tree** or from the 3D view area and then by holding **CTRL** key, select face of the second object from 3D view area by which the first object will cut; refer to Figure-161.

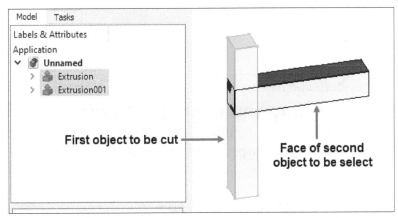

Figure-161. Selecting the objects to be cut

- Click on the **Cut Plane** tool from **Toolbar** in the **Arch** workbench; refer to Figure-162. The **Cut Plane** dialog will be displayed in the **Tasks** panel of **Combo View** along with the cut selection box; refer to Figure-163.

Figure-162. Cut Plane tool

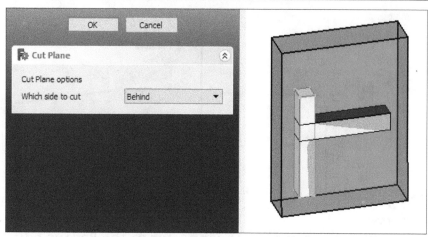

Figure-163. Cut Plane dialog with cut selection box

- Select **Behind** option to cut the object behind the normal face and select **Front** option to cut the object in front of normal face from **Which side to cut** drop-down in **Cut Plane options** section of the dialog.
- Click on **OK** button from the dialog. The object will be cut; refer to Figure-164.

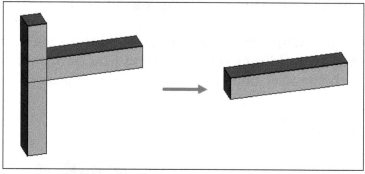

Figure-164. Object cut

Cutting the Object with Line

The **Cut with line** tool allows you to cut a solid Arch object like a wall or structure using a line that is crossing the object. The procedure to use this tool is discussed next.

- Select the object which you want to cut from the 3D view area and not from the **Model Tree** and then select the line or edge to be used to cut the object; refer to Figure-165.

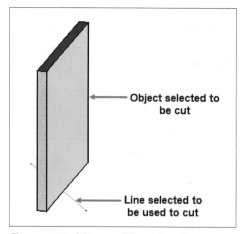

Figure-165. Object and line selected

- Click on the **Cut with a line** tool from **Toolbar** in the **Arch** workbench; refer to Figure-166. The **Cut Plane** dialog will be displayed as discussed in previous tool along with the cut selection box; refer to Figure-167.

Figure-166. Cut with a line tool

Figure-167. Cut selection box created

- Specify desired parameters as discussed in the previous tool and click on **OK** button from the dialog. The object will be cut; refer to Figure-168.

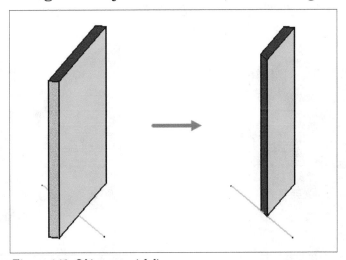

Figure-168. Object cut with line

Adding the Component

The **Add Component** tool allows you to add shape-based objects to an Arch component, such as a **Wall** or **Structure**. These objects then make part of the Arch component and allow you to modify its shape but keeping its base properties such as height and width. The procedure to use this tool is discussed next.

- First, select the objects from the **Model Tree** or from the 3D view area which you want to add to the Arch component and then by holding **CTRL** key, select the Arch component, such as a wall or structure; refer to Figure-169.

Figure-169. Selecting the object to add

- Click on the **Add component** tool from **Toolbar** in the **Arch** workbench; refer to Figure-170. The object will be added to the Arch component; refer to Figure-171.

Figure-170. Add component tool

Figure-171. Object added

Removing the Component

The **Remove Component** tool allows you to remove a subcomponent from an Arch component, for example, remove a box that has been added to a wall. The procedure to use this tool is discussed next.

- First, select the object from the **Model Tree** or from the 3D view area which you want to remove from the Arch component and then by holding **CTRL** key, select the Arch component from the **Model Tree** or from the 3D view area; refer to Figure-172.

Figure-172. Selecting the object to remove

- Click on the **Remove component** tool from **Toolbar** in the **Arch** workbench; refer to Figure-173. The object will be removed; refer to Figure-174.

Figure-173. Remove component tool

Figure-174. Object removed

Survey

The **Survey** tool enters a special surveying mode which allows you to quickly grab measurements and information from a model and transfer that information to other applications. The procedure to use this tool is discussed next.

- Click on the **Survey** tool from **Toolbar** in the **Arch** workbench; refer to Figure-175. The **Survey** dialog will be displayed in the **Tasks** panel of **Combo View**; refer to Figure-176.

Figure-175. Survey tool

Figure-176. Survey dialog

- Click on the vertices, edges, faces of the object, or double-click on the whole object from the 3D view area to get the measurement. The label will be displayed showing the measurement of selected object in the 3D view area as well as in the **Survey** dialog; refer to Figure-177.

Figure-177. Measurement displayed

- If you want to add description for the measurement then select the measurement created from the dialog and specify desired description in the box available next to the **Set description** button.
- After specifying the description, click on the **Set description** button from the dialog. The description for that measurement will be added; refer to Figure-178.

Figure-178. Description added

- Click on the **Export CSV** button from the dialog to export the contents of the dialog in csv file which can be opened in any spreadsheet application.
- Click on **Close** button to close the dialog.

PRACTICAL

Create the model as shown in Figure-179.

Figure-179. Practical 01

Creating Wall

- Start FreeCAD application using desktop icon if not started yet.
- Create a new document and switch to **Arch** workbench.
- Select the **Top** plane and create the sketch using **Line** tool as shown in Figure-180.

Figure-180. Creating first sketch

- Select all the lines created from the **Model Tree** and click on the **Wall** tool from the **Toolbar**. The wall will be created; refer to Figure-181.

Figure-181. Walls created

- Select all the walls created from the **Model Tree** and click on the **Add component** tool from the **Toolbar**. A single wall will be created; refer to Figure-182.

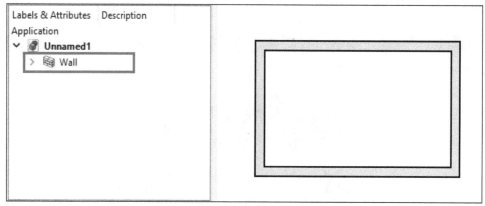

Figure-182. Single wall created

Creating Structure Object

- Click on the **Rectangle** tool from the **Toolbar** and create the rectangle on the **Top** plane as shown in Figure-183.

Figure-183. Rectangle created on wall

- Select the rectangle from the **Model Tree** and click on the **Structure** tool from the **Toolbar**. The structure object will be created; refer to Figure-184.

Figure-184. Structure object created

Creating Door

- Click on the **Window** tool from **Toolbar** and create a simple door as shown in Figure-185.

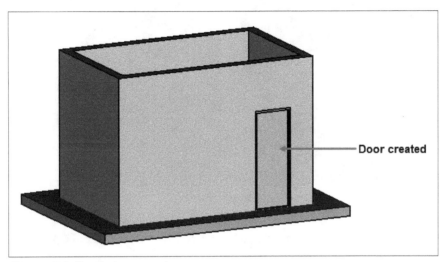

Figure-185. Simple door created

Creating Second Structure

- Click on the **Structure** tool from the **Toolbar** and create the structure as shown in Figure-186.

Figure-186. Second structure created

Creating Array

- Select the structure recently created, click on the **Array** tool from the **Toolbar** and create an array of the structure as shown in Figure-187.

Figure-187. Array of the structure created

Creating Third Structure

- Click on the **Rectangle** tool and create a rectangle on the **Top** plane as shown in Figure-188.

Figure-188. Rectangle created for third structure

- Select the recently created rectangle and create structure object using **Structure** tool. The model of house will be created; refer to Figure-189.

Figure-189. Model of house created

Chapter 8

Path Workbench

Topics Covered

The major topics covered in this chapter are:

- *Introduction to Path Workbench*
- *Create Jobs*
- *Tool Manager*
- *G-Code Generation*
- *Basic Path Operations*
- *Path Modification Tools*
- *Path Utilities*
- *Post-Processing*

INTRODUCTION

The **Path** workbench is used to generate machine codes for CNC machines using a FreeCAD 3D model; refer to Figure-1. You can use these codes on CNC machines such as mills, lathes, laser cutters, and so on to perform real machining on workpiece. These machine codes are also called G-codes.

Figure-1. Path Workbench Overview

The FreeCAD Path workbench creates these machine instructions as follows:

- A 3D model of the part to be achieved after performing machining is created in modeling workbenches of FreeCAD.
- A Job is created in Path Workbench. This job contains all the information required to generate necessary G-Code for a CNC machine like Stock material, list of cutting tools to be used for machining with cutting parameters, machine characteristics, and so on.
- Tools are selected as required by the Job Operations.
- Machining paths are created using operations like contour and pocket. These path objects use internal FreeCAD G-Code dialect independent of the CNC machine.
- Machining paths are exported to G-codes matching to your machine. This step is called post processing which uses a set of conversion codes called Post-processor specific to your machine.

STARTING PATH WORKBENCH

The **Path** workbench generates G-Code defining the paths required to mill the Project represented by the 3D model on the target mill in the Path Job Operations FreeCAD G-Code dialect which is later translated to the appropriate dialect for the target CNC controller by selecting the appropriate postprocessor.

- To start a new file in **Path** workbench, click on the **New** button from **File** menu and select **Path** workbench from **Switch between workbenches** drop-down in the **Toolbar**; refer to Figure-2. The tools related to **Path** workbench will be displayed in the **Toolbar**; refer to Figure-3.

Figure-2. Path workbench

Figure-3. Path workbench tools

JOBS, TOOLS, AND GENERATING G-CODE

After creating model, the first important step for generating CNC program is setting up a job. In FreeCAD, job means defining machine, tool controllers, geometry orientation, and so on. The tools to perform job setup are available in the **Toolbar** of **Path** workbench; refer to Figure-4. These tools are discussed next.

Figure-4. Tools for creating jobs and generating G codes

Creating Job

The **Job** tool is used to create a new machining job in the active document. The Job contains following information:

- A list of Tool-Controller definitions, specifying the geometry, Feeds, and Speeds for the Path Operations Tools.
- A Workflow sequential list of Path Operations.
- A Base Body—a clone used for offset.
- A Stock, representing the raw material that will be milled to get final model.
- A SetupSheet containing inputs used by the Path Operations, including static values and formulas.
- Configuration parameters specifying the output G-Code job's destination path, file name and extension, the Postprocessor—used to generate the appropriate dialect for the target CNC Controller, and so on.

The procedure to use this tool is discussed next.

- Open or create the 3D model for which you want to create the job; refer to Figure-5.

Figure-5. Model created

- Click on the **Job** tool from **Toolbar** in the **Path** workbench; refer to Figure-6. The **Create Job** dialog box will be displayed; refer to Figure-7. You will be asked to select the object.

Figure-6. Job tool

Figure-7. Create Job dialog box

- Select the object from **Model** list box in the dialog box for which you want to create a job and click on **OK** button from the dialog box. The **Job Edit** dialog will be displayed in the **Tasks** panel of **Combo View** with the **Setup** tab opened by default and the model will be enclosed by a bounding box; refer to Figure-8.

Figure-8. Job Edit dialog with the model

Defining Stock

- Select desired option from the drop-down in the **Stock** area of **Layout** section to define shape and size of the raw material to be machined. Select the **Create Box** option to create stock in box shape using parameters specified in **Length**, **Width**, and **Height** edit boxes; refer to Figure-9.

Figure-9. Create Box option for stock

- Select the **Create Cylinder** option from the drop-down if you want to create a cylindrical stock using specified radius and height parameters; refer to Figure-10.

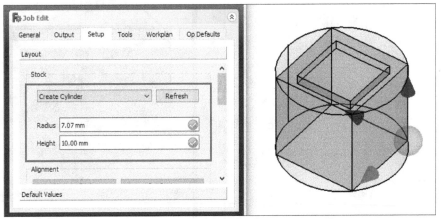

Figure-10. Create Cylinder option for stock

- Select the **Extend Model's Bound Box** option from the drop-down if you want to create stock by extending X, Y, and Z sides of base model by specified values; refer to Figure-11.

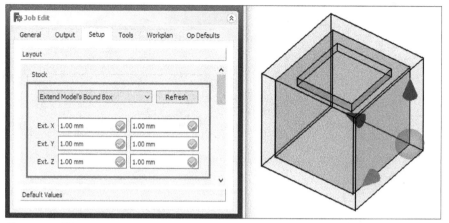

Figure-11. Extend model bound box option for stock

- Select the **Use Existing Solid** option from the drop-down to use existing solid currently available in the drawing area. On selecting this option, you can select any existing solid feature from the second drop-down of **Stock** area; refer to Figure-12.

Figure-12. Using existing solid body

- Select a face of the model to specify alignment of model, the options in **Alignment**, **Set**, and other areas of dialog will become active; refer to Figure-13.

Figure-13. Alignment options for stock

- Select the **Center in Stock** button from the **Alignment** area to place the model in the center of stock. Select the **XY in Stock** button from the **Alignment** area to place the base of model within stock.
- Select desired button from the **Set** area to set X, Y, and Z as 0.
- Using the buttons in the **Move - XY** and **Rotate - XY** areas, you can move or rotate the model with respect to stock.

General Tab

- Click on the **General** tab from the dialog. The options in the **General** tab will be displayed; refer to Figure-14.

Figure-14. General tab of Job Edit dialog

- Specify desired name of job in the **Label** edit box from **Job** section of **General** tab.
- The **Model** edit box displays the base object which defines final shape of model for generating the paths of the job.
- If you want to add more models for creating a job then click on the **Edit** button below **Model** edit box. The **Model Selection** dialog box will be displayed; refer to Figure-15.

Figure-15. Model Selection dialog box

- Select desired model and click on **OK** button from the **Model Selection** dialog box. The model will be added in the **Job Edit** dialog.
- You can add some notes to the job in the **Description** edit box of the **General** tab of the dialog.

- Click on the **Template Export** button at the bottom in the dialog to export current specified parameters as a template. The options will be displayed as shown in Figure-16.

Figure-16. Template Export options

- Select check boxes from the dialog to specify which parameters are to be exported in the template and click on the **Export** button. The **Path - Job Template** dialog box will be displayed; refer to Figure-17.

Figure-17. Path-Job Template dialog box

- Specify desired name for template in the **File name** edit box and save the template file at desired location.

Output Tab

- Click on the **Output** tab from **Job Edit** dialog. The options related to output tab will be displayed; refer to Figure-18.

Figure-18. Output tab of Job Edit dialog

- Set desired name, extension, and the file path of the G-Code output in the **Output File** edit box of the **Output** tab.
- Select desired post processor for your machine from **Processor** drop-down.
- You can add desired arguments for the post processor as needed in the **Arguments** edit box of **Output** tab in the dialog.
- Select desired check boxes from **Systems** area of **Work Coordinate Systems** section in the **Output** tab of the dialog to specify the G-Code output for defining coordinate systems.
- Select desired option from **Order By** drop-down of **Work Coordinate Systems** section of the **Output** tab. Select **Fixture** option to perform all the operations in the first coordinate system before switching to the second. Select **Tool** option to minimize the tool changes. A tool change will be done, then all operations will be performed in all coordinate systems before changing tools. Select **Operation** option to perform each operation in all coordinate systems before moving to the next operation.
- Select the **Split Output** check box to write the G-code to multiple output files as controlled by the 'order by' property. For example, if ordering by Fixture, the first output file will be for the first fixture and separate file for the second.

Tools Tab

- Click on the **Tools** tab from **Job Edit** dialog to specify parameters related to cutting tools displayed in the dialog; refer to Figure-19.

Figure-19. Tools tab of Job Edit dialog

- You can add the tool which you need for the operations of this job from **Tool** dialog box displayed by clicking on the **Add** button from **Tools** section of **Tools** tab; refer to Figure-20.

Figure-20. Tool dialog box

- The added tool will be displayed in the tool list box of **Tools** section in **Tools** tab of the **Job Edit** dialog.
- If you want to edit the tool then select the tool from the tool list and click on the **Edit** button from **Tools** section. The **Tool Controller Editor** dialog box will be displayed; refer to Figure-21.
- After adding a tool, you can set/change the feed rate and spindle speed if you need a different feed rate in this job.
- Similarly, specify desired parameters in **Workplan** and **Op Defaults** tab of the **Job Edit** dialog.
- Click on **OK** button from **Job Edit** dialog. The Job will be created and displayed in the **Model Tree**; refer to Figure-22.

Figure-22. Job created

Figure-21. Tool Controller Editor dialog box

ToolBit Library Editor

The **ToolBit Library editor** tool is used to create and manage cutting tools used in performing machining operations in the Path workbench. The procedure to use or manage cutting tools is discussed next.

- Click on the **ToolBit Library editor** tool from **Path** menu in the **Toolbar**; refer to Figure-23. If you are using this tool for the first time then you will be asked to specify the location of library.

Figure-23. ToolBit Library editor tool

- Click on the **Yes** button from the information box. The **Choose a writeable location for your toolbits** dialog box will be displayed; refer to Figure-24.

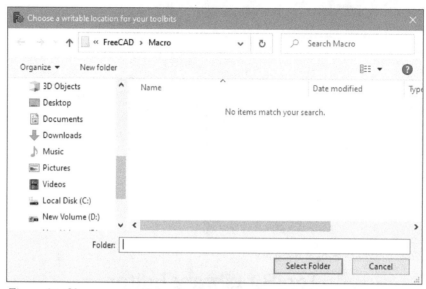

Figure-24. Choose a writeable location for your toolbits dialog box

- Select desired location/folder to specify location where toolbit files will be saved for library and click on the **Select Folder** button. A message box will be displayed for creating subdirectories like Library and Shape.
- Click on the **Yes** button from the message box. A message box to copy all example files of toolbits will be displayed.
- Click on the **Yes** button from the message box. The **Library** dialog box will be displayed; refer to Figure-25. Note that the **ToolBit Library editor** tool will be active in **Toolbar** only after you have created a job setup as discussed earlier.

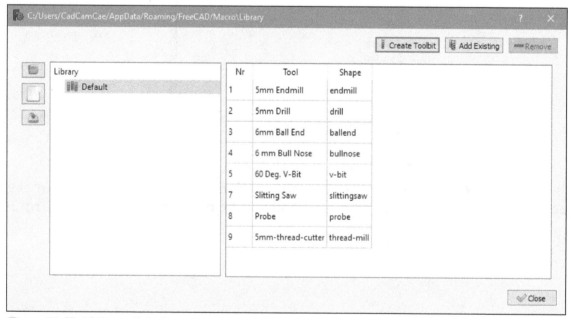

Figure-25. Tool Library dialog box

- Click on the **Add New Tool Table** ⬚ button from the dialog box to import a cutting tool table from an XML-file. The **Save toolbit library** dialog box will be displayed; refer to Figure-26.

Figure-26. Save toolbit library dialog box

- Specify desired name of library file in the **File name** edit box and click on the **Save** button. The new library will get added in the library list of dialog box.
- You can also import library files from selected folder. To do so, click on **Open** button from the dialog box. The **Tool Library Path** dialog box will be displayed; refer to Figure-27.

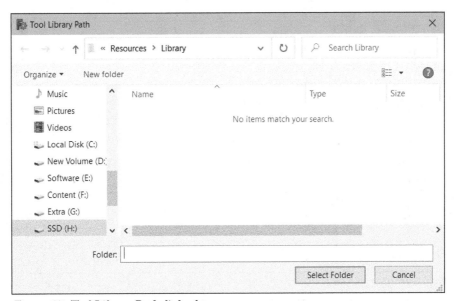

Figure-27. Tool Library Path dialog box

- Select the folder in which library files are available and click on the **Select Folder** button. The libraries of selected folder will be displayed in the dialog box.
- Click on the **Save** button from dialog box to save the toolbit library at desired location. The **Save toolbit library** dialog box will be displayed as discussed earlier.
- If you want to delete a tool from library then select the tool which you want to delete from the tools list box and click on **Remove** button.
- To create a cutting tool, click on **Create Toolbit** button. The **Select Tool Shape** dialog box will be displayed; refer to Figure-28.

Figure-28. Select Tool Shape dialog box

- Select desired tool shape template from the dialog box and click on the **Open** button. The **Tool** dialog box will be displayed for saving new toolbit; refer to Figure-29.

Figure-29. Tool dialog box

- Specify desired name for toolbit in the **File name** edit box and click on the **Save** button. The new tool will be added in the list.
- Select the newly added tool and double-click on it under **Tool** column. The options to modify cutting tool will be displayed; refer to Figure-30.

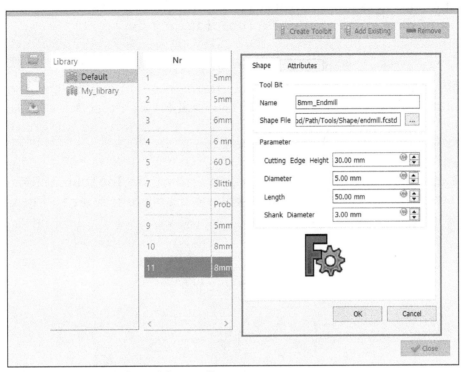

Figure-30. Parameters for modifying toolbit

- Specify desired parameters for cutting tool like diameter, length, cutting edge height, and so on in the edit boxes of **Shape** tab.
- Click on the **Attributes** tab in the dialog box to specify parameters like chipload, flutes, and material; refer to Figure-31.

Figure-31. Attributes tab

- Specify desired parameters for the tool in the dialog box and click on **OK** button from the dialog box. The parameters specified for the tool will be applied and displayed in the library.

- Click on the **Add Existing** button to add already existing tools in the library.
- Click on the **Close** button from the **Tool Library** dialog box.

Exporting Template

Exporting Job templates provide a convenient mechanism to save commonly used job definitions from within an existing Job. This facilitates the setup of future jobs that are largely similar by allowing job template import during the job creation process. The procedure to use this tool is discussed next.

- Click on **Export Template** tool from **Path** menu in the **Toolbar**; refer to Figure-32. The **Job Template Export** dialog box will be displayed; refer to Figure-33.

Figure-32. Export Template tool

Figure-33. Job Template Export dialog box

- Select **Post Processing** check box from the dialog box to include all post processing settings in the template.
- Select the **Tools** check box to store tool controller definitions in the template.
- Select the check boxes of all the tool controllers which should be included in the template from **Tools** area of the dialog box.

- Select **Setup Sheet** check box to include values of the Setup Sheet in the template.
- Select **Operation Heights** check box from **Setup Sheet** area to include the default heights for operations in the template.
- Select **Operation Depths** check box from **Setup Sheet** area to include values of the setup sheet in the template.
- Select **Tool Rapid Speeds** check box to include the default rapid tool speeds in the template.
- Select **Coolant Mode** check box to include the default coolant mode in the template.
- Select **Stock** check box to include the creation of stock in the template. If a template does not include a stock definition, the default stock creation algorithm will be used.
- Select **Extent** check box to include the current size settings for the stock object in the template.
- Select **Placement** check box to store the current placement of the stock solid in the template.
- Click on **OK** button from the dialog box. The **Path-Job Template** dialog box will be displayed to save the template; refer to Figure-34.

Figure-34. Path-Job Template dialog box

- Specify name for the template file in the **File name** edit box and specify desired location to save the file.
- Click on **Save** button from the dialog box. The template will be saved.

BASIC PATH OPERATIONS

Basic path operations are Contour, Profile Faces, and Profile Edges operations used to remove material from the stock. These tools are available in the **Toolbar** of **Path** workbench; refer to Figure-35. These tools are discussed next.

Figure-35. Basic Path Operations tool

Creating Contour Operation

The **Profile** tool creates a contour operation based on selected features of the model. It creates a simple external contour cut for 3D parts. The procedure to use this tool is discussed next.

- Click on the **Profile** tool from **Toolbar** in the **Path** workbench; refer to Figure-36. The **Choose a Tool Controller** dialog box will be displayed if you have not specified tool controller earlier for this operation while setting up the job; refer to Figure-37.

Figure-36. Profile tool

Figure-37. Choose a Tool Controller dialog box

- Select desired tool from the drop-down and click on the **OK** button. The **Profile** dialog will be displayed in the **Tasks** panel of **Combo View** and the model will be enclosed within the bounding box; refer to Figure-38.

Figure-38. Profile dialog and the model enclosed within bounding box

- The tool controller specified while creating the Job for this operation will be displayed in the **Tool Controller** drop-down in the **Operation** tab of **Profile** dialog.
- Select desired coolant mode to be used for this operation from **Coolant Mode** drop-down.
- Select desired option from **Cut Side** drop-down to specify whether the profile should be performed inside or outside the base geometry features.
- Specify desired direction in which the profile is to be machined from **Direction** drop-down.

- Enter desired value in the **Extra Offset** edit box to specify the amount of extra material left by this operation in relation to the target shape.
- Select desired option from **Enable Rotation** drop-down to gain access to pockets or areas not normal to Z axis. This allows to perform 4-axis and 5-axis machining.
- Select the **Use Start Point** check box if you want to use specified point as starting point for operation. Select **Use Compensation** check box to offset the profile operation by tool radius. The offset direction is determined by the cut side.
- Select one or more features in the 3D view area and click on **Add** button from **Base Geometry** tab of the dialog to add them as the base items for this operation.
- Enter desired values in **Start Depth** and **Final Depth** edit boxes of **Depths** tab in the dialog to specify start and final depth of tool for the operation, respectively.
- Enter desired value in **Step Down** edit box from **Depths** tab of the dialog to specify incremental step down of tool for each cutting pass of operation.
- Enter desired value in **Safe Height** edit box from **Height** tab of the dialog to specify the height above which Rapid motions are allowed.
- Enter desired value in **Clearance Height** edit box from **Height** tab of the dialog to specify the height needed to clear clamps and obstructions.
- Click on **Apply** button and then **OK** button from the dialog. The Contour operation path will be generated; refer to Figure-39.

Figure-39. Contour operation path generated

Creating Pocket Object

The **Pocket Shape** tool creates pocket machine operation for selected bottom faces or walls of one or more pockets in the model. The procedure to use this tool is discussed next.

- Click on the **Pocket Shape** tool from **Toolbar** in the **Path** workbench; refer to Figure-40. The **Pocket Shape** dialog will be displayed in the **Tasks** panel of **Combo View** and the model will be enclosed within the bounding box; refer to Figure-41.

Figure-40. Pocket Shape tool

Figure–41. Pocket Shape dialog and the model enclosed within bounding box

- Select the bottom face or the walls of an object in the 3D view area for which you want to create pocket machining operation and click on **Add** button from **Base Geometry** tab of the dialog to add them to the list as the base geometries for this operation; refer to Figure-42.

Figure–42. Selecting bottom face of the model

- Specify desired depth of final cut of the operation in the **Finish Depth** edit box from **Depths** tab of the dialog. Specify the value 0 to perform finishing operation.
- Select desired cutting mode from **Cut Mode** drop-down in the **Operation** tab of the dialog. The difference between conventional and climb cutting modes is shown in Figure-43.

Figure–43. Conventional and climb cutting modes

Note: Conventional Milling V/S Climb Milling
Characteristics of Conventional Milling:
- The width of the chip starts from zero and increases as the cutter finishes slicing.
- The tooth meets the workpiece at the bottom of the cut.
- Upward forces are created that tend to lift the workpiece during face milling.
- More power is required to conventional mill than climb mill.
- Surface finish is worse because chips are carried upward by teeth and dropped in front of cutter. There's a lot of chip recutting. Flood cooling can help!
- Tools wear faster than with climb milling.
- Conventional milling is preferred for rough surfaces.
- Tool deflection during Conventional milling will tend to be parallel to the cut.

Characteristics of climb milling:
- The width of the chip starts at maximum and decreases.
- The tooth meets the workpiece at the top of the cut.
- Chips are dropped behind the cutter--less recutting.
- Less wear with tools lasting up to 50% longer.
- Improved surface finish because of less recutting.
- Less power required.
- Climb milling exerts a down force during face milling which makes work-holding and fixtures simpler. The down force may also help to reduce chatter in thin floors because it helps brace them against the surface beneath.
- Climb milling reduces work hardening.
- It can, however, cause chipping when milling hot rolled materials due to the hardened layer on the surface.
- Tool deflection during Climb milling will tend to be perpendicular to the cut, so it may increase or decrease the width of cut and affect accuracy.
- There is a problem with climb milling which is that it can get into trouble with backlash if cutter forces are great enough. The issue is that the table will tend to be pulled into the cutter when climb milling. If there is any backlash, this allows leeway for the pulling in the amount of the backlash. If there is enough backlash, and the cutter is operating at capacity, this can lead to breakage and potentially injury due to flying shrapnel.

Some worthwhile rules of thumb:

- When cutting half the cutter diameter or less, you should definitely climb mill (assuming your machine has low or no backlash and it is safe to do so!).

- Up to 3/4 of the cutter diameter, it doesn't matter which way you cut.

- When cutting from 3/4 to 1x the cutter diameter, you should prefer conventional milling.

- Select desired option from the **Pattern** drop-down in the **Operation** tab to specify the pattern in which the tool bit will move to clear the material. It is always better to select the pattern which resembles with walls of pocket.
- Enter desired value in the **Angle** edit box from **Operation** tab to specify the angle in which the pattern is applied.

- Enter desired value in **Step Over Percent** edit box to specify the amount by which the cutting tool will move laterally on each cutting pass in the pattern (specified in percent of the tool diameter).
- Specify desired distance value in the **Pass Extension** edit box from **Operation** tab by which, the facing operation will be extended beyond the boundary shape.
- Select **Use Outline** check box from the **Operation** tab if the operation uses the outline of the selected base geometry and ignores all holes and islands.
- Select **Min Travel** check box from the **Operation** tab to use 3D Sorting of Path (when multiple base geometries used).
- Select **Show All** check box from **Extensions** tab to visualize all potential extensions. Enabled extensions are in purple and not enabled extensions are in yellow.
- Select **Extend Corners** check box from **Extensions** tab to extend the corner between two edges of a pocket.
- Specify desired extent of the dimension in the **Default Length** edit box from **Extensions** tab.
- A list box in the **Extensions** tab shows the tree of existing edges and their potential extensions.
- Other parameters in the **Pocket Shape** dialog have been discussed earlier.
- Click on **Apply** button and then **OK** button from the dialog. The Pocket shape operation path will be generated; refer to Figure-44.

Figure-44. Pocket shape operation path generated

Performing Drill Operation

The **Drilling** tool generates a drilling operation in the job. The procedure to use this tool is discussed next.

- Click on the **Drilling** tool from **Toolbar** in the **Path** workbench; refer to Figure-45. The **Choose a Tool Controller** dialog box will be displayed if no controller has been selected for this operation during job setup; refer to Figure-46.

Figure-45. Drilling tool

Figure-46. Choose a Tool Controller selection window

- Select desired drilling tool from **Tool Controller** drop-down and click on **OK** button from the selection window. The **Drilling** dialog will be displayed in the **Tasks** panel of **Combo View** and the model will be enclosed within the bounding box; refer to Figure-47.

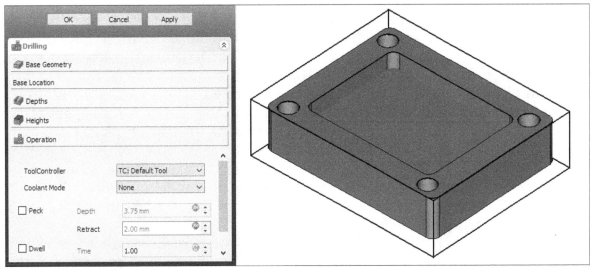

Figure-47. Drilling dialog and the model enclosed within bounding box

- Select the edges or/and faces of the hole feature of the model in the 3D view area on which you want to create the drilling operation; refer to Figure-48.

Figure-48. Selecting the faces and edges of hole feature

- After selecting the elements of the hole feature, click on **Add** button from the **Base Geometry** tab of dialog to add them in the selection list box.
- If you want to remove the element then select the element from the list box and click on **Remove** button from the **Base Geometry** tab.
- Click on the **Reset** button from the **Base Geometry** tab to delete all the current elements from the list and fills the list with all circular holes eligible for the drilling operation.

- Click on **Add** button from **Base Location** tab of the dialog. The direction edit boxes will be displayed in the tab and you will be asked to specify the locations.
- Click in the 3D view area at desired locations to specify the locations where you want to drill on the model or enter desired values in **Global X**, **Global Y**, and **Global Z** edit boxes to specify the drilling locations on the model.
- After specifying values in the edit boxes, click on **Save** button to save the drilling location in the list box or click on **Close** button to close the direction edit boxes.
- To remove the specified location, select the location from the list and click on **Remove** button from **Base Location** tab.
- Click on **Edit** button from **Base Location** tab to edit the specified location.
- Select **Peck** check box and enter desired value in the edit box from **Operation** tab to specify incremental drill depth before retracting to clear chips in Peck cycle of drilling.
- Enter desired value in **Retract** edit box from **Operation** tab of the dialog to specify the incremental height where feed starts above the model surface.
- Select **Dwell** check box and enter desired value in the edit box to specify the time of pause between peck cycles.
- Other parameters in the **Drilling** dialog have been discussed earlier. Click on **Apply** button and then **OK** button from the dialog. The drilling operation path will be generated; refer to Figure-49.

Figure-49. Drilling operation path generated

Creating Face Milling

The **Face Mill** tool creates toolpath to remove material from top flat face of the model. The procedure to use this tool is discussed next.

- Click on the **Face Mill** tool from **Toolbar** in the **Path** workbench; refer to Figure-50. The **MillFace** dialog will be displayed in the **Tasks** panel of **Combo View** and the model will be enclosed within the bounding box; refer to Figure-51.

Figure-50. Face Mill tool

Figure-51. MillFace dialog

- Select one or more features of the model in the 3D view area and click on **Add** button from **Base Geometry** tab of the dialog to add them to the list as the base geometries for this operation; refer to Figure-52.

Figure-52. Selecting face of the model

- Select desired option from **Boundary Shape** drop-down in the **Operation** tab whether the facing should be restricted by the actual shape of the selected face or the bounding box should be faced off.
- Enter desired value in **Material Allowance** edit box from **Operation** tab of the dialog to specify the amount of material that should be left by this operation in relation to the target shape.
- Other parameters in the **MillFace** dialog have been discussed earlier. Click on **Apply** button and then **OK** button from the dialog. The Face Mill operation path will be generated; refer to Figure-53.

Figure-53. Face Mill operation path generated

Creating Helical Operation

The **Helix** tool is used to create a helical path. This operation is useful to machine circular vertical surfaces of the model. The helix path gets appended automatically to a helical clearing operation in the Job. The procedure to use this tool is discussed next.

- Click on the **Helix** tool from **Toolbar** in the **Path** workbench; refer to Figure-54. The **Helix** dialog will be displayed in the **Tasks** panel of **Combo View** and the model will be enclosed within the boundary box; refer to Figure-55.

Figure-54. Helix tool

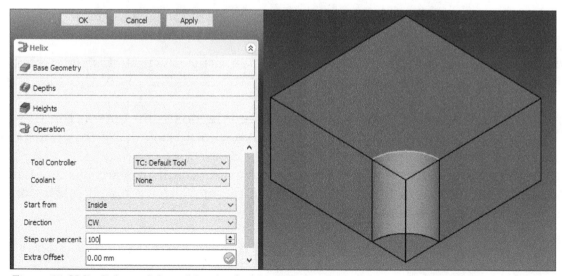

Figure-55. Helix dialog and the model enclosed within boundary box

- Select the edges of features to be machined by the helical operation and click on **Add** button from **Base Geometry** tab of the dialog to add them in the list as the base geometries for this operation.
- Specify appropriate values in the **Depths** section of dialog; refer to Figure-59. Note that you can activate the files by clicking on the arrow buttons next to edit boxes.

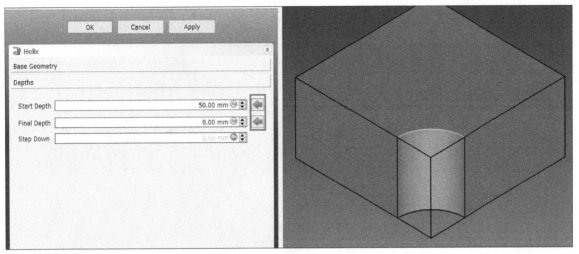

Figure-56. Specifying depths for helical toolpath

- Specify all the other parameters in the **Helix** dialog as discussed earlier.
- Click on **Apply** button and then **OK** button from the dialog. The helical operation path will be generated; refer to Figure-57.

Figure-57. Helical operation path generated

Creating Adaptive Operation

The **Adaptive** tool uses an adaptive algorithm to create clearing and profiling paths that manage cutter engagement so that engagement and material removal never exceed a maximum value. This toolpath uses all the possible cutting strategies available in FreeCAD for removing large stock of material. The procedure to use this tool is discussed next.

- Click on the **Adaptive** tool from **Toolbar** in the **Path** workbench; refer to Figure-58. The **Adaptive** dialog will be displayed in the **Tasks** panel of **Combo View** and the model will be enclosed within the bounding box; refer to Figure-59.

Figure-58. Adaptive tool

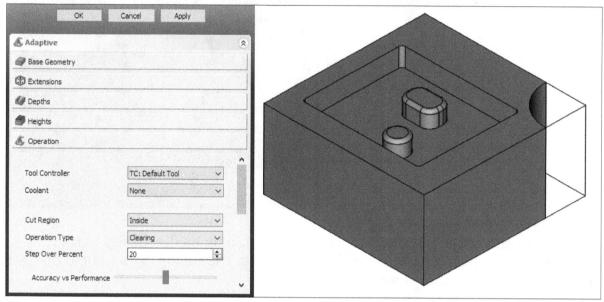

Figure-59. Adaptive dialog and the model enclosed within bounding box

- Select one or more faces of the model in the 3D view area to be machined (generally base face) and click on **Add** button from **Base Geometry** tab of the dialog to add them to the list as the base geometries for this operation; refer to Figure-60.

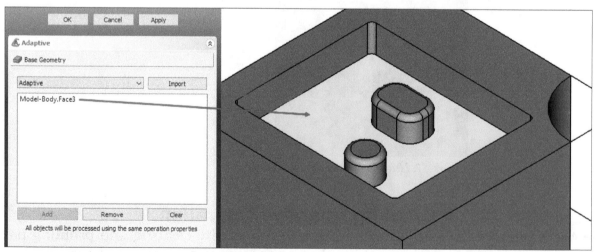

Figure-60. Selecting the face for adaptive operation

- Select desired option from **Cut Region** drop-down in the **Operation** tab to specify the cutting of model inside or outside of the selected shape.
- Select desired type of adaptive operation from **Operation Type** drop-down in the **Operation** tab of the dialog.

- Specify the percent of cutter diameter to step over on each pass in the **Step Over Percent** edit box of **Operation** tab.
- Slide the **Accuracy vs Performance** slider from **Operation** tab at desired position to influence the calculation performance vs stability and accuracy.
- Specify desired angle of the helix ramp entry in the **Helix Ramp Angle** edit box of **Operation** tab.
- Specify desired angle of the helix entry cone in the **Helix Cone Angle** edit box of the **Operation** tab.
- Enter desired value in **Helix Max Diameter** edit box from **Operation** tab to specify the limit helix entry diameter. If limit larger than tool diameter or 0, tool diameter is used.
- Enter desired value in **Lift Distance** edit box from **Operation** tab to specify how much to lift the tool up during the rapid linking moves over cleared regions. If linking path is not clear, tool is raised to clearance height.
- Enter desired value in **Keep Tool Down Ratio** edit box from **Operation** tab to specify maximum length of keep tool down linking path compared to direct distance between point. If exceeded, link will be done by raising the tool to clearance height.
- Enter desired value in **Stock to Leave** edit box to specify how much material to leave (i.e. for finishing operation).
- Select the **Force Clearing Inside-Out** check box from **Operation** tab to specify the force plunging into material inside and clearing towards the edges.
- Other parameters in the **Adaptive** dialog have been discussed earlier.
- Click on **Apply** button and then **OK** button from the dialog. The Adaptive operation path will be generated; refer to Figure-61.

Figure-61. Adaptive operation path generated

Creating an Engraving Path

The **Engrave** tool is primarily used for engraving a **Draft ShapeString** onto a part like writing name on part. However, it may be useful for other kinds of 2D toolpaths. Note that if you want to engrave string on face of model then you will need a new job setup for string. The procedure to perform engraving of a string is discussed next.

- Create the string of text as desired on the face of model in **Draft** workbench.
- Click on the **Path Job** tool from the **Toolbar** in **Path** workbench. The **Create Job** dialog box will be displayed.

- Select the check box for **ShapeString** in **2D** node of list in the dialog box; refer to Figure-62 and click on the **OK** button. The **Job Edit** dialog will be displayed in the **Tasks** panel of **Combo View**.

Figure-62. Shapestring selected for creating job

- Set the stock, cutting tool, and other parameters as discussed earlier and click on the **OK** button from the dialog box. A new job with name Job001 will be added in the **Model Tree**.

- Click on the **Engrave** tool from **Engraving Operations** drop-down in the **Toolbar** of **Path** workbench; refer to Figure-63. The **Choose a Path Job** dialog box will be displayed.

Figure-63. Engrave tool

- Select the newly created job from the drop-down and click on the **OK** button. The **Choose a Tool Controller** dialog box will be displayed if there are multiple tools available. Select desired engraving tool from the drop-down and click on the **OK** button. The **Engrave** dialog will be displayed in the **Tasks** panel of **Combo View** and the model will be enclosed within the bounding box; refer to Figure-64.

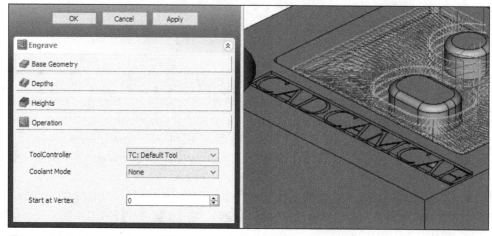

Figure-64. Engrave dialog and the model enclosed within bounding box

- Select faces of strings or every edge of the strings from the model in the 3D view area and click on **Add** button from **Base Geometry** tab of the dialog to add them to the list as the base geometries for this operation; refer to Figure-65.

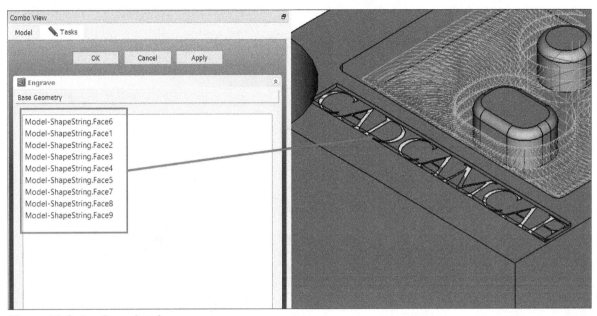

Figure-65. String faces selected

- Specify desired tool to be used for this operation from **ToolController** drop-down in the **Operation** tab of the dialog.
- Specify desired vertex number of the underlying shape string at which engraving should start in the **Start at Vertex** edit box of **Operation** tab in the dialog.
- Other parameters in the **Engrave** dialog have been discussed earlier.
- Click on **Apply** button and then **OK** button from the dialog. The Engraving operation path will be generated at specified depth; refer to Figure-66.

Figure-66. Engraving operation path generated

Creating Deburring Operation

The **Deburr** tool creates a deburr path along edges or around faces. The procedure to use this tool is discussed next.

- Click on the **Deburr** tool from **Engraving Operations** drop-down in the **Toolbar** of **Path** workbench; refer to Figure-67. The **Deburr** dialog will be displayed in the **Tasks** panel of **Combo View** and the model will be enclosed within the boundary box; refer to Figure-68.

Figure-67. Deburr tool

Figure-68. Deburr dialog and the model enclosed within bounding box

- Select the edges and/or faces of the model in the 3D view area on which you want to create the deburring operation and click on **Add** button from **Base Geometry** tab of the dialog to add them to the list as the base geometries for this operation; refer to Figure-69.

Figure-69. Selecting the faces of model for deburring

- Select desired tool to be used for this operation from **Tool Controller** drop-down in the **Operation** tab of dialog.
- Select desired direction in which the profile is to be performed from **Direction** drop-down in the **Operation** tab.

- Specify desired width of chamfer cut in the **W** edit box from **Operation** tab of the dialog.
- Specify desired extra depth of tool immersion in the **h** edit box from **Operation** tab of the dialog.
- Other parameters in the **Deburr** dialog have been discussed earlier.
- Click on **Apply** button and then **OK** button from the dialog. The Deburring operation path will be generated; refer to Figure-70.

Figure-70. Deburr operation path generated

Creating Vcarve

The **Vcarve** tool is primarily for center-line engraving a Draft ShapeString onto a part. The procedure to use this tool is same as discussed for **Engrave** tool. Use the V -bit cutting tool to perform the operation.

Creating 3D Pocket Operation

The **3D Pocket** tool inserts a path 3D Pocket object into the job. A 3D pocket takes into account the bottom surface of the pocket. The procedure to use this tool is discussed next.

- Click on the **3D Pocket** tool from **Toolbar** in the **Path** workbench; refer to Figure-71. The **Pocket 3D** dialog will be displayed in the **Tasks** panel of **Combo View** and the model will be enclosed within the bounding box; refer to Figure-72.

Figure-71. 3D Pocket tool

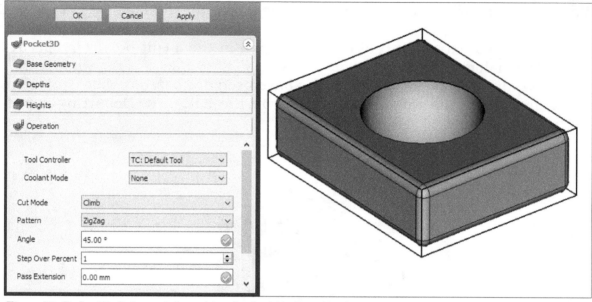

Figure-72. Pocket 3D dialog and the model enclosed within bounding box

- Select one or more faces from the model in the 3D view area and click on **Add** button from **Base Geometry** tab of the dialog to add them to the list as the base geometries for this operation; refer to Figure-73.

Figure-73. Selecting the face of model for 3D pocketing

- Specify all the parameters in the **Pocket 3D** dialog as discussed earlier.
- Click on **Apply** button and then **OK** button from the dialog. The 3D Pocket operation path will be generated; refer to Figure-74.

Figure-74. 3D Pocket operation path generated

PATH MODIFICATION TOOLS

There are three tools in **Toolbar** of **Path** workbench used to create multiple copies of the selected toolpaths. These tools are discussed next.

Creating Copy

The **Copy** tool creates a copy of selected path operation. The procedure to use this tool is discussed next.

- Select desired path operation from the **Model Tree** for which you want to create the copy or multiple copies; refer to Figure-75.

Figure-75. Selecting the path operation

- Click on the **Copy** tool from **Toolbar** in the **Path** workbench; refer to Figure-76. The copy of selected path operation will be created; refer to Figure-77.

Figure-76. Copy tool

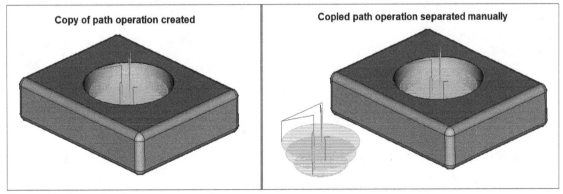

Figure-77. Copy of path operation created

- You can edit the properties of copied path operation using **Property editor** dialog as discussed earlier.

Creating Array

The **Array** tool creates a new path by duplicating another path several times at a certain interval distance. The procedure to use this tool is discussed next.

- Select desired path operation created from the **Model Tree** of which you want to create array; refer to Figure-78.

Figure-78. Selecting the path operation

- Click on the **Array** tool from **Toolbar** in the **Path** workbench; refer to Figure-79. The array of path operation will be created; refer to Figure-80.

Figure-79. Array tool

Figure-80. Array of path operation created

- You can edit the properties of array of path operation created from **Property editor** dialog as discussed earlier.

Creating Simple Copy

The **Simple Copy** tool creates a non-parametric copy of a given path. The procedure to use this tool is similar to discussed for **Copy** tool.

PATH UTILITIES

The tools in **Path Utilities** section of **Toolbar** are used to inspect, manage, and simulate toolpaths. These tools are discussed next.

Inspection of G-Code

The **Inspect Path Commands** tool allows inspection of the internal FreeCAD G-code dialect contents of a Path Operation Object. The procedure to use this tool is discussed next.

- Select desired path operation from the **Model Tree** which you want to inspect for G-code; refer to Figure-81.

Figure-81. Selecting the path operation

- Click on the **Inspect Path Command** tool from **Toolbar** in the **Path** workbench; refer to Figure-82. The **FreeCAD** dialog box will be displayed showing the G-code of path operation; refer to Figure-83.

Figure-82. Inspect path command tool

Figure-83. FreeCAD dialog box showing the G code

- Make any changes in the dialog box as desired and click on **OK** button from the dialog box to apply the changes.

CAM Simulator

The **CAM simulator** tool is used to simulate selected operations on 3D model along the G-Code paths, subtracting material from the stock, where the stock and tool overlap, providing visualization of the Job. This allows detection and isolation of errors prior to running the job on a mill. The procedure to use this tool is discussed next.

- Click on the **CAM simulator** tool from **Toolbar** in the **Path** workbench; refer to Figure-84. The **Path Simulator** dialog will be displayed in the **Tasks** panel of **Combo View** and the model will be enclosed within the stock, ready to be simulate; refer to Figure-85.

Figure-84. Cam simulator tool

Figure-85. Path Simulator dialog and the model enclosed within stock

- All the operations created that are to be simulated will be selected by default in the **Operation list** area of the dialog.
- Deselect the operations which you do not want to be simulate from the **Operation list**.
- Select desired Job which you want to use as the basis of the simulation from **Job** drop-down in the dialog.
- Press the ▶ **Play** button to play or playback an animation of the operations.
- Press the ■ **Stop** button to stop the animation.
- Press the ❚❚ **Pause** button to pause animation for troubleshooting purposes.
- Press the ❚▶ **Single-Step** button for slowing down the animation, this functionality helps troubleshooting and resolving specific cuts and/or movements.
- Press the ▶▶ **Fast-Forward** button to increase the speed substantially.
- Tune the speed of simulation by sliding the **Speed** slider from the dialog.
- Tune the accuracy of simulation by sliding the **Accuracy** slider from the dialog.
- After performing the simulation, the object will be displayed as shown in Figure-86.

Figure-86. Object simulated

- Click on the **OK** button from the dialog to finish the simulation process.

FINISH SELECTING LOOP

The **Finish Selecting Loop** tool is used to complete a loop from two selected edges. The procedure to use this tool is discussed next.

• Select two edges that share a common vertex as shown in the Figure-87.

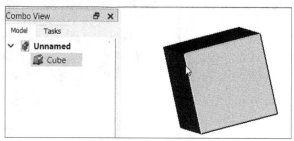

Figure-87. Select loop

• Click on the **Finish Selecting Loop** tool from the **Toolbar** in the path workbench; refer to Figure-88. Full loop of selected edges will be selected; refer to Figure-89.

Figure-88. Finish selecting tool

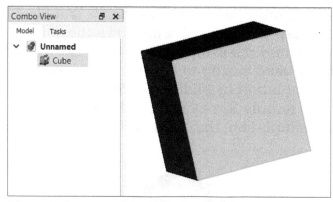

Figure-89. Loop created

POST-PROCESSING

The **Post Process** command exports the selected Path Jobs to a G-code file. Each CNC controller speaks a specific G-Code dialect-correct Postprocessor to translate the final output from the agnostic internal FreeCAD G-Code dialect. The procedure to use this tool is discussed next.

• Click on the **Edit** menu and select the **Preferences** option. The **Preferences** dialog box will be displayed; refer to Figure-90.

Figure-90. post processing preference dialog box

- Select desired option from **Default Post Process** drop-down in the **Preferences** tab of **Path** option available in the left side of the dialog box. After specifying desired parameters click on the **Ok** button from the dialog box.

- Select the Path Job created from the **Model Tree** which you want to export; refer to Figure-91.

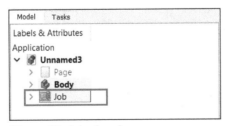

Figure-91. Job selected

- Click on the **Post Process** tool from **Toolbar** in the **Path** workbench; refer to Figure-92. The **Output File** dialog box will be displayed asking you to specify the file name and directory; refer to Figure-93.

Figure-92. Post Process tool

Figure-93. Output File dialog box

- Specify desired name of output file in the **File name** edit box and specify desired directory to save the file.
- Click on **Save** button to save the output file. The **FreeCAD** dialog box will be displayed showing the postprocessor properties of the job; refer to Figure-94.

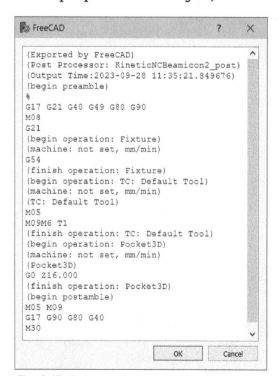

Figure-94. FreeCAD dialog box showing the post processor properties

PRACTICAL

Generate the toolpath of the given model shown in Figure-95.

Figure-95. Model for generating toolpaths

Starting a New File

- Start FreeCAD if not started yet and open the model file for this practical from resources.
- Select the **Path** option from the **Switch between workbenches** drop-down in the **Toolbar**. The path environment will be displayed.

Creating Stock

- Click on the **Path Job** tool from the **Toolbar**. The **Create Job** dialog box will be displayed.
- Select the object which you want to create a job and click on **OK** button from the dialog box. The **Job Edit** dialog will be displayed along with the model enclosed by a bounding box; refer to Figure-96.

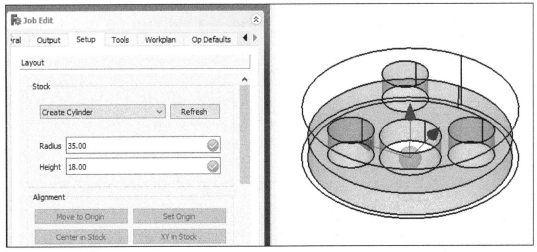

Figure-96. Model enclosed within bounding box

- Specify the parameters in the dialog as shown in Figure-96 and click on **OK** button. The path job will be created; refer to Figure-97.

Figure-97. Path job created

Performing Facing Operation

- Removing material from the top face of stock is the first step in machining. Click on the **Face** tool from the **Path** menu. The **MillFace** dialog will be displayed.
- Select top face of model (any of the boss feature) and click on the **Add** button from **Base Geometry** section in the dialog; refer to Figure-98.

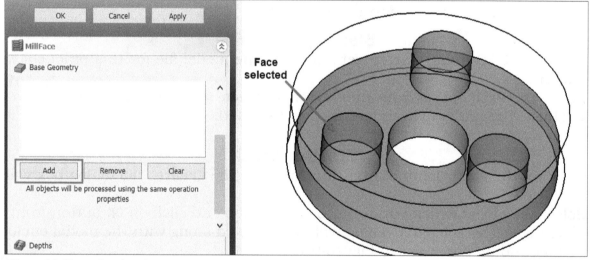

Figure-98. Face to be selected

- Expand the **Operations** section and select desired cutting tool from the **Tool Controller** drop-down.
- Select the **Stock** option from the **Boundary Shape** drop-down to machine up to stock boundaries for facing operation.
- Set the other parameters as desired and click on the **OK** button to create facing operation; refer to Figure-99.

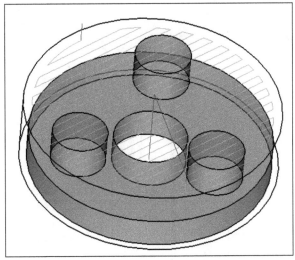

Figure-99. After facing operation

- Select the **MillFace** option from **Job->Operations** category of **Combo View** and click on **Toggle the Active State of the Operation** tool from **Toolbar** or **Path** menu. This will hide the toolpaths from graphics area.

Creating Pocket Operation

- Click on the **Pocket Shape** tool from the **Path** menu and select face of the model as base geometry for pocket operation as shown in Figure-100. Click on the **Add** button from **Base Geometry** section in the dialog to add it to operation.

Figure-100. Selecting face of model as base geometry

- Specify other parameters for pocket operation in the dialog as shown in Figure-101 and click on the **OK** button. The pocket operation path of the model will be created; refer to Figure-102.

Figure-101. Parameters for pocket operation

Figure-102. Pocket operation of model created

Simulating the Path

- Click on the **Simulator** tool from the **Toolbar**. The **Path Simulator** dialog will be displayed along with the pocket shape tool; refer to Figure-103.

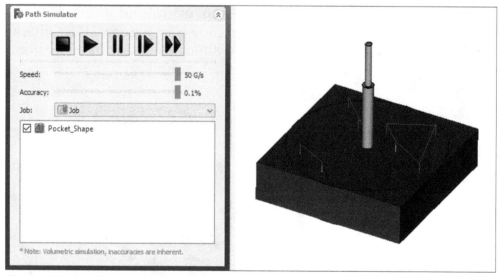

Figure-103. Path Simulator dialog to simulate pocket operation

- Click on the **Activate/resume simulation** button to simulate the pocket shape operation. The pocket operation will be simulated; refer to Figure-104.

Figure-104. Pocket operation simulated

Chapter 9

FEM Workbench

Topics Covered

The major topics covered in this chapter are:

- *Introduction to FEM Workbench*
- *Starting FEM Workbench*
- *Creating Analysis Container*
- *Analysis Types in FreeCAD*
- *Defining Material*
- *Applying Constraints*
- *Creating Geometry Elements*
- *Meshing*
- *Solvers*
- *Post Processing*

INTRODUCTION TO FEM

The **FEM** Workbench provides a modern finite element analysis (FEA) workflow for FreeCAD. This means all the tools to perform an analysis are combined into one graphical user interface (GUI); refer to Figure-1.

Figure-1. FEM Workbench Overview

The workflow for FEM is shown in Figure-2. The steps to carry out a finite element analysis are given next.

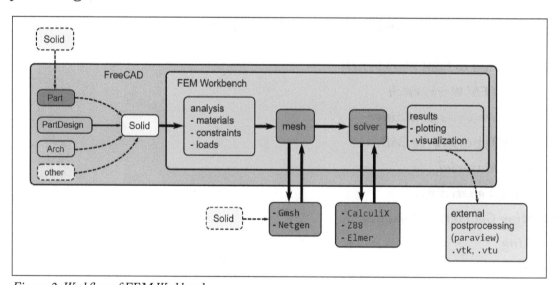

Figure-2. Workflow of FEM Workbench

1. **Preprocessing**: Setting up the analysis problem mainly on plain paper or in your mind.
2. **Modeling the geometry**: Creating the geometry with FreeCAD, or importing it from a different application.
3. **Creating an analysis**: This step is also called Defining Boundary Conditions. It involves adding simulation constraints such as loads and fixed supports to the geometric model. Adding materials to the parts of the geometric model. Creating a finite element mesh for the geometrical model, or importing it from a different application.

4. **Solving**: Running an external solver from within FreeCAD.
5. **Postprocessing**: Visualizing the analysis results from within FreeCAD, or exporting the results so they can be post processed with another application.

STARTING FEM WORKBENCH

The **FEM** Workbench calls two external programs to perform meshing of a solid object and perform the actual solution of the finite element problem.

*   To start a new **FEM** file, click on the **New** button from **File** menu and select **FEM** workbench from **Switch between workbenches** drop-down in the **Toolbar**; refer to Figure-3. The tools related to **FEM** workbench will be displayed in the **Toolbar**; refer to Figure-4.

Figure-3. FEM Workbench

Figure-4. FEM Workbench tools

CREATING ANALYSIS CONTAINER

The **FEM** Analysis in FreeCAD could be seen as a Container that holds all objects of a Finite Element Analysis. It is mandatory to have an analysis container which holds all the needed objects. At least one of the following objects is needed for a mechanical analysis: material, fixed constraint, force constraint, or pressure constraint. The procedure to use this tool is discussed next.

*   Click on the **Analysis Container** tool from **Toolbar** in the **FEM** workbench; refer to Figure-5. A new Analysis will be created and displayed in the **Model Tree**; refer to Figure-6.

Figure-5. Analysis Container tool

Figure-6. New Analysis created

Some details about common analyses available in FreeCAD are given next.

Static Analysis

This is the most common type of analysis we perform. In this analysis, loads are applied to a body due to which the body deforms and the effects of the loads are transmitted throughout the body. To absorb the effect of loads, the body generates internal forces and reactions at the supports to balance the applied external loads. These internal forces and reactions cause stress and strain in the body. Static analysis refers to the calculation of displacements, strains, and stresses under the effect of external loads, based on some assumptions. The assumptions are as follows.

1. All loads are applied slowly and gradually until they reach their full magnitudes. After reaching their full magnitudes, load will remain constant (i.e. load will not vary against time).
2. Linearity assumption: The relationship between loads and resulting responses is linear. For example, if you double the magnitude of loads, the response of the model (displacements, strains, and stresses) will also double. You can make linearity assumption if:

• All materials in the model comply with Hooke's Law that is, stress is directly proportional to strain.
• The induced displacements are small enough to ignore the change in stiffness caused by loading.
• Boundary conditions do not vary during the application of loads. Loads must be constant in magnitude, direction, and distribution. They should not change while the model is deforming.

If the above assumptions are valid for your analysis then you can perform **Linear Static Analysis**. For example, a cantilever beam fixed at one end and force applied on other end; refer to Figure-7.

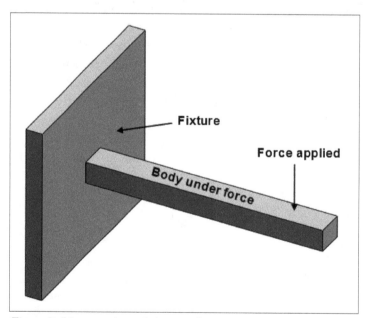

Figure-7. Linear static analysis example

If the above assumptions are not valid then you need to perform the **Non-Linear Static analysis**. For example, an object attached with a spring being applied under forces.

Frequency Analysis (Modal Analysis)

By its very nature, vibration involves repetitive motion. Each occurrence of a complete motion sequence is called a "cycle." Frequency is defined as so many cycles in a given time period. "Cycles per seconds" or "Hertz". Individual parts have what engineers call "natural" frequencies. For example, a violin string at a certain tension will vibrate only at a set number of frequencies, which is why you can produce specific musical tones. There is a base frequency in which the entire string is going back and forth in a simple bow shape.

Harmonics and overtones occur because individual sections of the string can vibrate independently within the larger vibration. These various shapes are called "modes". The base frequency is said to vibrate in the first mode and so on up the ladder. Each mode shape will have an associated frequency. Higher mode shapes have higher frequencies.

The most disastrous kinds of consequences occur when a power-driven device such as a motor for example, produces a frequency at which an attached structure naturally vibrates. This event is called "resonance." If sufficient power is applied, the attached structure will be destroyed. Note that ancient armies, which normally marched "in step," were taken out of step when crossing bridges. Should the beat of the marching feet align with a natural frequency of the bridge, it could fall down. Engineers must design machines so that resonance does not occur during regular operation of machines. This is a major purpose of Modal Analysis. Ideally, the first mode has a frequency higher than any potential driving frequency. Frequently, resonance cannot be avoided, especially for short periods of time. For example, when a motor comes up to speed, it produces a variety of frequencies. So, it may pass through a resonant frequency.

Thermal analysis

There are three mechanisms of heat transfer. These mechanisms are Conduction, Convection, and Radiation. Thermal analysis calculates the temperature distribution in a body due to some or all of these mechanisms. In all three mechanisms, heat flows from a higher-temperature medium to a lower temperature one. Heat transfer by conduction and convection requires the presence of an intervening medium while heat transfer by radiation does not.

There are two modes of heat transfer analysis.

Steady State Thermal Analysis

In this type of analysis, we are only interested in the thermal conditions of the body when it reaches thermal equilibrium, but we are not interested in the time it takes to reach this status. The temperature of each point in the model will remain unchanged until a change occurs in the system. At equilibrium, the thermal energy entering the system is equal to the thermal energy leaving it. Generally, the only material property that is needed for steady state analysis is the thermal conductivity.

Transient Thermal Analysis

In this type of analysis, we are interested in knowing the thermal status of the model at different instances of time. A thermos designer, for example, knows that the temperature of the fluid inside will eventually be equal to the room temperature(steady state), but he is interested in finding out the temperature of the fluid as a function of time. In addition to the thermal conductivity, we also need to specify density, specific heat, initial temperature profile, and the period of time for which solutions are desired.

DEFINING MATERIAL FOR SOLID

The **Material for Solid** tool specifies material properties for the model. The procedure to use this tool is discussed next.

- Click on the **Material for Solid** tool from **Toolbar** in the **FEM** workbench; refer to Figure-8. The **FEM material** and **Geometry reference selector** dialog will be displayed in the **Tasks** panel of **Combo View**; refer to Figure-9 and Figure-10, respectively.

Figure-8. Material for Solid tool

Figure-9. FEM material dialog

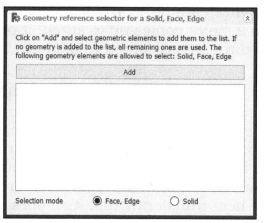

Figure-10. Geometry reference selector for a Solid Face Edge dialog

- Select desired material option from the **Material card** drop-down in **FEM material** dialog to be used as material for the model during analysis.

- Select desired radio button from **Selection mode** section of the **Geometry reference selector for a Solid, Face, Edge** dialog to define whether you want to apply material to single face/edge or solid object. Generally, the **Solid** radio button is used from this section. After selecting desired radio button, click on the **Add** button from the dialog and select the object to which you want to apply the material. The object will be added in the list; refer to Figure-11.

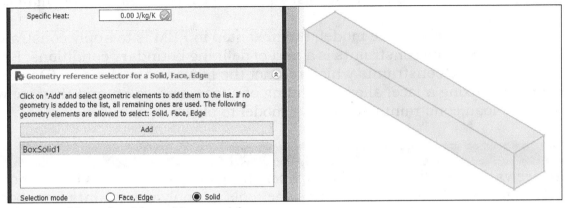

Figure-11. Object selected for applying material

Editing Material

The **use FreeCAD material editor** button in **FEM material** dialog is used to modify the parameters of material like density, Young's Modulus, and so on. The procedure to edit material is given next.

- Click on the **use FreeCAD material editor** button from the **FEM material** dialog in the **Tasks** panel of **Combo View**. The **Material Editor** dialog box will be displayed; refer to Figure-12.

Figure-12. Material Editor dialog box

- Select desired material from the **Material card** drop-down to modify its parameter.
- Double-click in desired field of material to modify its value.
- If you want to add a new property for material then specify the name of property in the edit box of **Add/Remove parameter** section in the dialog box and click on the **Add property** button. If you want to delete a property then select desired property from the dialog box and click on the **Delete property** button.
- You can open a new material by selecting the **Open** button and save current modified material properties by clicking on the **Save as** button.

- After setting desired parameters, click on the **OK** button from the dialog box to exit.

Note that you can modify the properties of material directly in **FEM material** dialog if you have selected the **use this task panel** check box in the dialog.

- Click on the **OK** button from the **Tasks** panel in **Combo View** to apply material.

After applying material to the model, the next step in FEM is to apply constraints to the model. Applying constraints is a part of defining boundary conditions. First, we will discuss the constraints which restrict the movements of model (Degree of Freedom) like fixing a face, allowing sliding of a face, and so on and then we will discuss the load constraints applied on model to check their effect.

APPLYING CONSTRAINTS

There are various toolbars in FEM workbench to apply constraints to the model; refer to Figure-13. You can also access the tools to apply constraints from **Model** menu in the FEM workbench; refer to Figure-14. The tools in these toolbars are discussed next.

Figure-13. Constraint Toolbars

Figure-14. Constraint tools in Model menu

Mechanical Constraints

Mechanical constraints are used to stop physical movements of selected objects in specified directions. The tools to apply mechanical constraints are available in the **Mechanical Constraints Toolbar** and **Mechanical Constraints** cascading menu of **Model** menu; refer to Figure-15. These tools are discussed next.

Figure-15. Mechanical Constraints cascading menu

Applying Fix Constraint

The **Fix Constraint** tool in **Mechanical Constraint Toolbar** is used to stop movement of selected object in specified directions. The procedure to apply this constraint is given next.

- Click on the **Fix Constraint** tool from the **Mechanical Constraints Toolbar** or **Model** menu. The **FEM constraint parameters** dialog will be displayed in the **Tasks** panel of **Combo View**; refer to Figure-16.

Figure-16. FEM constraint parameters dialog

- Select the face(s) whose movements are to be fixed and click on the **Add** button from the dialog box. Red boxes will be displayed attached to the selected faces showing that they are fixed; refer to Figure-17.

Figure-17. Fixed constraint applied to face

- After selecting desired faces, click on the **OK** button from the dialog to apply constraint.

Applying Displacement Constraint

The **Constraint Displacement** tool is used to specify the degree of freedom for selected objects in 3 linear and 3 rotation directions. The procedure to apply displacement constraint is given next.

- Click on the **Constraint Displacement** tool from the **Mechanical Constraints Toolbar** or **Model** menu. The **FEM constraint parameters** dialog will be displayed as shown in Figure-18.

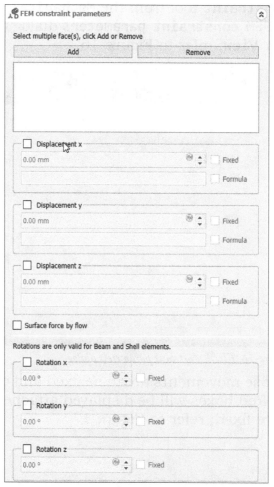

Figure-18. FEM constraint parameters dialog for displacement constraint

- Select the face(s) on which you want to apply constraint and click on the **Add** button from the dialog.

- Below the selection list box, there are six sections in the dialog box : `Displacement x`, `Displacement y`, `Displacement z`, `Rotation x`, `Rotation y`, and `Rotation z`. Select the **Free** check box for a section if you want the movement for respective direction to be free. You can also specify the limit up to which the displacement or rotation is allowed in edit box for that section. If you select the **Fixed** check box for a section then respective displacement or rotation will not be allowed. Note that rotation constraint can be applied for beam and shell elements only.
- After setting desired parameters, click on the **OK** button from the dialog to apply the constraint. The annotation for displacement constraint will be shown on the model; refer to Figure-19.

Figure-19. Applying displacement constraint

Applying Contact Constraint

The Contact constraint is applied to the model for making two selected faces in contact during the analysis. You can also specify the contact stiffness and friction coefficient between the faces when applying this constraint. This constraint is useful when you are working on an assembly of components. In assembly, you can define the parameters for contact between faces of two components. The procedure to apply this constraint is given next.

- Click on the **Constraint Contact** tool from the **Mechanical Constraints Toolbar** or **Model** menu in **FEM** workbench. The **FEM constraint parameters** dialog will be displayed as shown in Figure-20.

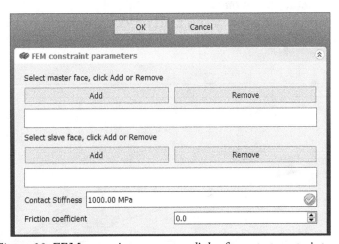

Figure-20. FEM constraint parameters dialog for contact constraint

- Select the face of first body to be used as master face and click on the **Add** button for **Select master face** section in the dialog; refer to Figure-21.

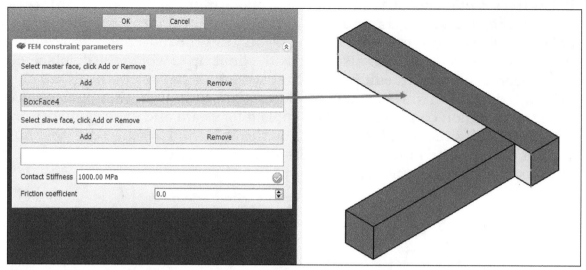

Figure-21. Master face selected

- Now, hide the master body by selecting any of its face and then pressing **SPACEBAR** from keyboard, so that you can select other face touching the master body.
- Select the slave face from other body and click on the **Add** button from **Select slave face** section of the dialog; refer to Figure-22.

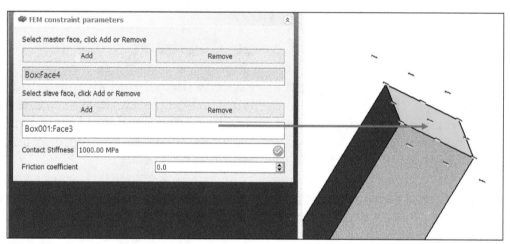

Figure-22. Slave face selected for constraint

- Open the **Model** panel in **Combo View** and select the hidden body to be displayed again and press **SPACEBAR**. The hidden body will be displayed again; refer to Figure-23.

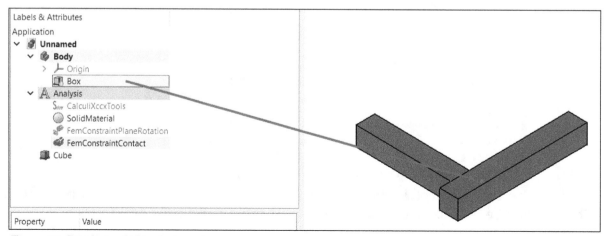

Figure-23. Displaying hidden body again

- Switch back to **Tasks** panel in the **Combo View** and specify desired value in **Contact Stiffness** edit box to define stiffness of contact point up to which the two faces will remain in contact under load.
- Click in the **Friction coefficient** edit box and specify desired value of friction coefficient between two faces.
- After setting desired parameters, click on the **OK** button from the dialog.

Applying Tie Constraint

The **Constraint Tie** tool defines a tie constraint that connects two selected surfaces in such a way that they cannot separate or slide on each other throughout the analysis. It is similar two bonded contact between two faces. The procedure to use this tool is discussed next.

- Click on the **Constraint Tie** tool from **Toolbar** or **Mechanical Constraints** cascading menu of **Model** menu in the **FEM** workbench. The **Tie parameter** and **Geometry reference selector for a Face** dialogs will be displayed in the **Tasks** panel of **Combo View**; refer to Figure-24.

Figure-24. Tie parameter and Geometry reference selector for a Face dialogs

- Click on the **Add** button from **Geometry reference selector for a Face** dialog and select the surface of the model in the 3D view area which you want to add to define tie constraint. The selected surface will be displayed in the list area of the dialog.
- Click on the **Add** button again and select the another surface between which you want to define tie constraint.
- Specify desired tolerance value in the **Tolerance** edit box from **Parameter** area in the **Tie parameter** dialog.
- Click on the **OK** button from the dialog. The tie constraint will be applied to the model; refer to Figure-25.

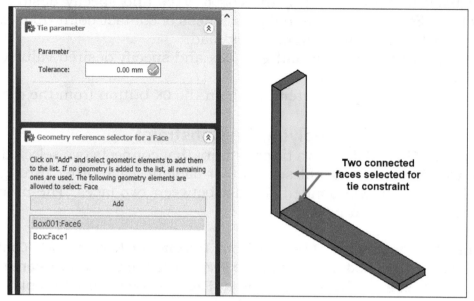

Figure-25. Tie constraint applied to the model

Applying Spring Constraint

The **Constraint spring** tool is used to apply constraint for a spring acting on a face. You can use this constraint to represent a soft base of the part that can absorb energy like soil or rubber mat. The procedure to use this constraint is given next.

- Click on the **Constraint spring** tool from the **Toolbar** or **Mechanical Constraints** cascading menu of **Model** menu in **FEM** workbench. The **FEM constraint parameters** dialog will be displayed; refer to Figure-26.

Figure-26. FEM constraint parameters dialog for spring constraint

- Click on the **Add** button from the dialog and select the face model you want to be connected by a spring; refer to Figure-27.

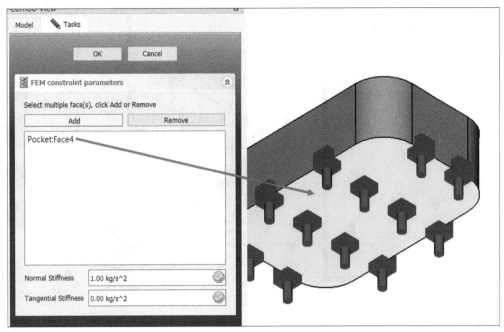

Figure-27. Selecting the face for spring constraint

- Specify desired value in the **Normal Stiffness** edit box.
- Specify desired value in the **Tangential Stiffness** edit box.
- After specifying desired parameters, click on the **OK** button from the dialog.

Applying Force Constraint

The **Constraint force** is used to apply force load on selected face of specified value. The procedure to use this constraint is given next.

- Click on the **Constraint force** tool from the **Toolbar** or **Mechanical Constraints** cascading menu of **Model** menu in **FEM** workbench. The **FEM constraint parameters** dialog will be displayed; refer to Figure-28.

Figure-28. FEM constraint parameters dialog for force constraint

- Select desired face from the freeCAD model on which you want to apply the load and specify desired load value in the **Area load** edit box.
- Select the **Reverse direction** check box to reverse direction of load.
- If you want to define desired direction for load then click on the **Direction** button and select desired reference; refer to Figure-29.

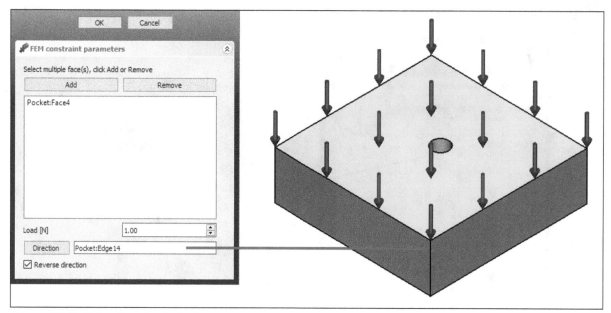

Figure-29. Defining direction for load

- After setting desired parameters, click on the **OK** button from the dialog.

Applying Pressure Load

The **Constraint pressure** tool is used to define pressure load on selected face/ surface. The procedure to use this tool is given next.

- Click on the **Constraint pressure** tool from the **Toolbar** or **Model** menu in **FEM** workbench. The **FEM constraint parameters** dialog will be displayed; refer to Figure-30.

Figure-30. FEM constraint parameters dialog for pressure constraint

- Select the face(s) on which you want to specify pressure load and specify desired value in the **Pressure** edit box of the dialog.
- Specify the other parameters as discussed earlier and click on the **OK** button to apply load.

Applying Centrif Constraint

The **Centrif Constraint** tool is used to create constraint of centrifugal force. The procedure to use this tool is given next.

- Click on the **Centrif constraint** tool from the **Model** menu or from the **Toolbar**. The **Centrif parameter**, **Geometry reference selector for a Solid**, and **Geometry reference selector for a Edge** dialogs will be displayed; refer to Figure-31, Figure-32, and Figure-33.

Figure-31. Centrif parameter dialog

Figure-32. Geometry reference selector for Solid dialog   *Figure-33. Geometry reference selector for a Edge dialog*

- Specify desired revolutions per second in the **Rotation frequency f$_{rot}$ [rps]** edit box from **Centrif parameter** dialog.
- Click on the **Add** button from **Geometry reference selector for a Solid** dialog and select desired solid object on which you want to apply the constraint; refer to Figure-34.

Figure-34. Selecting the elements for centrif constraint

- Click on the **Add** button from **Geometry reference selector for a Edge** dialog and select desired edge of the solid object; refer to Figure-34.
- After specifying desired parameters, click on the **OK** button from the dialog. The centrif constraint will be applied; refer to Figure-35.

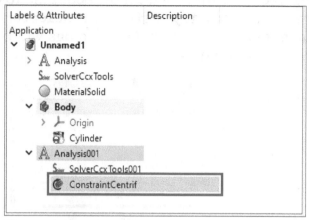

Figure-35. Centrif constraint applied

Applying Self Weight Constraint

The **Self Weight** constraint is applied on the model to check the effect of weight of the model in various situations. The procedure to use this tool is given next.

- Click on the **Constraint self weight** tool from the **Model** menu or from the **Toolbar**. The self weight constraint will be created and displayed in the **Model Tree**.
- If you want to modify the parameters of self weight then select it from the **Model Tree** and set desired parameters in the **Properties** box; refer to Figure-36.

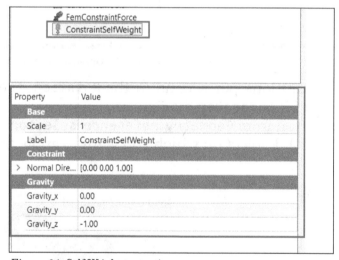

Figure-36. Self Weight properties

Geometrical Constraints

The tools to apply geometrical constraints are available in the **Geometrical Constraints Toolbar** and **Geometrical Constraints** cascading menu of **Model** menu; refer to Figure-37. These tools are discussed next.

Figure-37. Geometrical Constraints cascading menu

Applying Plane Rotation Constraint

The **Constraint plane rotation** tool is applied on selected faces to allow rotation of model in plane parallel to selected face but restrict translation in any direction. This type of constraint is useful for fixing gears in an analysis so that they are free to rotate. The procedure to apply this constraint is given next.

- Click on the **Constraint plane rotation** tool from the **Toolbar** or **Geometrical Constraints** cascading menu in the **Model** menu. The **FEM constraint parameters** dialog will be displayed in the **Tasks** panel; refer to Figure-38.

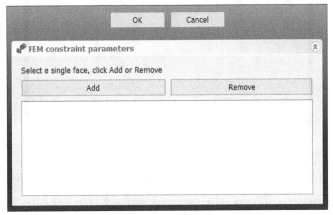

Figure-38. FEM constraint parameters dialog for plane rotation constraint

- Select the face on which you want to apply plane rotation constraint and click on the **Add** button from the dialog. The annotation of constraint will be displayed on the model face; refer to Figure-39.

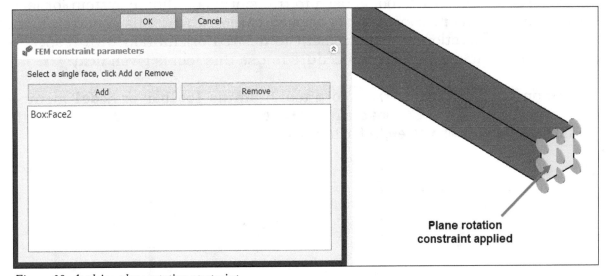

Figure-39. Applying plane rotation constraint

- After adding faces, click on the **OK** button from the dialog to apply constraints and exit the dialog.

Applying Section Print Constraint

The **Constraint sectionprint** tool is used to print internal section forces at nodes of section surface. The procedure to use this tool is discussed next.

- Click on the **Constraint sectionprint** tool from the **Toolbar** or **Geometrical Constraint** cascading menu in the **Model** menu. The **SectionPrint parameter** and **Geometry reference selector for a Face** dialogs will be displayed in the **Tasks** panel of **Combo View**; refer to Figure-40.

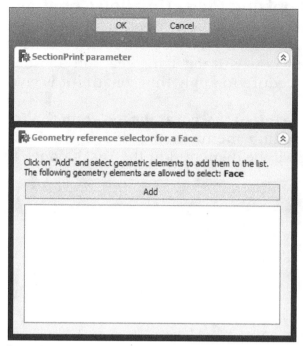

Figure-40. SectionPrint parameter and Geometry reference selector for a Face dialogs

- Click on the **Add** button from dialog and select the face to be used for checking internal forces.
- Click on the **OK** button to apply constraint.

Applying Transform Constraint

The **Constraint transform** tool is used to transform coordinate system for faces on which displacement constraint is applied earlier. Using this constraint, you can modify the direction of displacement to desired orientation with respect to original coordinate system. The procedure to use this tool is given next.

- Click on the **Constraint transform** tool from the **Toolbar** or **Model** menu in **FEM** workbench. The **FEM constraint parameters** dialog will be displayed in the **Tasks** panel of **Combo View**; refer to Figure-41.

Figure-41. FEM constraint parameters dialog for transform constraint

- The surfaces/faces that can be used for applying transform constraint are available in the **Transformable surfaces** list box at the bottom in the dialog. Note that these are the faces on which displacement constraint has been applied.
- Select the face from model on which transform constraint is to be applied and click on the **Add** button from the dialog. The current orientation of coordinate system will be displayed on the face; refer to Figure-42.

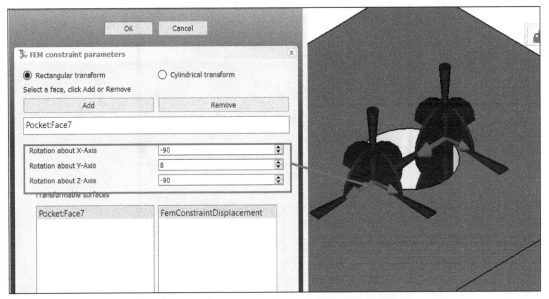

Figure-42. Orientation of coordinate system

- By default, the **Rectangular transform** radio button is selected, so you can apply rotation to each axis in Cartesian coordinate system. You can select the **Cylindrical** radio button to use Polar coordinate system for transformation.
- Set desired parameters in the **Rotation about X-Axis**, **Rotation about Y-Axis**, and **Rotation about Z-Axis** edit boxes. The coordinate system will orient according to the values specified in these edit boxes.
- Click on the **OK** button from the dialog to apply constraint.

Constraints Without Solver

The tools to apply these constraints are available in the **Constraints without solver** cascading menu of **Model** menu; refer to Figure-43. Note that at present the solver used for performing analysis related to these constraints is not available in FreeCAD. These tools are discussed next.

Figure-43. Constraints without solver cascading menu

Defining Fluid Boundary Conditions

The **Fluid boundary condition** tool is used to define boundary conditions for performing fluid flow analysis. The procedure to apply fluid boundary conditions is given next.

- Click on the **Fluid boundary condition** tool from the **Constraints without solver** cascading menu of the **Model** menu. The **FEM constraint parameters** dialog will be displayed; refer to Figure-44.

Figure-44. FEM constraint parameters dialog for fluid boundary conditions

- Select desired option from the **Boundary** drop-down in the dialog. Select the **inlet** option from the drop-down if you want to specify properties of fluid like pressure, velocity, flow rate, and so on at the inlet of fluid. Select the **wall** option from the drop-down if you want to specify physical conditions for wall in fluid domain like wall is fixed, slipping, partial slipping with specified slip ratio, moving at specified speed, rough, and so on. Select the **outlet** option from the drop-down if you want to specify fluid properties at the outlet like total pressure, static pressure, uniform velocity, or outflow parameter for fluid. Select the **interface** option from the drop-down if you want to specify symmetry parameters for selected face(s). Select the **freestream** option from the drop-down if you want to apply free fluid flow condition at selected face. The freestream condition is applicable at both inlet as well as outlet.
- After selecting desired option from the **Boundary** drop-down, select the related option from the **Subtype** drop-down below it.
- Select desired face(s) on which you want to apply boundary conditions.
- Specify the other parameters as desired and click on the **OK** button.

Applying Bearing Constraint

The Bearing constraint is used to apply constraints to selected faces for giving them bearing type load during analysis. This tool is not available by default in the Toolbar, so you need to access this tool from **Model** menu. The procedure to use this tool is given next.

- Click on the **Constraint bearing** tool from the **Constraints without solver** cascading menu of the **Model** menu in **FEM** workbench. The **FEM constraint parameters** dialog will be displayed as shown in Figure-45.

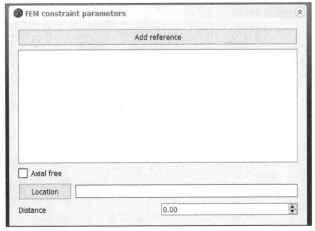

Figure-45. FEM constraint parameters dialog for bearing constraint

- Select the cylindrical face to which you want to apply bearing constraint.
- Select the **Axial free** check box if you want to allow translation of faces in direction parallel to cylindrical axis.
- Click on the **Location** button and select a point or line to define reference to be used for placing constraint.
- You can set the distance of constraint from selected point/line in the **Distance** spinner.
- After setting desired parameters, click on the **OK** button from the dialog. The annotation of constraint will be displayed on the model; refer to Figure-46.

Figure-46. Bearing constraint

Applying Gear Constraint

The **Gear** constraint is used to restrict the rotation of selected face as if it is attached to a gear which needs specified load to be applied before it can rotate. In other words, a load equivalent to specified gear parameters will be applied on selected face/surface. The procedure to use this constraint is given next.

- Click on the **Constraint gear** tool from the **Constraints without solver** cascading menu of the **Model** menu. The **FEM constraint parameters** dialog will be displayed; refer to Figure-47.

Figure-47. FEM constraint parameters dialog for gear constraint

- Select the cylindrical face to which you want to apply gear constraint. Preview of constraint will be displayed; refer to Figure-48.

Figure-48. Preview of gear constraint

- Specify desired value in **Gear diameter** edit box to specify the diameter of gear to be used for constraining. This should represent actual gear attached to selected face.
- Specify desired parameters in the **Force** and **Force location[deg]** edit boxes to specify value and direction of force load on gear.
- Set the other parameters as discussed earlier and click on the **OK** button to apply the constraint.

Applying Pulley Constraint

The **Pulley** constraint is used to apply load on selected cylindrical face/surface equivalent to pulley load based on specified parameters. The procedure to apply this constraint is given next.

- Click on the **Constraint pulley** tool from the **Constraints without solver** cascading menu of **Model** menu in **FEM** workbench. The **FEM constraint parameters** dialog will be displayed as shown in Figure-49.

Figure-49. FEM constraint parameters dialog for pulley constraint

- Select the cylindrical face on which you want to apply pulley constraint (load). Preview of pulley constraint will be displayed; refer to Figure-50.

Figure-50. Preview of pulley constraint

- Set the parameters related to pulley in the edit boxes of dialog. Note that by default, selected cylindrical face is assigned as driving pulley. If you want to constraint selected face as driven pulley then select the **Driven pulley** check box from the dialog.
- After setting desired parameters, click on the **OK** button from dialog to apply constraint.

Thermal Constraints

The **Thermal** constraints are used to define various thermal conditions like temperature, heat flux, body heat source, and so on. The tools to apply thermal constraints are available in the **Thermal Constraints Toolbar** and **Thermal Constraints** cascading menu of the **Model** menu; refer to Figure-51. These tools are discussed next.

Figure-51. Thermal Constraints cascading menu

Defining Initial Temperature Constraint

The **Constraint initial temperature** tool is used to apply initial temperature to the model for thermal analysis. The procedure to use this tool is given next.

- Click on the **Constraint initial temperature** tool from the **Thermal Constraints Toolbar**. The **FEM constraint parameter** dialog will be displayed; refer to Figure-52.

Figure-52. FEM constraint parameters dialog for initial temperature constraint

- Specify desired value of initial temperature in the edit box of dialog. Note that you need to specify temperature in Kelvin (K) unit.
- After specifying desired value, click on the **OK** button to apply constraint.

Applying Heat Flux Constraint

Heat flux is the rate of heat energy transfer from selected face(s). The **Constraint heatflux** tool in **Thermal Constraints** cascading menu of **Model** menu is used to define the heat flux for selected faces. The procedure to apply this constraint is given next.

- Click on the **Constraint heatflux** tool from **Thermal Constraints** cascading menu of **Model** menu. The **FEM constraint parameters** dialog will be displayed as shown in Figure-53.

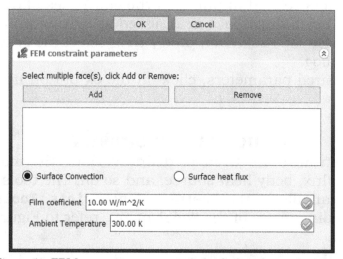

Figure-53. FEM constraint parameters dialog for heat flux constraint

- Select the **Surface Convection** radio button if you want to define heat convection rate for selected face/surface at specified temperature. Heat convection is the amount of heat passing through selected face/surface. Select the **Surface heat flux** radio button if you want to specify heat energy transfer rate.
- Specify the parameters as desired in the edit boxes of the dialog.
- Select the face on which you want to apply heat flux constraint and click on the **Add** button from the dialog. If you want to apply heat flux to multiple faces then you can select multiple faces while holding the **CTRL** key.
- After setting desired parameters, click on the **OK** button from the dialog. The constraint will be applied; refer to Figure-54.

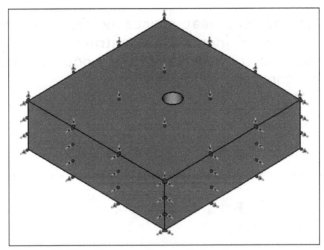

Figure-54. Heat flux constraint applied on faces

Applying Temperature Constraint

The **Constraint temperature** tool is used to apply fix temperature to selected faces. Applying fix temperature to a face makes that face a heat source for analysis. The procedure to apply this constraint is given next.

- Click on the **Constraint temperature** tool from the **Thermal Constraints** cascading menu in the **Model** menu. The **FEM constraint parameters** dialog will be displayed; refer to Figure-55.

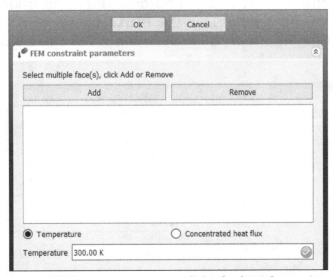

Figure-55. FEM constraint parameters dialog for thermal constraint

- Select the face(s) to which you want to apply thermal constraint and click on the **Add** button.
- Select the **Temperature** radio button to specify temperature of selected face(s).

- Select the **Concentrated heat flux** radio button from the bottom in the dialog if you want to specify total heat energy of the selected face.
- Specify desired value of temperature/heat flux in the edit box below the radio buttons.
- After setting desired parameters, click on the **OK** button from the dialog to apply the constraint.

Applying Body Heat Source Constraint

The **Body Heat Source** constraint is used to define the heat generated by model. The procedure to apply this constraint is given next.

- Click on the **Constraint body heat source** tool from the **Thermal Constraints** cascading menu of the **Model** menu. The constraint will be applied to the model.
- Select the constraint from the **Model Tree** and modify the parameters as desired in the **Property editor**; refer to Figure-56.

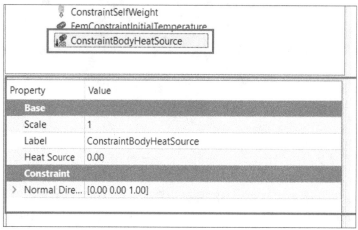

Figure-56. Property editor for body heat source constraint

- Specify desired value of heat in the **Heat Source** edit box. Click anywhere in the model area to exit the editing mode.

Fluid Constraints

The **Fluid** constraints are used to define boundary conditions for fluid flow in the model. These tools are available in the **Fluid Constraints** Toolbar and **Fluid Constraints** cascading menu of the **Model** menu; refer to Figure-57. These tools are discussed next.

Figure-57. Fluid Constraints cascading menu

Applying Initial Flow Velocity

The **Constraint initial flow velocity** tool is used to specify initial velocity for flow of fluid in the model. This constraint is useful for fluid flow analysis. The procedure to use this tool is given next.

- Click on the **Constraint initial flow velocity** tool from the **Fluid Constraints** cascading menu of the **Model** menu. The **Constraint Properties** dialog will be displayed; refer to Figure-58.

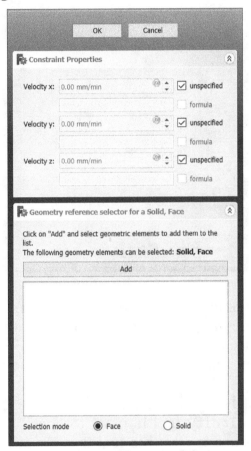

Figure-58. Constraint Properties dialog

- Clear the **unspecified** check box(es) for desired velocity direction(s) and specify the value(s) in respective edit box(es).
- Click on the **Add button from Geometry reference selector for a Solid, Face** dialog to add the selected geometric elements to the list. Select **Face** radio button from **Selection mode** area to select the face as a geometric element. Select **Solid** radio button to select the solid body as a geometric element.
- After setting desired parameters, click on the **OK** button from the dialog to apply constraint.

Constraint Initial Pressure

The **Constraint initial pressure** tool is used to create an initial pressure constraints for a fluid flow analysis. The procedure to use this tool is given next.

- Click on the **Constraints initial pressure** tool from the **Fluid Constraints** cascading menu of the **Model** menu. The **Constaint Properties** dialog will be displayed; refer to Figure-59.

Figure-59. Constraint Properties dialog

- Specify desired value of initial pressure in the **Pressure** edit box.
- Specify desired parameters in the **Geometry reference selector for a Solid, Face** dialog as discussed earlier.
- Click on the **Ok** button from the dialog to apply constaints.

Applying Constraint Flow Velocity

The **Constraint Flow Velocity** tool is used to specify velocity of fluid through selected face. The procedure to use this tool is given next.

- Click on the **Constraint Flow Velocity** tool from the **Fluid Constraints** cascading menu of the **Model** menu. The **Select Faces/Edges/Vertexes** and **Constraint Properties** dialogs will be displayed in the **Tasks** panel of **Combo View**; refer to Figure-62.

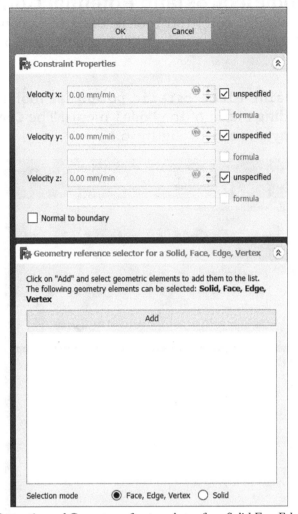

Figure-60. Constraint Properties and Geometry reference selector for a Solid Face Edge Vertex dialogs

- Clear the unspecified check box(es) for desired velocity direction(s) and specify the value(s) in respective edit box(es).
- Select the faces, edges, or vertices for which you want to define flow velocity and click on the **Add** button.
- Set the other parameters as discussed earlier and click on the **OK** button.

Electromagnetic Constraints

The **Electromagnetic Constraints** is an icon button in the FEM constraints toolbar that groups electromagnetic constraints. These tools are available in the **Fluid Constraints Toolbar** and **Electromagnetic Constraints** cascading menu of the **Model** menu; refer to Figure-61.

Figure-61. Electromagnetic Constraints cascading menu

Applying Electrostatic Potential Constraint

The Constraint electrostatic potential tool is used to specify electrical potential (voltage) at selected faces/edges/vertices. The procedure to use this tool is given next.

- Click on the **Constraint electrostatic potential** tool from **Electromagnetic Constraints** cascading menu of the **Model** menu. The **Constraint Properties** dialog will be displayed; refer to Figure-62.

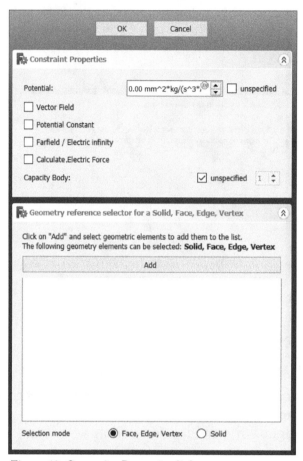

Figure-62. Constraint Properties dialog

- Select the faces/edges/vertices on which you want to apply electrostatic potential constraint and click on the **Add** button from the dialog.
- Clear the unspecified check box for **Potential** edit box and specify desired value of voltage.
- Select the **Potential Constant** check box from the dialog to keep voltage constant in analysis.
- After setting desired parameters, click on the **OK** button from the dialog to apply the constraint.

Constraint current density

The **Constraint current density** tool is used to create an FEM constraint for defining the current density. The procedure to use this tool is given next.

- Click on the **Constraint current density** tool from **Electromagnetic Constraints** cascading menu of the **Model** menu. The **Constraint Properties** dialog will be displayed; refer to Figure-63.

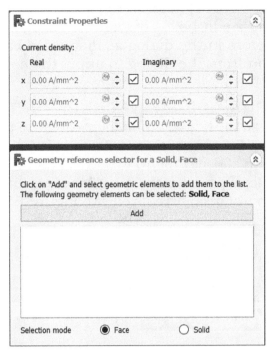

Figure-63. Constraint Properties dialog

- Clear the check boxes from **Current density** area of the dialog to specify desired values in the **x**, **y**, and **z** edit boxes of **Real** and **Imaginary** sections.
- Specify other parameters in the dialog as discussed earlier.
- Click on the **Ok** button from the dialog to apply the constraint.

Constraint magnetization

The **Constraint magnetization** tool is used to restrict the magnetism value of selected solid/face at specified value. The procedure to use this constraint is given next.

- Click on the **Constraint magnetization** tool from **Electromagnetic constraints** cascading menu of the **Model** menu. The **Constraint Properties** dialog will be displayed; refer to Figure-64.

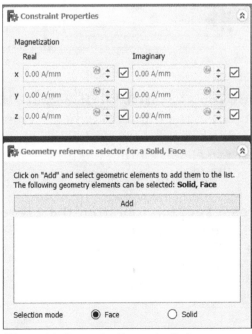

Figure-64. Constraint Properties dialog

- Clear the check boxes from **Magnetization** area of the dialog to specify desired values in the **x**, **y**, and **z** edit boxes of **Real** and **Imaginary** sections.
- Specify other parameters in the dialog as discussed earlier.
- Click on the **Ok** button from the dialog to apply the constraint.

Applying Constant Vacuum Permittivity

Vacuum permittivity is used to define ability of medium to pass current in dielectric environment. Click on the **Constant Vacuum Permittivity** tool from the **Overwrite Constants** cascading menu of **Model** menu to add the parameter in current analysis. You can modify the value of parameter using **Property** dialog after selecting this constant from **Model** tab in **Combo View**; refer to Figure-65.

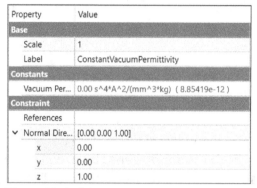

Property	Value
Base	
Scale	1
Label	ConstantVacuumPermittivity
Constants	
Vacuum Per...	0.00 s^4*A^2/(mm^3*kg) (8.85419e-12)
Constraint	
References	
∨ Normal Dire...	[0.00 0.00 1.00]
x	0.00
y	0.00
z	1.00

Figure-65. Applying constant vacuum permittivity

CREATING GEOMETRY ELEMENTS

In FreeCAD FEM, you can create 2D elements of specified thickness by using the tools in **Element Geometry** toolbar or **Element Geometry** cascading menu of the **Model** menu; refer to Figure-66. These tools are discussed next.

Figure-66. Element Geometry tools

Creating Beam Cross-section

The **Beam cross section** tool is used to create beam elements with different type of cross sections like circular beam, rectangular beam, and pipe beam. To create beam elements, you need to select reference edges from the model. These edges can be of a surface or a solid. The procedure to use this tool is given next.

- Click on the **Beam cross section** tool from the **Element Geometry** cascading menu of **Model** menu. The **Beam section parameter** and **Geometry reference selector for a Edge** dialogs will be displayed; refer to Figure-67.

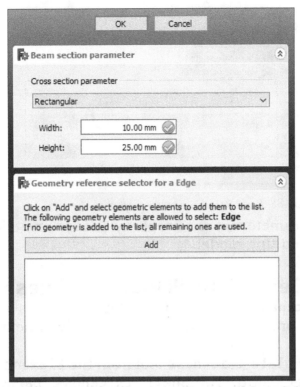

Figure-67. Beam section parameters and geometry selector dialogs

- Select desired option from the **Cross section parameter** drop-down in the dialog. Select the **Rectangular** option from the drop-down to create rectangular beam element. Select the **Circular** option to create circular rod elements. Select the **Pipe** option from the drop-down to create tube like elements.
- After selecting desired option from the drop-down, specify the related parameters in edit box(es) below the drop-down.
- After setting desired parameters in the dialog, click on the **Add** button from the **Geometry reference selector for a Edge** dialog and select the edge on which you want to create the element. Note that you need to click on **Add** button each time before you select an edge for creating element.
- Click on the **OK** button from the dialog to create the elements.

Applying Rotation to Beam Elements

You can change the default orientation of beam elements by specified rotation value using the **Beam rotation** tool. The procedure to use this tool is given next.

- Click on the **Beam rotation** tool from the **Element Geometry** cascading menu of the **Model** menu. The **ElementRotation1D** feature will be added in the **Model Tree**.
- Select the **ElementRotation1D** feature from **Model Tree**. The **Property Editor** dialog will be displayed; refer to Figure-68.

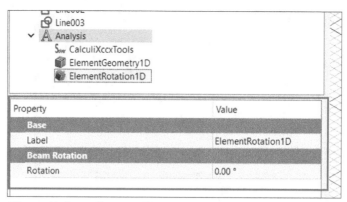

Figure-68. Beam Rotation Element properties

- Specify desired parameters in the **Property Editor** and click in the empty area of drawing to exit editing model.

Defining Shell Plate Thickness

The **Shell plate thickness** tool is used to define thickness of shell element based on selected face. The procedure to use this tool is given next.

- Click on the **Shell plate thickness** tool from the **Element Geometry** cascading menu of the **Model** menu. The **Shell thickness parameter** and **Geometry reference selector for a Face** dialogs will be displayed in the **Tasks** panel of **Combo View**; refer to Figure-69.

Figure-69. Dialogs for shell element

- Specify desired value of thickness for shell element in the **Thickness** edit box of the dialog.
- Click on the **Add** button from the dialog and select the face on which you want to create the shell element of specified thickness.
- After setting desired parameters, click on the **OK** button from the dialog to create the geometry.

Creating Fluid Section for 1D Flow

The **Fluid section for 1D flow** tool is used to create section in the model for fluid flow. The procedure to use this tool is given next.

- Click on the **Fluid section for 1D flow** tool from the **Element Geometry** cascading menu of the **Model** menu. The **Form** and **Geometry reference selector for a Edge** dialogs will be displayed; refer to Figure-70.

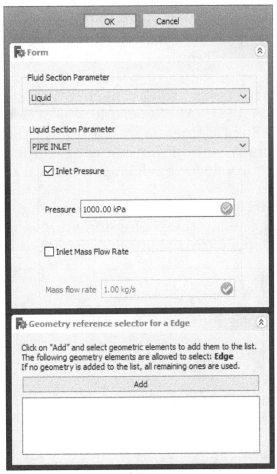

Figure-70. Dialogs for 1D fluid section

- Select desired option from the **Fluid Section Parameter** drop-down to define whether fluid section is for liquid, gas, or it is open for both liquid and gas. The other parameters will be displayed based on selected option.
- If you have selected **Liquid** option for fluid section then select desired option from the **Liquid Section Parameter** drop-down to specify whether the section is at inlet, outlet, diaphragm, or any other section of fluid domain.
- Specify the other parameter as needed and click on the **Add** button from the dialog. You will be asked to select an edge to define location of section.
- Select desired edge and click on the **OK** button from the dialog to create the section.

Note that the functionality to work with Beam elements has not been added yet up to version 0.21, so you will not be able to work with them in software.

MESHING

Meshing is the base of FEM (Finite Element Method) which is one of the methods used for FEA. Meshing divides the solid/shell models into elements of finite size and shape. These elements are joined at some common points called nodes. These nodes define the load transfer from one element to other element. Meshing is a very crucial step in design analysis because it directly affects the accuracy of analysis. The automatic mesher in the software generates a mesh based on a global element size, tolerance, and local mesh control specifications. Mesh control lets you specify different sizes of elements for components, faces, edges, and vertices.

The software estimates a global element size for the model taking into consideration its volume, surface area, and other geometric details. The size of the generated mesh (number of nodes and elements) depends on the geometry and dimensions of the model, element size, mesh tolerance, mesh control, and contact specifications. In the early stages of design analysis where approximate results may suffice, you can specify a larger element size for a faster solution. For a more accurate solution, a smaller element size may be required.

Meshing generates 3D tetrahedral solid elements, 2D triangular shell elements, and 1D beam elements. A mesh consists of one type of elements unless the mixed mesh type is specified. Solid elements are naturally suitable for bulky models. Shell elements are naturally suitable for modeling thin parts (sheet metals), and beams and trusses are suitable for modeling structural members.

There are two plug-ins in FreeCAD to perform meshing: Netgen and Gmsh. The tools to perform meshing using these plug-ins are available in the **Mesh** menu; refer to Figure-71. The tools in this menu are discussed next.

Figure-71. Mesh menu

Creating FEM mesh using Netgen Mesher

The **FEM mesh from shape by Netgen** tool is used to create mesh from selected shape(body) using Netgen plug-in. The procedure to use this tool is given next.

* Click on the **FEM mesh from shape by Netgen** tool from the **Mesh** menu after selecting the body. The **Tet Parameter** dialog will be displayed; refer to Figure-72.

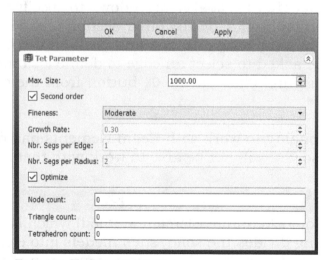

Figure-72. Tet Parameter dialog

* Specify desired value in **Max. Size** edit box to define maximum size of element that can be generated in current mesh. Note that this size can be length, width, or height of the element.
* Select the **Second order** check box if you want to generate elements in mesh with additional nodes that define curvature of segments; refer to Figure-73.

Figure-73. Difference between first order and second order mesh

- Select desired option from the **Fineness** drop-down to define smoothness in mesh elements at edges. A finer mesh gives better results of analysis but it also increases the computational time for analysis. Figure-74 shows same object mesh with both very coarse and very find mesh types.

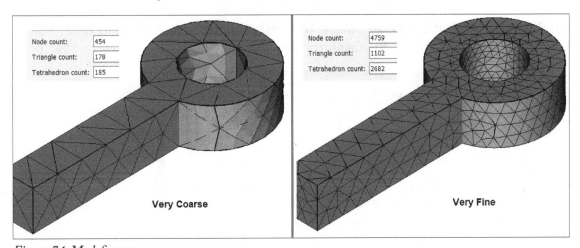

Figure-74. Mesh fineness

Note that a simple move from very coarse to very fine mesh can exponentially increase the number of elements as in our case. There is a certain threshold in meshing fineness after which you get a marginal accuracy at the loss of a lot of computing power and time. So, you should start with a moderate fineness before moving to finer side when performing analysis.

- If you want to manually specify the smoothness of your mesh then select the **UserDefined** option from the **Fineness** drop-down. The edit boxes below the drop-down will become active.
- Specify desired value in **Growth Rate** edit box to define increase in edge length of element when moving away from edges or complex regions of the model.
- Specify desired value in **Nbr. Segs per Edge** edit box to define number of segments in which the edges of model will be divided in mesh. Note that the mesh will be a compromise between maximum element size and number of segments per edge values. So, you may or may not get the definite number of segments per edge depending on the model.
- Specify desired value in **Nbr. Segs per Radius** edit box to define number of segments in circular/arc type edges.
- Select the **Optimize** check box to automatically increase number of elements near edges and complex regions.
- After setting desired parameters, click on the **OK** button from the dialog. The mesh will be generated; refer to Figure-75.

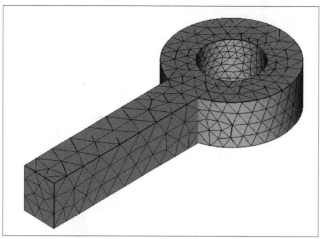

Figure-75. Mesh generated

Creating FEM mesh using Gmsh Mesher

The **FEM mesh from shape by Gmsh** tool is used to create mesh using the Gmsh plug-in. The procedure to use this tool is given next.

- Click on the **FEM mesh from shape by Gmsh** tool from **Mesh** menu after selecting the body. The **FEM Mesh by Gmsh** dialog will be displayed; refer to Figure-76.

Figure-76. FEM Mesh by Gmsh dialog

- Select desired option from the **Element dimension** drop-down to define what type of mesh elements are to be created. By default, **From Shape** option is selected in this drop-down, so dimension of element is automatically decided based on selected object. For example, for a 3D solid body, 3D elements will be generated in the mesh. You can manually specify the dimensions of element by selecting 1D, 2D, or 3D option from the drop-down.

The element types available in FreeCAD FEM are as follows:

1D Elements:

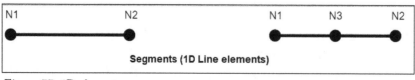

Figure-77. 1D elements

2D Elements:

Figure-78. 2D elements

3D Elements:

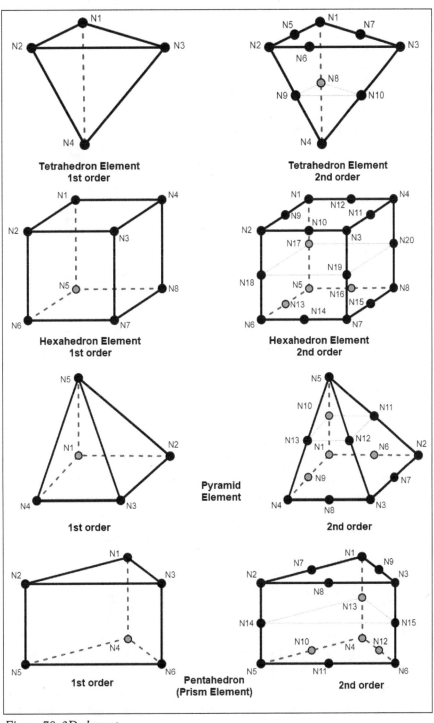

Figure-79. 3D elements

- Specify desired values in **Max element size** and **Min element size** edit boxes to define the maximum and minimum element size in mesh.
- After setting desired parameters, click on the **Apply** button to generate mesh.

Creating Boundary Mesh Layer

The **FEM mesh boundary layer** tool is used to create a layer of mesh with different element size from main mesh at the boundary of model. The procedure to use this tool is given next.

- Click on the **FEM mesh boundary layer** tool from the **Mesh** menu after selecting the **FEMMeshGmsh** feature (mesh created by Gmsh) from **Model Tree**. The **Mesh boundary layer settings** and **Geometry reference selector for a Solid, Face, Edge, Vertex** dialogs will be displayed; refer to Figure-80.

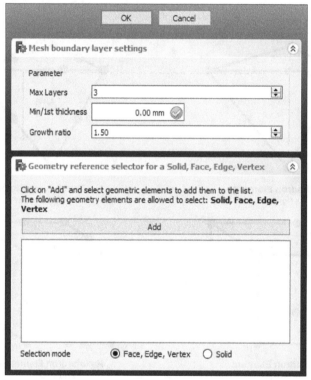

Figure-80. Mesh boundary layer settings and Geometry reference selector dialogs

- Specify desired value in the **Max Layers** edit box to define number of layers to be meshed by boundary layer settings.
- Specify desired value in the **Min/1st thickness** edit box to define the thickness of 1st layer in the mesh boundary.
- Specify desired value in **Growth ratio** edit box to specify the rate of increase/decrease in element size while moving away from boundary regions.
- Click on the **Add** button and select the solid/face/edge/vertex at which you want to apply boundary layer mesh.
- After setting desired parameters, click on the **OK** button from the dialog. The mesh will be created.

Create FEM mesh region

The **FEM mesh region** tool is used to create meshing for specific region of model with different element size from main mesh. This tool is useful for creating finer mesh in complex regions of model. The procedure to use this tool is given next.

- Click on the **FEM mesh region** tool from the **Mesh** menu. The **Mesh region** and **Geometry reference selector for a Solid, Face, Edge, Vertex** dialogs will be displayed; refer to Figure-81.

Figure-81. Mesh region and Geometry reference selector dialogs

- Specify desired value of maximum element size for refined mesh in the **Max element size** edit box.
- Select desired radio button for **Selection mode** and add the geometry region to be refined.
- After setting desired parameters, click on the **OK** button from the dialog.

Creating FEM mesh group

The **FEM mesh group** tool is used to create a group of mesh for exporting to other applications. The procedure to use this tool is given next.

- Click on the **FEM mesh group** tool from the **Mesh** menu. The **Mesh group** and **Geometry reference selector for a Solid, Face, Edge, Vertex** dialogs will be displayed; refer to Figure-82.

Figure-82. Mesh group dialogs

- Select desired radio button from **Mesh group** dialog to define whether name is used for exporting mesh or label.

- Select desired face, edge, vertex, or solid object to be added in mesh group. If no object is selected then all the unused objects will be selected for mesh group.
- After setting desired parameters, click on the **OK** button to create mesh group.

Creating Nodes set

The **Nodes set** tool is used to create a set of selected nodes for checking results of analysis later on those nodes. The procedure to use this tool is given next.

- Click on the **Nodes set** tool from the **Mesh** menu. The **TaskObjectName** and **Nodes set** dialogs will be displayed; refer to Figure-83.

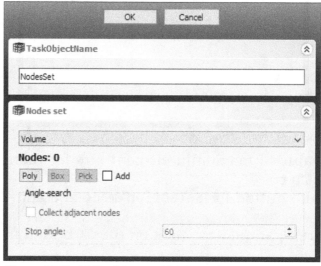

Figure-83. TaskObjectName and Nodes set dialogs

- Specify desired name of node set in the **TaskObjectName** edit box.
- Click on the **Poly** button from the **Nodes set** dialog if the **Volume** option is selected and click on the **Pick** button if the **Surface** option is selected in dropdown of this dialog. You will be asked to select the volume or surface based on your selection.
- Select desired volume or surface using the polygon or pick selection method respectively; refer to Figure-84.

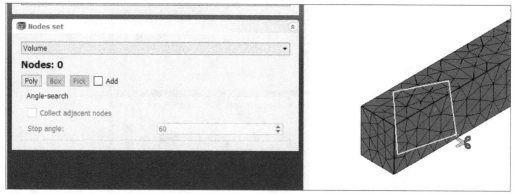

Figure-84. Polygon created for selection

- If you are using polygon selection then after drawing polygon, right-click in the drawing area. A shortcut menu will be displayed as shown in Figure-85. Select the **Inner** option to select all the nodes falling inside a polygon and select the **Outer** option if you want to select the nodes outside the polygon. Selected nodes will be displayed in green color; refer to Figure-86.

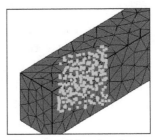

Figure-86. Selected nodes

Figure-85. Shortcut menu for polygon selection

- Click on the **OK** button from the dialog to create node set.

FEM Mesh to Polygonal Mesh Model Conversion

The Mesh created in FEM workbench is not a model mesh. It is representation of original model in the form of integration of small elements. You can convert the FEM mesh to a model mesh by using the **FEM mesh to mesh** tool in **Model** menu. The procedure to use this tool is given next.

- Click on the **FEM mesh to mesh** tool from the **Model** menu after selecting **FEMMeshGmsh** feature from the **Model Tree**. The FEM mesh will be converted to mesh model; refer to Figure-87.

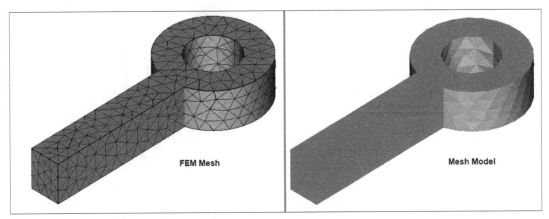

Figure-87. Mesh model conversion

SOLVERS

FreeCAD FEM has many analysis solvers available in **Solve** menu; refer to Figure-88. Some of these solvers are in experimental stage and some are specific to analysis type. Various solvers available in FreeCAD are discussed next.

Figure-88. Solve menu

Calculix Standard Solver

CalculiX is a package designed to solve field problems. The method used by this solver is the finite element method. With CalculiX, Finite Element Models can be built, calculated, and post-processed. The solver is able to do linear and non-linear calculations. Static, dynamic, and thermal solutions can be obtained by this solver. Because the solver makes use of the abaqus input format, it is possible to use commercial pre-processors as well with this solver. The pre-processor of this solver is able to write mesh related data for nastran, abaqus, ansys, code-aster and for the free-cfd codes like duns, ISAAC and OpenFOAM. The procedure to use this solver is given next.

- Click on the **Solver CalculiX Standard** option from the **Solve** menu. The **CalculiXccxTools** feature will be added in the **Model Tree**. Note that when you start a mechanical analysis then **CalculiXccxTools** feature is added in the **Model Tree** by default.
- Double-click on the **CalculiXccxTools** feature. The **Mechanical analysis** dialog will be displayed in **Tasks** panel of **Combo View**; refer to Figure-89.

Figure-89. Mechanical analysis dialog

- Click on the **Write .inp file** button to create an input file for solver if you have specified all the other parameters of analysis like constraints, material, and so on in the model. This input file will tell the solver, what is to be calculated and what parameters are known.
- On clicking the **Write .inp file** button, the **Run CalculiX** button will become active and overview of input file writing process will be displayed; refer to Figure-90. Note that if there are parameters like material of body, constraints applied on the model, and loads which have not been defined yet then you will get an error message. Also, you can perform analysis on single body with one mesh feature. Multiple meshes are not allowed till writing this book in FreeCAD. Most of the time, mesh is major cause of error in solving the analysis, so make sure the mesh is created without error. Easiest way in FreeCAD to create a fine mesh is by using the **FEM mesh from shape by Netgen** tool if you are not familiar with advanced mesh terms.

Figure-90. Input file writing process overview

- Click on the **Run CalculiX** button from the dialog. The resulting mesh will be generated if solver successfully runs. You may not get results if model is inconsistent, parameters specified are not feasible, or all the parameters are not defined.
- After running the analysis, click on the **Close** button to exit the dialog. The **Result mesh** feature will be added in the **Model Tree**; refer to Figure-91.

Figure-91. Result mesh feature generated

- To view results of analysis, select the CalculiX static results (or CCX results in some cases) feature from the **Model Tree** and click on the **Show Results** option from the **Results** menu; refer to Figure-92. The **Show result** dialog will be displayed result mesh preview; refer to Figure-93.

Figure-92. Show result option

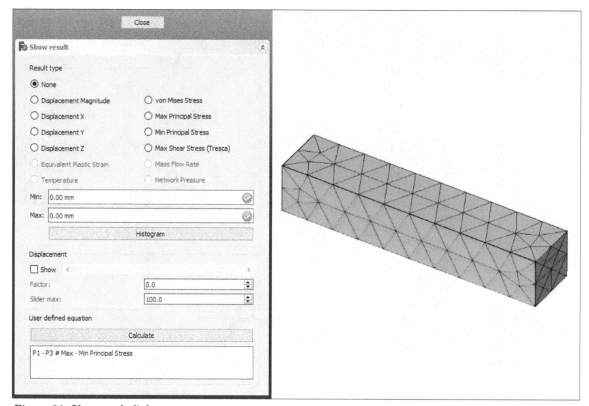

Figure-93. Show result dialog

- Select desired radio button from the **Result type** area of the dialog to check respective result.
- Select the **Show** check box from **Displacement** area of dialog and set desired magnification factor using slider to check displacement cause due to load.
- You can check the Minimum and Maximum value of selected result type from the **Min** and **Max** edit boxes in the dialog.
- Click on the **Close** button from the top in the dialog to exit.

Calculix New Framework Solver

CalculiX New Framework solver is based on standard solver with some enhancements under development. The procedure to use this solver is similar to standard solver.

Elmer Solver

Elmer is an open source multi-physics simulation software mainly developed by CSC - IT Center for Science (CSC). Elmer development was started 1995 in collaboration with Finnish Universities, research institutes, and industry. After it's open source publication in 2005, the use and development of Elmer has become international.

Elmer includes physical models of fluid dynamics, structural mechanics, electromagnetism, heat transfer, and acoustics. These are described by partial differential equations which Elmer solves by the Finite Element Method (FEM). **Note that in FreeCAD, you can use the Elmer Solver with mesh generated by using Gmsh Mesher only.** Before using this solver, you need to install the Elmer solver with its binary files which will be linked to FreeCAD GUI. Elmer requires two components to be interfaced with FreeCAD:

- ElmerGrid is the interface handling meshes.
- ElmerSolver is handling the computation.

The procedure to install this solver is given next.

- Go to the CSC binaries resources for Elmer: `https://www.nic.funet.fi/pub/sci/physics/elmer/bin/` OR `https://www.csc.fi/web/elmer/binaries`
- Download and install the version best suited to your Operating System (Windows 64 bits or Linux); refer to Figure-94.

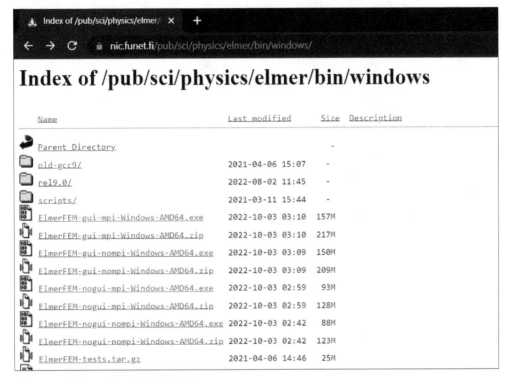

Figure-94. Elmer binaries for Windows

- After installing the Elmer solver, go to **Edit → Preferences → FEM → Elmer** in menu and link the correct path for both ElmerGrid and ElmerSolver; refer to Figure-95.

Figure-95. Linking Elmer with FreeCAD GUI

Now, we are ready to use the solver for performing analysis. The procedure to use this solver is given next.

- Create the analysis setup including mesh created by Gmsh, thermal/ mechanical/electrostatic/fluid loads applied on model, and material applied to model which has properties related to analysis; refer to Figure-96. After that, click on the **Solver Elmer** tool from the **Solve** menu. The **SolverElmer** feature will be added in the **Model Tree**.

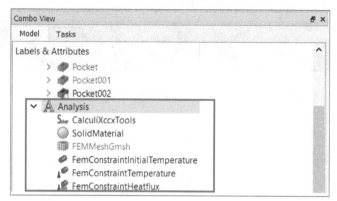

Figure-96. Features created before using Elmer solver for thermal analysis

For using Elmer solver, you need to create equations related to the analysis. There are six types of equations available in FreeCAD for different type of analyses; refer to Figure-97.

Figure-97. Equations available for Elmer solver

- Select the **Elasticity equation** option from **Mechanical equations** cascading menu of the **Solve** menu to perform linear mechanical analyses. Select the **Deformation equation** option to perform nonlinear mechanical analyses (deformations).
- Select the **Electrostatic equation** option from **Electromagnetic equations** cascading menu to add electrostatic analysis equation for finding force between two charged static objects. Select the **Electricforce equation** option to add equation for finding electric force between two moving charged objects/ particles. Select the **Magnetodynamic equation** option to create a FEM Equation for magnetodynamic forces. Select the **Magnetodynamic2D equation** option to create a FEM equation for 2D magnetodynamic forces.
- Select the **Flow equation** option to add fluid dynamics analysis equation.
- Select the **Flux equation** option to add equation for finding out flux of any selected scalar field variable.
- Select the **Heat equation** option to add thermal analysis equation. In our case, we are using **Heat equation** option. On selecting this option, **Heat** feature will be added below **Elmer** solver in **Model Tree**.
- Select the **Solver job control** option to open the menu to adjust and start the selected solver.

- Select the **Run Solver Calculations** option to run the selected solver of the active analysis.
- Select the feature from **Model Tree**. Properties of the equation will be displayed in the **Property Editor** dialog; refer to Figure-98.

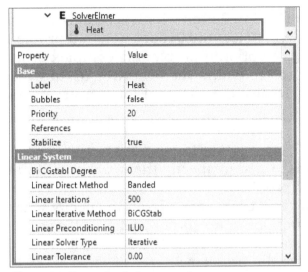

Figure-98. Properties for heat equation

- Specify desired parameters in the **Property Editor** dialog. Various parameters for equation are given next:

--Base Section--

Label : Used to specify name of the equation.

Bubbles : Used to stabilize solution of equation depending on compressibility of model. If you are working on compressible fluids then you should set this value to True.

Priority : Used to define priority level of current equation with respect to other equations to be solved by Elmer solver. A higher value of priority means respective equation will be solved first.

References : Used to change the linked object.

Stabilize : Used to specify whether Stabilized finite element method will be used or RFB (Residual Free Bubble) stabilization will be used.

--Linear System--

BiCGstabl Degree : In numerical linear algebra, the biconjugate gradient stabilized method, often abbreviated as BiCGSTAB, is an iterative method developed by H. A. van der Vorst for the numerical solution of nonsymmetric linear systems. Used to solve iterative linear equations for 3D analysis cases in FreeCAD. Set desired value to define stabilization order for the solver.

Linear Direct Method : Select desired type of linear direct method for solving linear FEM equations. For stress analysis, a direct solver is used instead of an iterative solver. It is often difficult for the iterative solver to find a solution for a structure that contains parts with varying stiffness properties. Use the **Banded** option when matrix to be formed for equation is dense (In simple words, a small region is to be analyzed with minimum void areas). Use the **umfpack** option when problem is complex with load free matrices in equation. If you do not know what to select then go with **umfpack**, although it will take more resources but it will do the trick.

Linear Iterative Method : Select desired option from the drop-down to define which iterative method is to be used for solving linear equations if the problem becomes complex or the matrix of equation is large. Each iterative method has its pros and cons which you can browse on Internet.

Linear Preconditioning : Linear Preconditioning is used to form a matrix easily solvable by selected iteration method. Preconditioning is typically related to reducing a condition number of the problem. The options available in FreeCAD try to form Incomplete Lower unitriangular and Upper triangular (ILU) matrices for preconditioning.

Linear Solver Type : Used to define which type of solver will be used for solving current analysis problem. Select the **Direct** option if problem is small scale linear and select the **Iterative** option if problem is complex or large.

Linear Tolerance : Used to define the accuracy up to which solution is needed. The value specified here also acts as goal for convergence if true solution is not available.

--Nonlinear System--

Nonlinear Iterations : Used to specify the number of iterations to be performed for finding solution of problem.

Nonlinear Newton After Iterations : Used to specify when Newton method will be used for solving equations in terms of iterations. If a certain number of specified iterations have been solved then their solutions will be used as input for Newton method to further solve the problem.

Nonlinear Newton After Tolerance : Used to specify when Newton method will be introduced for solving equation in terms of tolerance. If specified tolerance in solution has been achieved then Newton method will further solve the equation for convergence.

Nonlinear Tolerance : Used to specify the tolerance allowed in final solution of problem.

Relaxation Factor : The relaxation factor is used to speed up convergence of iteration. You can assume relaxation factor as a value which removes the amount of error in solution of each iteration due to numerous unknown real conditions.

--Steady State--

Steady State Tolerance : Used to define the accuracy in results to be achieved after performing the analysis.

- After setting desired parameters in the **Property Editor** dialog, select the **SolverElmer** feature from the **Model Tree**. The options in **Property Editor** will be displayed as shown in Figure-99.

Figure-99. SolverElmer properties

- Specify the name and group for feature in the **Label** and **Group** fields, respectively.
- Specify desired values in the **Steady State Max Iterations** and **Steady State Min Iterations** fields to define maximum and minimum number of iterations to be solved when you are running coupled analysis (Coupled analysis involve multiple variables which are interdependent like temperature of body is function

of current in analysis and current value changes with temperature of body as well). Generally, default value is **1** for max and **0** for min when you are solving for one variable only. If your analysis is coupled then specify desired value for iterations up to which you will be able to achieve solution of coupled equations.

- After setting desired solver parameters, double-click on the **SolverElmer** feature from **Model Tree**. The **Solver Control** dialog will be displayed in **Tasks** panel; refer to Figure-100.

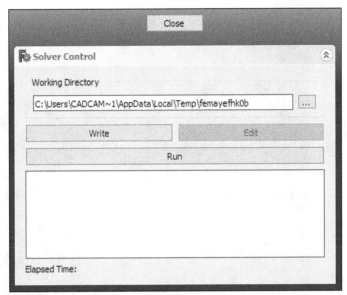

Figure-100. Solver Control dialog

- Click on the **Write** button from dialog to generate input file. Since, we are using external solver, so input file parameters will not be displayed in the dialog. If you want to check the input file then browse to the location of **Working Directory** edit box in the dialog and open the **case.sif** file in any Word processor program.
- Click on the **Run** button from the dialog after writing input file. If all the parameters specified are feasible then results will be generated with Output file; refer to Figure-101. You can also check the results of analysis by double-clicking on the **SolverElmerResult** node in **Model** tab of **Combo View** dialog.

Figure-101. Elmer solver results generated

Note that sometimes, you may not get the results displayed in FreeCAD GUI. In that case, reach to the result file (*.vtu) of analysis in temporary folder of Elmer solver analysis and open it in ElmerGUI application which we have installed earlier. ParaView is also a good post processor for analyzing results of analysis. You can download ParaView from https://www.paraview.org/download web page. Figure-102 shows the result of our current analysis in ParaView.

Figure-102. Result of Elmer analysis in ParaView

Similarly, you can use the other equations of Elmer Solver for performing related analyses.

Z88 Solver

The software was developed by Frank Rieg, a professor for engineering design and CAD at the University of Bayreuth. Originally written in FORTRAN 77, the program was ported to the programming language C in the early 1990s.

There are two programs for finite element analysis:

Z88OS is available as free software including the source code under the GNU General Public License. Due to the modular structure of the program and the open availability of the source code, it is possible to develop customized extensions, add-ons and several special case 2D and 3D continuum elements (e.g. anisotropic shell element).

Z88Aurora originally described the user interface of the Z88 finite element analysis program. After several additions and further development, it now comprises a significantly larger range of functionality than Z88OS. Z88Aurora is freeware, however the source code is not publicly available. Z88Aurora's current version contains several computation modules:

In the case of linear static analyses, it is assumed that the result is proportional to the applied forces.
Nonlinear analyses are used for nonlinear geometries and nonlinear materials.
Using thermal and thermomechanical analyses, it is possible to not only compute results about temperature or heat currents, but also thermomechanical displacements and stresses.

By utilizing natural frequency simulation, natural frequencies and the resulting oscillations can be determined.

A contact module makes it possible to simulate interacting parts and assemblies. An integrated part management tool enables an effective handling of assemblies. There are options to simulate a glued connection or a friction-free connection and the contact discretization (type of contact: node-surface-contact, or surface-surface-contact), the mathematical imposition method (lagrange method, perturbed lagrange method, or penalty method), and the direction of contact stiffness (normal or tangential direction) can be changed via the contact settings. This module only supports tetrahedrons and hexahedrons with linear or quadratic shape functions. Additionally, the module is only available for linear mechanical strength analyses.

The procedure to use Z88 solver in FreeCAD is similar to using CalculiX solver which has been discussed earlier.

Mystran Solver

Mystran is an open source general purpose solver used to perform linear finite element analysis. The primary author of this solver Dr. Bill Case has been working on development of this solver for approximately 30 years till he died in 2021. In a linear analysis, displacement, forces, and stresses are directly proportional to applied load. Mystran is acronym of **My Structural Analysis** which signifies it can solve wide variety of FEA problems. Note that most of the structural analyses modeled for Nastran can be solved in Mystran. There is a large community of developers working on development of this solver. You can know more about the solver from website https://www.mystran.com/.

POST PROCESSING IN FREECAD FEM

The tools in **Results** menu of **FEM** workbench are used to perform post processing, so that you can analyze the results. The tools of this menu are active only when you performed the analysis and result of solver is selected in the **Model Tree**; refer to Figure-103. Various tools of this menu are discussed next.

Figure-103. Results menu

Purge Results

The **Purge results** tool is used to delete the results generated by using **Show result** tool. Note that this tool is active only after you have used the **Show result** tool.

Show Result

The **Show result** tool from the **Results** menu to generate analysis results. The procedure to use this tool has been discussed earlier.

Creating Post Processing Pipeline for Results

The **Post pipeline from result** tool is used to create post processing pipelines from the results. The procedure to use this tool is given next.

- Click on the **Post pipeline from result** tool from the **Results** menu after selecting solver's result feature from the **Model Tree**. The **Pipeline** feature will be added in the **Model Tree**.
- Double-click on this feature to modify its parameters. The **Result display options** dialog will be displayed; refer to Figure-104.

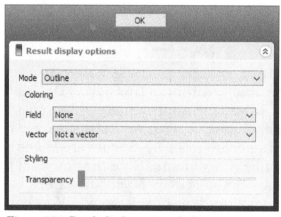

Figure-104. Result display options dialog

- Select desired option from the **Mode** drop-down to define what type of pipeline plot you want to generate using the result parameters. Select the **Outline** option to generate only border lines of the model. Select the **Nodes** option to generate colored nodes for analyzing results. Select the **Surface** option to generate result surface with multicolored annotations. Select the **Surface with Edges** option to generate result surface with wireframe edges. Select the **Wireframe** option to generate wireframe model of results. Select the **Wireframe (surface only)** option to generate wireframe model over the surface of model for analyzing results.
- Select desired option from the **Field** drop-down to define result parameter to be displayed over the pipeline plot. Preview of result will be displayed in the modeling area; refer to Figure-105.

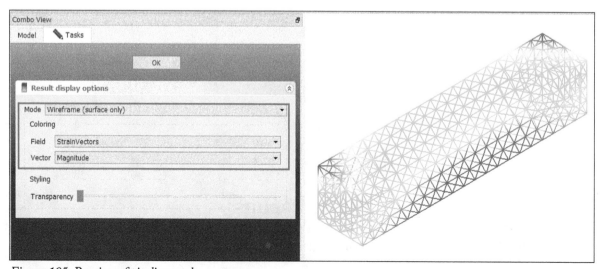

Figure-105. Preview of pipeline result

- Select desired option from the **Vector** drop-down to define which component of vector is to be used for generating result model.
- Click on the **OK** button from the dialog to generate pipeline result.

Applying Warp Filter

The **Warp filter** tool is used to distort the result model, so that higher values are further away from base model as compared to lower result values. The procedure to use this tool is given next.

- Click on the **Warp filter** tool from the **Results** menu after selecting **Pipeline** feature from the **Model Tree**. The **Warp options** and **Result display options** dialogs will be displayed; refer to Figure-106.

Figure-106. Warp options and Result display options dialogs

- Select desired option from the **Vector** drop-down to specify which result vector is to be displayed in warped model.
- Set the other parameters as discussed earlier in the **Result display options** dialog.
- Use the warp slider and warp related parameters in the **Warp options** dialog to check warped model of results; refer to Figure-107.

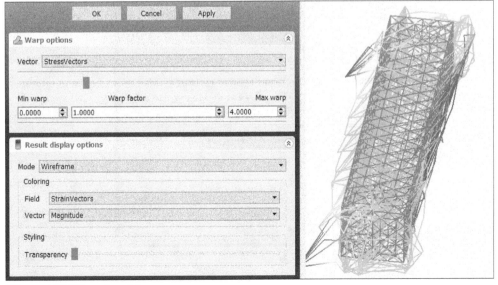

Figure-107. Warped model

- After setting desired parameters, click on the **OK** button from the dialog.

Applying Scalar Clip Filter

The **Scalar clip filter** tool is used to check internal section of model for result parameters. The procedure to use this tool is given next.

- Click on the **Scalar clip filter** tool from the **Results** menu after selecting **Pipeline** feature from **Model Tree**. The **Scalar clip options** and **Result display options** dialogs will be displayed.
- Set the parameters as discussed earlier in the dialogs and use the clipping slider for checking internal regions of model for analysis results; refer to Figure-108.

Figure-108. Pipeline results with scalar clipping

- Click on the **OK** button from the dialog to create the result model and exit dialog.

Applying Function Cut Filter

The **Function cut filter** tool is used to check the pipeline results of analysis along a plane or over the surface of a sphere. The procedure to use this tool is given next.

- Click on the **Function cut filter** tool from the **Results** menu after selecting the **Pipeline** feature from **Model Tree**. The **Function cut, choose implicit function** dialog will be displayed along with **Result display options** dialog; refer to Figure-109.

Figure-109. Function cut dialog

- Click on the **Create** button from the dialog and select desired option to define scope of result output geometrically. Select the **Plane** option if you want to display result on a plane intersecting with the model. Select the **Sphere** option if you want to check results on surface of a sphere intersecting with the model. We have selected **Plane** option in our case for tutorial.
- On selecting the **Plane** option, parameters related to location of plane will be displayed in the dialog; refer to Figure-110.

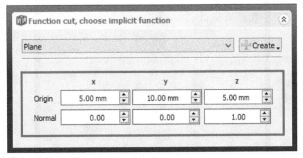

Figure-110. Parameters for plane option

- Set desired location and orientation for plane using the **Origin** and **Normal** edit boxes.
- Set desired options in the **Result display options** dialog as discussed earlier. Preview of the plot will be displayed; refer to Figure-111.

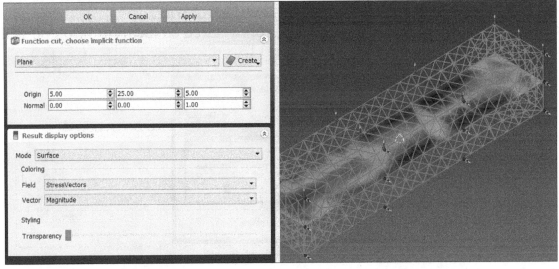

Figure-111. Preview of plot

- Click on the **OK** button from the dialog to create the plot and exit the dialog.

Region Clip Filter

The **Region Clip Filter** tool is used to clip a field using a sphere or a plane cutting through the model. The procedure to use this tool is given next.

- Click on the **Region clip filter** tool from the **Results** menu after selecting the **Pipeline** feature from **Model Tree**. The **Clip region, choose implicit function** dialog will be displayed along with **Result display options** dialog; refer to Figure-112.

Figure-112. Clip region dialog

- Click on the **Create** button from the dialog and select desired option to define scope of result output geometrically. Select the **Plane** option if you want to display result on a plane intersecting with the model. Select the **Sphere** option if you want to check results on surface of a sphere intersecting with the model. We have selected **Plane** option in our case for tutorial.
- On selecting the **Plane** option, parameters related to location of plane will be displayed in the dialog; refer to Figure-113.

Figure-113. Parameters for plane option

- Set desired location and orientation for plane using the **Origin** and **Normal** edit boxes.
- Select **Inside Out** check box to invert the cut.
- Select **Cut Cells** check box to smoothen the clipped region by removing parts of finite elements that are sticking out.
- Set desired options in the **Result display options** dialog as discussed earlier. Preview of the plot will be displayed; refer to Figure-114.

Figure-114. Preview of clip region

- Click on the **OK** button from the dialog to create the plot and exit the dialog.

Contours filter

The **Contours filter** tool is used to create iso-contours and iso-lines in the results mesh. The procedure to use this tool is discussed next.

- Click on the **Contours filter** tool from the **Results** menu after selecting the Pipeline feature. The **Contours filter options** dialog will be displayed; refer to Figure-115.

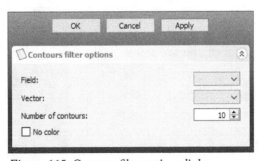

Figure-115. Contours filter options dialog

- Select desired results field to be drawn from **Field** drop-down.
- Select desired vector component to be used from **Vector** drop-down.
- Specify desired number of contours to be created in the **Number of contours** edit box.
- Select the **No color** check box to not color the contour lines.
- After specifying desired parameters, click on the **OK** button from the dialog.

Line Clip Filter

The **Line clip filter** tool is used to generate line plot of desired result parameter. The procedure to use this tool is given next.

- Click on the **Line clip filter** tool from the **Results** menu after selecting the **Pipeline** feature. The **Data along a line options** dialog will be displayed; refer to Figure-116.

Figure-116. Data along a line options dialog

- Click on the **Select Points** button from the dialog and one by one select two points from model to define result output limits.
- Set the parameters as discussed earlier in **Resolution**, **Mode**, **Field**, and **Vector** drop-downs.
- Click on the **Create Plot** button to generate the plot. The plot will be created and preview of plot will be displayed in the model; refer to Figure-117.

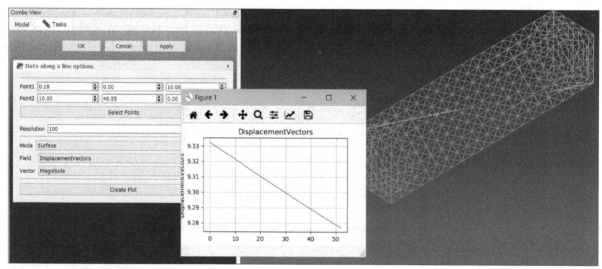

Figure-117. Line plot generated

- Using the **Zoom** button in the **Figure 1** dialog box, you can zoom into a specific region of plot by drawing a box.
- Using the **Pan** button you can move left/right/up/down in the plot to check values.
- The **Back** and **Forward** buttons are used to switch between previous and next views of plot.
- Click on the **Configure subplots** button to modify various spacings in the plot. On clicking this button, the **FreeCAD** dialog box will be displayed as shown in Figure-118. Set desired values and click on the **Close** button.
- Click on the **Edit axes, curve, and image parameters** button from the dialog box. The **Figure options** dialog will be displayed; refer to Figure-119.

Figure-118. FreeCAD dialog box

Figure-119. Figure options dialog box

- Set desired parameters in **Axes** and **Curves** tab of the dialog box and click on the **OK** button to exit the dialog box.
- Click on the **Save** button from the **Figure 1** dialog to save current plot as a png file.
- Close the dialog after checking plot and click on the **OK** button from the **Data along a line** dialog to exit the tool.

Similarly, you can use other post processing tools in **Results** menu. Sometimes, you may get errors in postprocessing due to some programming needs of software. In such cases, your other post processor ParaView is available to help.

PRACTICAL

Perform finite element analysis of the given model; refer to Figure-120.

Figure-120. Model for finite element analysis

Starting a New File

- Start FreeCAD if not started yet.
- Select the **FEM** option from the **Switch between workbenches** drop-down in the **Toolbar**. The path environment will be displayed.

Creating Analysis Container

- Click on the **Analysis Container** tool from the **Toolbar**. The Analysis container will be created in the **Model Tree**; refer to Figure-121.

Defining the Material

- Click on the **Material For Solid** tool from the **Toolbar**. The **FEM Material** dialog will be displayed.
- Select the **Calculix - Steel** material from the **Material card** drop-down and specify the parameters in the dialog as shown in Figure-122.

Figure-121. Analysis container created

Figure-122. Material defined

- Scroll down in dialog, select the **Solid** radio button and click on the **Add** button. You will be asked to select face of the body.
- Select the model to apply material and click on the **OK** button. The material will be defined.

Creating FEM Mesh from shape

- Select the body and click on the **FEM mesh from shape by Gmsh** tool from the **Toolbar**. The **FEM Mesh by Gmsh** dialog will be displayed.
- Click on the **Apply** button from the dialog. The FEM mesh will be created; refer to Figure-123.

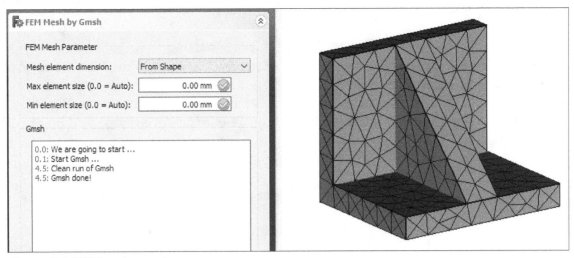

Figure-123. FEM mesh created

- Click on the **OK** button from the dialog. The FEM mesh will be displayed in the **Model Tree**.

Defining Fixed Constraint

- Click on the **Constraint fixed** tool from **Toolbar**. The **FEM constraint parameters** dialog will be displayed.
- Select the face of model to apply the fixed constraint as shown in Figure-124, and click on **Add** button from the dialog. The fixed constraint will be applied; refer to Figure-125.

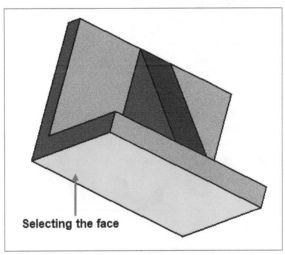

Figure-124. Selecting the face to apply fixed constraint

Figure-125. Fixed constraint applied

- Click on the **OK** button to close the dialog.

Defining Force Constraint

- Click on the **Constraint force** tool from the **Toolbar**. The **FEM constraint parameters** dialog will be displayed.
- Select the face of model to apply the force constraint as shown in Figure-126, and click on the **Add** button from the dialog. The force constraint will be applied; refer to Figure-127.

Figure-126. Selecting the face to apply force constraint

Figure-127. Force constraint applied

- Specify the load value as **100** in **Load** edit box and click on the **OK** button to close the dialog.

Running the Solver

- Select the solver from the **Model Tree** to run the calculations; refer to Figure-128, and click on the **Run solver calculations** tool from the **Toolbar** or **Solve** menu. The result will be displayed; refer to Figure-129.

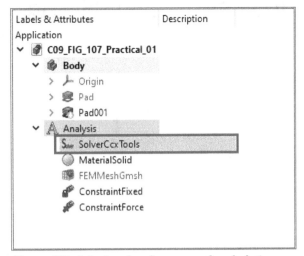

Figure-128. Selecting the solver to run the calculation

Figure-129. Results displayed

Tip: While applying constraints, you may get error like Face already selected when selecting a face to apply load/constraint. In such cases, save the file, close it and then reopen it to apply load/constraint. This can be software programming fault for retaining objects in cache memory.

FOR STUDENT NOTES

Chapter 10

Tech Drawing

Topics Covered

The major topics covered in this chapter are:

- *Introduction to TechDraw Workbench*
- *Starting New Drawing Page*
- *Inserting Views*
- *Creating Section Views and Detail Views*
- *Applying Annotations*
- *Clip Groups*
- *Linking Dimensions of 3D Geometry*
- *Exporting Drawing Pages*

Note: **Image** workbench has been removed in FreeCAD 0.21 and later versions as the functionality is now available in **Import** tool of **File** menu.

INTRODUCTION TO TECHDRAW WORKBENCH

The **TechDraw** workbench of FreeCAD is used to create technical drawings from the model earlier created in the software using other workbenches. To start creating technical drawings, select the **TechDraw** option from the **Switch between workbenches** drop-down in the **Toolbar**; refer to Figure-1. On selecting the TechDraw workbench, tools will be displayed similar to Figure-2.

Figure-1. TechDraw workbench option

Figure-2. TechDraw toolbar

Various tools of this workbench are discussed next.

STARTING A NEW DRAWING PAGE

The **Insert Default Page** tool is used to generate a drawing page using default drawing template. The procedure to use this tool is given next.

- Click on the **Insert Default Page** tool from the **Toolbar** in **TechDraw** workbench; refer to Figure-3. A new drawing page will be created and displayed in the interface; refer to Figure-4.

Figure-3. Insert Default Page tool

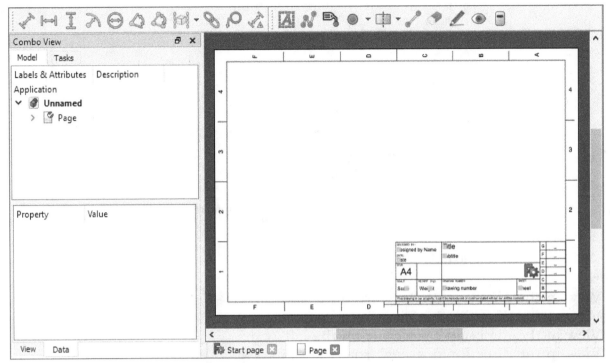

Figure-4. Drawing page generated

- Select the **Page** feature from the **Model Tree**. The properties of page will be displayed as shown in Figure-5.

Figure-5. Properties for page feature

- Select desired option from the **Projection Type** field in **Property Editor** dialog. Select the **First Angle** option from the field if you want to use First Angle projection for generating views. Select the **Third Angle** option from the field if you want to use Third Angle projection for generating views.

First Angle Projection and Third Angle Projection
Figure-6 shows an object with different view directions say, a, b, c, d, e, and f.

Figure-6. Object with view directions

Here,
1. View in the direction a = view from the front
2. View in the direction b = view from top
3. View in the direction c = view from the left
4. View in the direction d = view from the right
5. View in the direction e = view from bottom
6. View in the direction f = view from the back

In First Angle projection, these views are arranged as shown in Figure-7. In Third Angle projection, these views are arranged as shown in Figure-8.

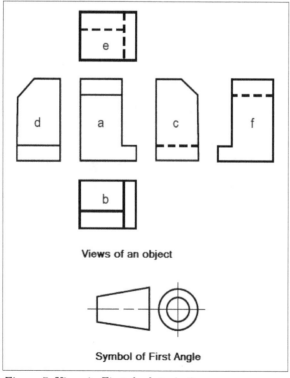

Figure-7. Views in First Angle projection

Figure-8. Views in Third Angle projection

• Specify desired name of page in the **Label** field of **Property Editor**.

- If you want to change template of drawing page and click on the **Template** link button from **Template** field of **Property Editor**. The **Template** feature will be selected in **Property Editor** and related options will be displayed; refer to Figure-9.

Figure-9. Properties of template

- Click on the **Browse** button in **Template** field of **Property Editor** and select desired template file.
- You can specify the scale of drawing views in the **Scale** field of **Property Editor** for **Page**.

INSERTING NEW PAGE OF SELECTED TEMPLATE

The **Insert Page using Template** tool is used to create new page using selected template. A template includes various features like size of page, orientation of page, scale of views, and so on. The procedure to use this tool is given next.

- Click on the **Insert Page using Template** tool from the **Toolbar** or **TechDraw** menu of **TechDraw** workbench; refer to Figure-10. The **Select a Template File** dialog box will be displayed; refer to Figure-11.

Figure-10. Insert Page using Template tool

Figure-11. Select a Template File dialog box

- Select desired template from the dialog box and click on the **Open** button. The page will be created and displayed in the application; refer to Figure-12.

Figure-12. Page created by template

REDRAW PAGE

The **Redraw Page** tool forces an update of the selected page. The procedure to use this tool is discussed next.

- Select desired page from the **Model Tree** which you want to update and click on the **Redraw Page** tool from the **Toolbar** or **TechDraw** menu in the **TechDraw** workbench; refer to Figure-13. The selected page will be updated.

Figure-13. Redraw page tool

Print All Pages

The **Print All Pages** tool is used to send all pages in a document to the printer. The procedure to use this tool is discussed next.

- Click on the **Print All Pages** tool from the **Toolbar** or **TechDraw** menu in the **TechDraw** workbench. The **Print** dialog box will be displayed; refer to Figure-14.

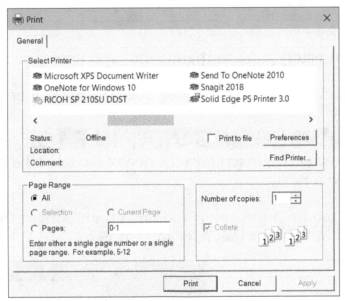

Figure-14. Print dialog box

- Specify desired parameters in the dialog box to print the page and click on the **Print** button. The **Save Print Output As** dialog box will be displayed; refer to Figure-15.

Figure-15. Save Print Output As dialog box

- Specify desired directory to save the file and specify desired name of file in the **File name** edit box.
- Click on the **Save** button from the dialog box. The file will be saved.

INSERTING VIEW IN PAGE

The **Insert View** tool is used to insert current view of model in the page. The procedure to use this tool is given next.

- Select the **Model** tab from the application menu and orient the model to desired view to be placed in the drawing; refer to Figure-16.

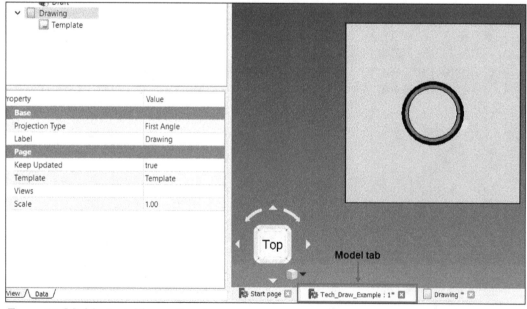

Figure-16. Model oriented for creating view

- Now, select the **Page** tab from the application window, select the model whose views are to be placed from **Model Tree**, and click on the **Insert View** tool from the **Toolbar** or **TechDraw** menu; refer to Figure-17. The current view of model will be placed in the page; refer to Figure-18. Note that the view generated has same orientation as defined in Figure-16.

Figure-17. Insert View tool

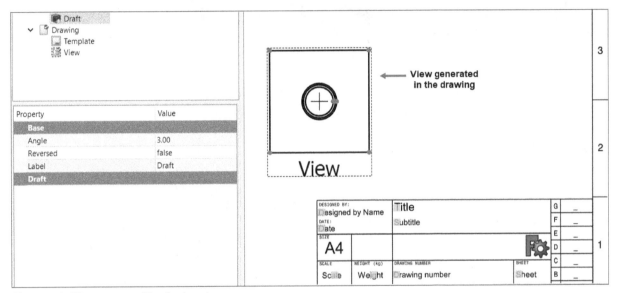

Figure-18. View generated in the drawing

- You can move this view by selecting the dotted boundary and dragging it to desired location.
- On selecting the view from **Model Tree**, various properties of the view are displayed in **Property Editor**; refer to Figure-19.

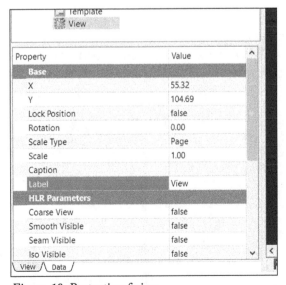

Figure-19. Properties of view

- Specify desired values in **X** and **Y** edit boxes to define location of view on page.

- Select **True** option for **Lock Position** field if you want to lock the position of view on page and do not want it to move.
- Specify desired value in **Rotation** field to rotate selected view and click on the **Recompute** button ⟳. The rotated view will be generated.
- Set desired (**True** or **False**) value for various display objects in **HLR Parameters** section of **Property Editor**.
- By default, orthographic views are generated from the model but if you want to generate perspective view of model then select the true option from **Perspective** field in the **Projection** section. You can also specify the Focus distance for perspective view in the **Focus** field.

INSERTING ACTIVE VIEW IN PAGE

The **Insert Active View (3D View)** tool inserts a copy of a 3D window into a drawing page. The procedure to use this tool is discussed next.

- Select the tab of model from the application menu and orient the model to desired view to be placed in the drawing; refer to Figure-20.

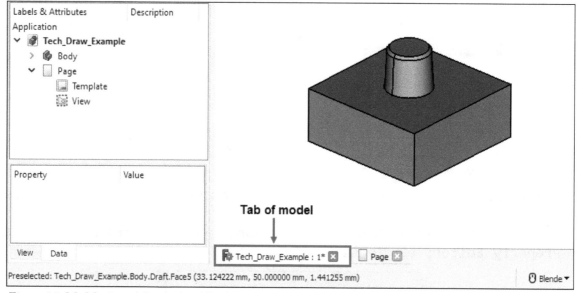

Figure-20. Model oriented for creating 3D view

- Click on the **Insert Active View (3D View)** tool from **Toolbar** or **TechDraw** menu in the **TechDraw** workbench; refer to Figure-21. The **ActiveView to TD View** dialog will be displayed in the **Tasks** panel of **Combo View**; refer to Figure-22.

Figure-21. Insert active view 3D view tool

Figure-22. ActiveView to TD View dialog

- Specify desired width and height of generated view in the **Width** and **Height** edit boxes of the dialog.
- Enter desired value in the **Border** edit box to specify minimal distance of the object from the top and left view border.
- Select the **Background** check box and click on the **Background color** button to specify the color for the background.
- Enter desired value in **Line Width** edit box to specify desired thickness of individual lines in the generated view.
- Select desired drawing style from **Render Mode** drop-down.
- After specifying desired parameters in the dialog, click on the **OK** button. The active 3D view will be created and displayed in the **Page** tab; refer to Figure-23.

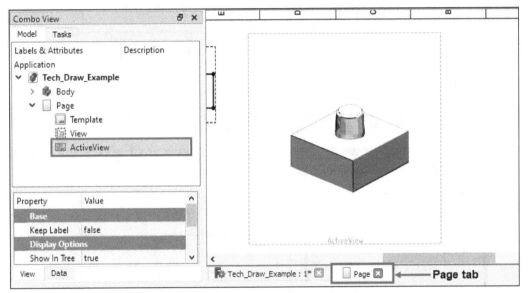

Figure-23. Active 3D view created

INSERTING PROJECTION GROUP

The **Insert projection group** tool is used to create multiple views (like front, left, and top views) of selected model object. The procedure to use this tool is given next.

- Click on the **Insert Projection Group** tool from the **Toolbar** or from the **TechDraw** menu after selecting the model from **Model Tree**; refer to Figure-24. The **Projection Group** dialog will be displayed with preview of base view; refer to Figure-25.

Figure-24. Insert multiple linked views tool

Figure-25. Projection Group dialog

- Select check boxes for views to be generated. Preview of views will be displayed accordingly in the drawing area; refer to Figure-26.

Figure-26. Preview of views

- Select desired option from the **Projection** drop-down to specify whether you want to use First Angle projection or Third Angle projection.
- Select desired option from the **Scale** drop-down to define whether you want to use Automatic scaling, Scale parameter set for page, or you want to specify custom scale values in the **Custom Scale** edit boxes. The views will be generated as per the scale option selected.
- Select the **Auto Distribute** check box to distribute the projections automatically using the given X/Y Spacing.
- Enter desired value in **X Spacing** and **Y Spacing** edit box to specify horizontal distance and vertical distance between border of projections, respectively.
- After setting desired parameters, click on the **OK** button from the dialog to create the views.

INSERT SECTION VIEWS

The **Section View** tool is used to insert section view using selected view in the drawing area. The procedure to use this tool is given next.

- Click on the **Section View** tool from the **Toolbar** or **TechDraw** menu after selecting a view from the drawing area; refer to Figure-27. The **Create Section View** dialog will be displayed; refer to Figure-28.

Figure-27. Section View tool

Figure-28. Create Section View dialog

- Select desired button from the dialog to specify the direction in which the section view will be generated. Preview of the section view will be displayed if **Live Update** check box is selected in the **Preview** area of the dialog; refer to Figure-29.
- You can change the section line angle by specifying desired value in the **View Direction as Angle** edit box of the dialog.
- Use the edit boxes in **Section Plane Location** area of dialog to move section line in desired direction.
- Click on the **OK** button from the dialog to create the section view.

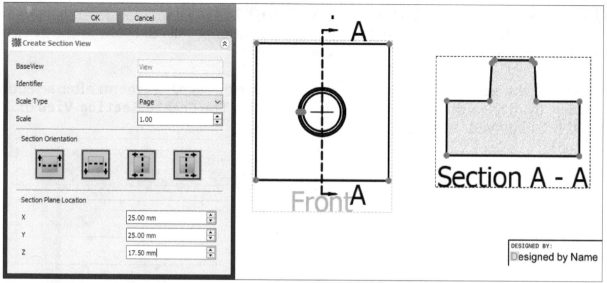

Figure-29. Preview of section view

Insert Complex Section View

The **Complex Section** tool is used to insert a complex cross section view based on selected open profile object (line/arc/multiline created in Draft workbench) and view. The procedure to use this tool is given next.

*   Select a profile object passing through model created in draft workbench on the model and view from which cross-section view is to be generated; refer to Figure-30.

Figure-30. Selecting objects for complex cross-section

*   Click on the **Complex Section** tool from the **Toolbar** or **TechDraw** menu after selecting a view from the drawing area; refer to Figure-31. The **New Complex Section** dialog will be displayed; refer to Figure-32

Figure-31. Complex Section tool

Figure-32. New complex section dialog box

- Click on the **OK** button from the dialog without modifying any parameters. The complex cross-section view will be generated; refer to Figure-33.

Figure-33. Complex section view generated

INSERTING DETAIL VIEW

The **Insert Detail View** tool is used to create a magnified view of the model for defined region. The procedure to use this tool is given next.

- Click on the **Insert Detail View** tool from the **Toolbar** or **TechDraw** menu after selecting a view; refer to Figure-34. The **New Detail View** dialog will be displayed in the **Tasks** panel of **Combo View** along with the preview of detail view; refer to Figure-35.

Figure-34. Insert Detail View tool

Figure-35. New Detail View dialog with detail view

- The **Base View** box displays the view on which the detail view is based.
- The **Detail View** box displays the view which is being created.
- Click on the **Drag Highlight** button to highlight the detail origin border bold and with the label drag.
- Enter desired value in **X** and **Y** edit boxes to specify the x and y position of detail highlight within view, respectively.
- Specify desired size of detail view in the **Radius** edit box.
- Select desired type of scale from **Scale Type** drop-down. Select **Page** option from the drop-down to use the scale factor of drawing page. On selecting the **Automatic** option, in case, the detail view would be larger than the page, it will be scaled down to fit into the page. Select **Custom** option from the drop-down to set the custom scale factor value in the **Scale Factor** edit box.
- Enter desired value in **Reference** edit box to specify an identifier to indicate the area of the base view that is displayed.
- After specifying desired parameters, click on the **OK** button from the dialog to create the detail view.
- If you want to edit the properties of detail view then select the detail view created from the **Model Tree**, the **Property Editor** dialog will be displayed in the **Model** panel of **Combo View**; refer to Figure-36.

Figure-36. Property editor dialog with detail view parameters

- Specify desired value of scale factor in the **Scale** field of **Property Editor** to increase or decrease the size of detail view.
- Expand the **Anchor Point** field and specify desired location parameters in **x**, **y**, and **z** fields of **Property Editor** to specify location whose detail view will be generated with respect to center of selected main view; refer to Figure-37.

Figure-37. Setting location of detail view circle

- Set the other parameters of **Property Editor** as discussed earlier and click in the empty area of drawing to exit.

INSERTING DRAFT WORKBENCH OBJECT

The **Insert Draft Workbench Object** tool is used to create a view of objects in Draft workbench. The procedure to use this tool is given next.

- Click on the **Insert Draft Workbench Object** tool from the **Toolbar** or **TechDraw** menu after selecting the **Draft** workbench objects to be inserted in the view; refer to Figure-38. The draft views will be generated for selected objects; refer to Figure-39.

Figure-38. Insert Draft Workbench Object tool

Figure-39. Draft views generated

- Set the parameters as discussed earlier in the **Property Editor** section of the **Combo View**.

Insert Arch Workbench Object

The **Insert Arch Workbench Object** tool is used to insert a section view of model created in Arch workbench.

- Select the section view from the architectural model; refer to Figure-40 and click on the **Insert Arch Workbench Object** tool from the **Toolbar** or **TechDraw** menu; refer to Figure-41. The architectural section view will be generated in the drawing; refer to Figure-42.

Figure-40. Section view selected

Figure-41. Insert arch workbench object

Figure-42. Arch view generated

Insert Spreadsheet View

The **Insert Spreadsheet view** tool is used to place view of selected spreadsheet in the drawing area. The procedure to use this tool is given next.

- Make sure you have already created a spreadsheet to be inserted in the drawing view using the Spreadsheet workbench; refer to Figure-43. Select the spreadsheet to be inserted in the drawing and click on the **Insert Spreadsheet View** tool from the **Toolbar** or **TechDraw** menu in the **TechDraw** workbench; refer to Figure-44. The sheet view will be placed in the drawing; refer to Figure-45.

Figure-43. Spreadsheet created

Figure-44. Insert spreadsheet view

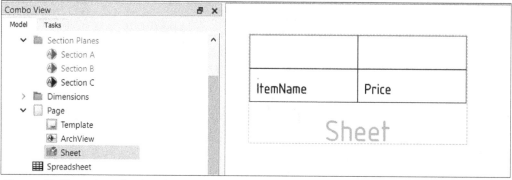

Figure-45. Sheet view placed in drawing

- Select the **Sheet** view from **Combo View** and modify the **Cell Start** & **Cell End** parameters in **Property Editor** to include desired cells in the sheet view; refer to Figure-46.

Figure-46. Sheet view after modifying Cell End parameter

MOVING THE VIEW

The **Move View** tool moves a view and all its dependents (Balloons, Dimension, etc.) to a different page. The procedure to use tool is discussed next.

- Click on the **Move View** tool from the **Toolbar** or **Techdraw** menu; refer to Figure-47. The **Move View to a different Page** dialog will be displayed; refer to Figure-48.

Figure-47. Move View tool

Figure-48. Move View to a different Page dialog

- Click on the button [...] next to **View to move** selection box. The **Select View** dialog box will be displayed; refer to Figure-49.

Figure-49. Select View dialog box

- Select desired view from the list box which you want to move to another page; refer to Figure-49 and click on the **OK** button. The selected view will be displayed in the **View to move** selection box of the dialog.
- Click on the [...] button next to **From Page** selection box. The **Select Page** dialog box will be displayed; refer to Figure-50.

Figure-50. Select Page dialog box

- Select desired page from the list box from which you want to move the view; refer to Figure-50 and click on the **OK** button. The selected page will be displayed in the **From Page** selection box of the dialog.
- Click on the button [...] next to **To Page** selection box. The **Select Page** dialog box will be displayed; refer to Figure-51.

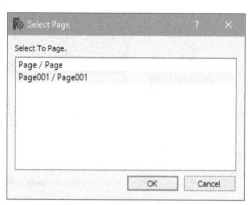

Figure-51. Select Page dialog box

- Select desired page from the list box to which you want to move the view; refer to Figure-51 and click on the **OK** button. The selected page will be displayed in the **To Page** selection box of the dialog.
- After specifying the parameters, click on the **OK** button from the dialog. The selected view will be move to the another page; refer to Figure-52.

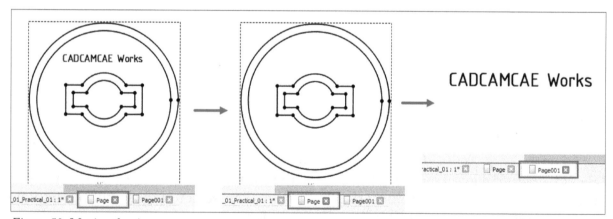

Figure-52. Moving the view

SHARING THE VIEW

The **Share View** tool makes a view and all its dependents (Balloons, Dimensions, etc.) visible on a second page. The procedure to use this tool is same as discussed for **Move View** tool.

PROJECTING THE SHAPE

The **Project Shape** tool creates projections of shapes. The projections are created in the 3D view, and not on a page. The procedure to use this tool is discussed next.

- Select desired feature from **Model Tree** or from the 3D view area which you want to project; refer to Figure-53.

Figure-53. Selecting the feature

- Click on the **Project Shape** tool from the **Toolbar** or **TechDraw** menu; refer to Figure-54. The **Project shapes** dialog will be displayed; refer to Figure-55.

Figure-54. Project Shape tool

Figure-55. Project shapes dialog

- Select desired check boxes in the dialog and click on the **OK** button. The selected feature will be projected; refer to Figure-56.

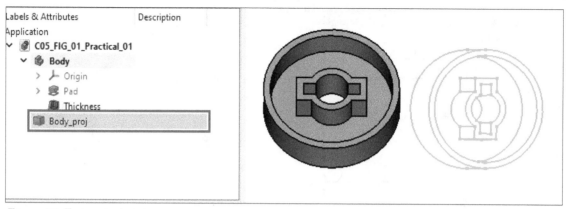

Figure-56. Feature projected

ADJUST STACKING ORDER OF VIEW
Stack Top View

The **Stack Top** tool is used to move view to the top of the stacking order. The stacking order controls the apparent depth of views on a page. For example if two views are overlapping then you can move selected view to top of another view.

- Select the view from model which you want to stack at top of other views and click on the **Stack Top** tool from the **Toolbar** or **TechDraw** menu; refer to Figure-57. The view will come at top; refer to Figure-58.

Figure-57. Stack top view

Figure-58. Stacking view at top

Stack Bottom View

The **Stack Bottom** tool is used to moves views to the bottom in the stacking order.

Stack UP View

The **Stack Up** tool is used to move view up one level in the stacking order.

Stack Down View

The **Stack Down** tool is used to move views down one level in the stacking order.

INSERTING ANNOTATIONS

The **Insert Annotation** tool is used to insert desired text in the drawing view. The procedure to use this tool is given next.

- Click on the **Insert Annotation** tool from the **Toolbar** or **TechDraw** menu; refer to Figure-59. The annotation will be placed in drawing area and properties of annotation will be displayed in the **Property Editor**; refer to Figure-60.

Figure-59. Insert Annotation tool

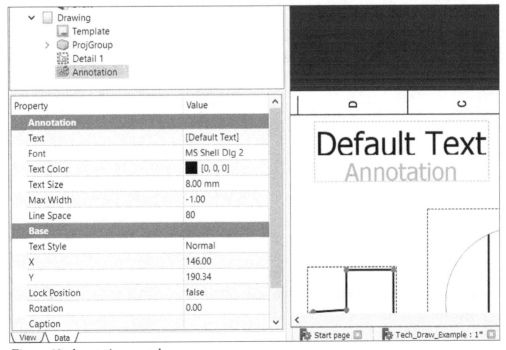

Figure-60. Annotation created

- Click in the **Text** field and then click on the **Browse** button. The **List** dialog box will be displayed; refer to Figure-61.

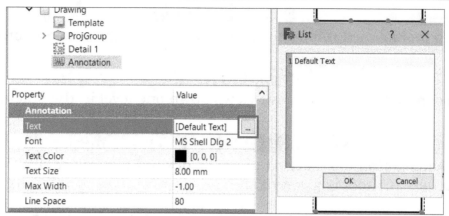

Figure-61. List dialog box

- Specify desired text in the edit box and click on the **OK** button from the dialog box. The annotation will be modified accordingly.
- Specify other parameters like font, text color, text size, and so on in the **Property Editor**.

ADDING LEADER LINE TO VIEW

The **Add Leaderline to View** tool adds a line to a view. Other annotation objects can be connected to the leaderline to form complex annotations. The procedure to use this tool is discussed next.

- Select desired view from the **Model Tree** or from the 3D view area to which you want to add leaderline; refer to Figure-62.

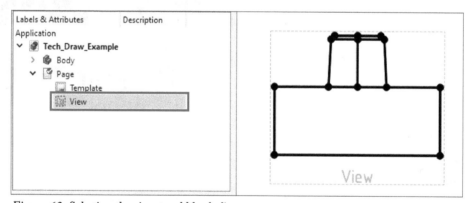

Figure-62. Selecting the view to add leaderline

- Click on the **Add Leaderline to View** tool from the **Toolbar** or **TechDraw** menu; refer to Figure-63. The **New Leader Line** dialog will be displayed in the **Tasks** panel of **Combo View**; refer to Figure-64.

Figure-63. Add Leaderline to View tool

Figure-64. New Leader Line dialog

- Click on the **Pick points** button from the dialog and click in the 3D view area at desired location to specify the starting point of leaderline. The line will be attached to the cursor; refer to Figure-65.

Figure-65. Leaderline attached to the cursor

- Move the cursor away and click at desired location to specify the another point of leaderline.
- Now, click on the **Save Points** button from the dialog to finish the line creation or move the cursor away and click at desired locations to create more line segments.
- Select desired symbol from **Start Symbol** and **End Symbol** drop-downs which you want at the start and end of line, respectively.
- Click on the **Color** button in the dialog to specify desired color for the leaderline.
- Specify desired width of the leaderline in the **Width** edit box.
- Select desired style for the leaderline from **Style** drop-down.
- After specifying desired parameters, click on the **OK** button from the dialog to create the leaderline; refer to Figure-66.

Figure-66. Leaderline added to the view

INSERTING RICH TEXT ANNOTATION

The **Insert Rich Text Annotation** tool adds a formatted annotation block to a leaderline or a view. The procedure to use this tool is discussed next.

- Click on the **Insert Rich Text Annotation** tool from **Toolbar** or **TechDraw** menu; refer to Figure-67. The **Rich text creator** dialog will be displayed in the **Tasks** panel of **Combo View**; refer to Figure-68.

Figure-67. Insert Rich Text Annotation tool

Figure-68. Rich text creator dialog

- Specify desired maximal width of the text in the **Max. Width** edit box of the dialog.
- Enter desired annotation text which you want to create in the **Input** box of the dialog.
- Click on the **Start Rich Text Editor** button from the dialog. The **Rich text editor** dialog will be displayed with the text written in the **Input** box; refer to Figure-69.

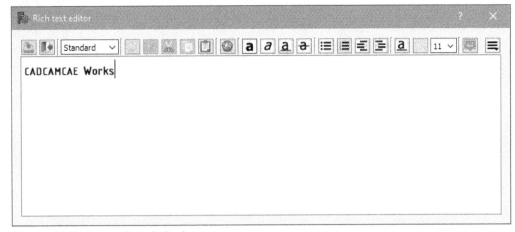

Figure-69. Rich text editor dialog box

- Specify desired parameters in the **Rich text editor** dialog and click on the **Save changes** button to save all the changes created in the dialog. The dialog will be closed.
- Select the **Show Frame** check box from **Rich text creator** dialog. The options below the check box will be activated.
- Click on the **Color** button to select desired color for the text.
- Specify desired width for the text in the **Width** edit box of the dialog.
- Select desired style for the text from the **Style** drop-down.
- After specifying desired parameters, click on the **OK** button from the dialog. The rich text annotation will be created; refer to Figure-70.

Figure-70. Rich text annotation created

ADDING COSMETIC VERTEX

The **Add Cosmetic Vertex** tool adds a vertex which is not part of the source geometry to a view. This vertex behaves like any other vertex and can be used for dimensioning. The procedure to use this tool is discussed next.

- Select desired view from the **Model Tree** or from the 3D view area to which you want to add the vertex; refer to Figure-71.

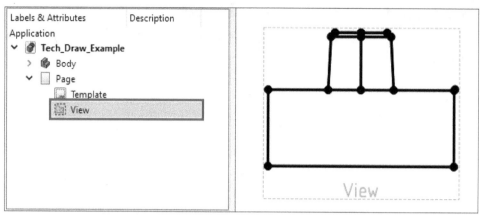

Figure-71. Selecting the view to add vertex

- Click on the **Add Cosmetic Vertex** tool from the drop-down in the **Toolbar** or **TechDraw** menu; refer to Figure-72. The **New Cosmetic Vertex** dialog will be displayed in the **Tasks** panel of **Combo View**; refer to Figure-73.

Figure-72. Add Cosmetic Vertex tool

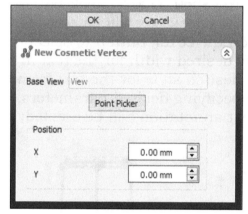

Figure-73. New Cosmetic Vertex dialog

- Click on the **Point Picker** button from the dialog and click at desired location in the 3D view area to place the point or enter desired values in the **X** and **Y** edit boxes from **Position** area of the dialog.
- After clicking at desired location for the point, click on the **OK** button from the dialog. The cosmetic vertex will be added to the view; refer to Figure-74.

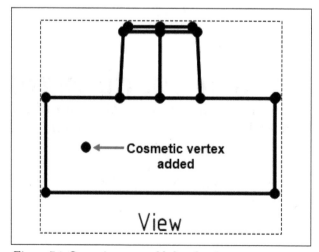

Figure-74. Cosmetic vertex added to the view

- If you want to remove the added cosmetic vertex then select the cosmetic vertex added to the view from the 3D view area and click on the **Remove Cosmetic Object** tool from the **Toolbar** or **TechDraw** menu; refer to Figure-75. The cosmetic vertex will be removed.

Figure-75. Remove Cosmetic Object tool

ADDING MIDPOINT VERTICES

The **Add Midpoint Vertices** tool adds cosmetic vertices at the midpoints of one or more edges. The procedure to use this tool is discussed next.

- Select one or more edges of desired view to which you want to add the midpoint vertex or vertices; refer to Figure-76.

Figure-76. Selecting the edges to add midpoint vertices

- Click on the **Add Midpoint Vertices** tool from the drop-down in the **Toolbar** or **TechDraw** menu; refer to Figure-77. The midpoint vertices will be added to the view; refer to Figure-78.

Figure-77. Add Midpoint Vertices tool

Figure-78. Midpoint vertices added to the view

• You can remove the midpoint vertices using **Remove Cosmetic Object** tool as discussed earlier.

ADDING QUADRANT VERTICES

The **Add Quadrant Vertices** tool adds cosmetic vertices at the 90/180/270° points of a circular edge. The 0° vertex should already be there as a geometric vertex. The procedure to use this tool is discussed next.

• Select one or more circular edges of desired view to which you want to add the quadrant vertices; refer to Figure-79.

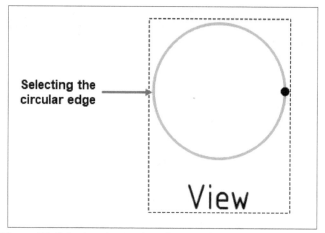

Figure-79. Selecting the circular edge to add quadrant vertices

• Click on the **Add Quadrant Vertices** tool from the drop-down in the **Toolbar** or **TechDraw** menu; refer to Figure-80. The quadrant vertices will be added to the view; refer to Figure-81.

Figure-80. Add Quadrant Vertices tool

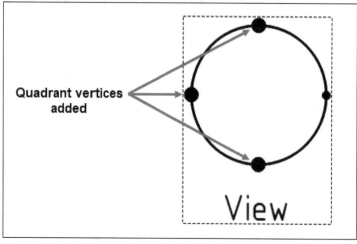

Figure-81. Quadrant vertices added to the view

- You can remove the quadrant vertices using **Remove Cosmetic Object** tool as discussed earlier.

ADDING CENTERLINE TO FACES

The **Add Centerline to Faces** tool adds a centerline to selected faces. The procedure to use this tool is discussed next.

- Select one or more faces of desired view to which you want to add centerline; refer to Figure-82.

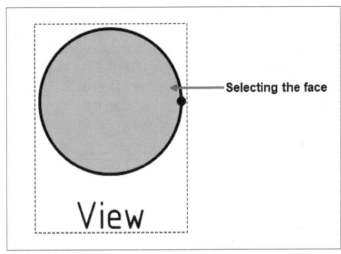

Figure-82. Selecting the face to add centerline

- Click on the **Add Centerline to Faces** tool from the drop-down in the **Toolbar** or **TechDraw** menu; refer to Figure-83. The **Create Center Line** dialog will be displayed in the **Tasks** panel of **Combo View**; refer to Figure-84.

Figure-83. Add Centerline to Faces tool

Figure-84. Create Center Line dialog

- The selected face will be displayed in the **Elements** area of the dialog.

- Select desired radio button from **Orientation** area of the dialog. Select **Vertical** radio button to orient the centerline from Top to Bottom line. Select **Horizontal** radio button to orient the centerline from left to right line. Select **Aligned** radio button to orient the centerline between lines or points.
- Specify desired value in **Shift Horizontal** edit box to move the centerline left or right of its normal position.
- Specify desired value in **Shift Vertical** edit box to move the centerline up or down from its normal position.
- Specify desired value in **Rotate** edit box to rotate the centerline around its center.
- Specify desired value in **Extend By** edit box to make the centerline longer by this amount.
- Click on the **Color** button to specify desired color for the centerline.
- Enter desired value in the **Weight** edit box to specify thickness of the centerline.
- Select desired style for the centerline from **Style** drop-down.
- After specifying desired parameters, click on the **OK** button from the dialog. The centerline will be added to the face; refer to Figure-85.

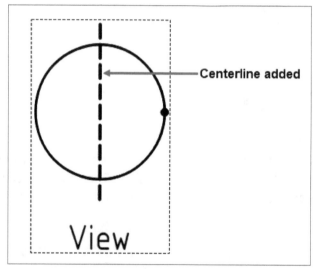

Figure-85. Centerline added to the face

- You can remove the centerline using **Remove Cosmetic Object** tool as discussed earlier.

ADDING CENTERLINE BETWEEN 2 LINES

The **Add Centerline between 2 Lines** tool adds a centerline between two edges. The procedure to use this tool is discussed next.

- Select the two edges of desired view between which you want to add the centerline; refer to Figure-86.

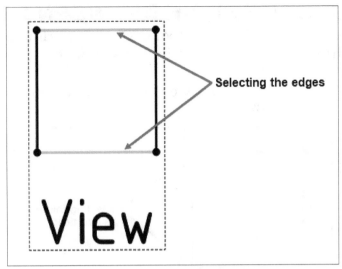

Figure-86. Selecting the edges to add centerline

- Click on the **Add Centerline between 2 Lines** tool from the drop-down in the **Toolbar** or **TechDraw** menu; refer to Figure-87. The **Create Center Line** dialog will be displayed in the **Tasks** panel of **Combo View** as discussed earlier.

Figure-87. Add Centerline between 2 Lines tool

- Specify desired parameters in the dialog as discussed earlier and click on **OK** button. The centerline between the two edges will be added; refer to Figure-88.

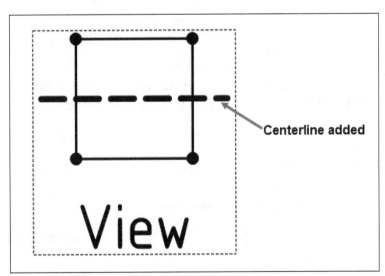

Figure-88. Centerline added between the edges

- You can remove the centerline using **Remove Cosmetic Object** tool as discussed earlier.

ADDING CENTERLINE BETWEEN 2 POINTS

The **Add Centerline between 2 Points** tool adds a centerline between two vertices (Points). The procedure to use this tool is discussed next.

- Select the two vertices of desired view between which you want to add the centerline; refer to Figure-89.

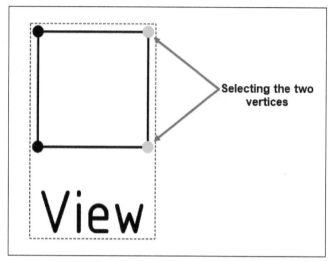

Figure-89. Selecting the vertices to add centerline

- Click on the **Add Centerline between 2 Points** tool from the drop-down in the **Toolbar** or **TechDraw** menu; refer to Figure-90. The **Create Center Line** dialog will be displayed in the **Tasks** panel of **Combo View** as discussed earlier.

Figure-90. Add Centerline between 2 Points tool

- Specify desired parameters in the dialog as discussed earlier and click on the **OK** button. The centerline between the two points will be added; refer to Figure-91.

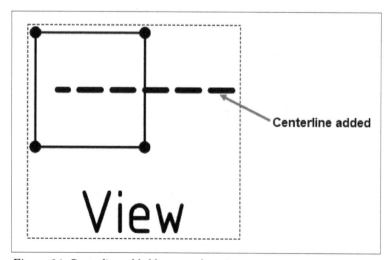

Figure-91. Centerline added between the points

- You can remove the centerline using **Remove Cosmetic Object** tool as discussed earlier.

ADDING COSMETIC LINE THROUGH 2 POINTS

The **Add Cosmetic Line Through 2 Points** tool adds a cosmetic line between two vertices (points). The vertices can be 2D or 3D. The resulting line can be used for dimensioning. The procedure to use this tool is discussed next.

- Select the two vertices of desired view or in the 3D view area between which you want to add the cosmetic line; refer to Figure-92.

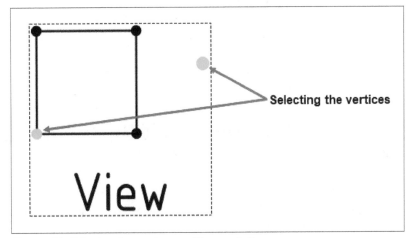

Figure-92. Selecting the vertices to add cosmetic line

- Click on the **Add Cosmetic Line Through 2 Points** tool from **Toolbar** or **TechDraw** menu; refer to Figure-93. The **Create Cosmetic Line** dialog will be displayed in the **Tasks** panel of **Combo View**; refer to Figure-94.

Figure-93. Add Cosmetic Line Through 2 Points tool

Figure-94. Create Cosmetic Line dialog

- Select desired radio button from **2D Point** and **3D Point** radio buttons in the dialog.
- Enter desired values in the **X**, **Y**, and **Z** edit boxes to specify the coordinates of line in the x, y, and z directions, respectively.
- After specifying desired parameters, click on the **OK** button from the dialog. The cosmetic line will be created between the points; refer to Figure-95.

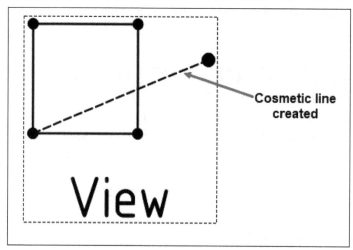

Figure-95. Cosmetic line between the points created

- You can remove the cosmetic line using **Remove Cosmetic Object** tool as discussed earlier.

CHANGING APPEARANCE OF LINE

The **Change Appearance of selected Lines** tool changes the appearance of edges. The procedure to use this tool is discussed next.

- Select one or more edges of desired view of which you want to change the appearance; refer to Figure-96.

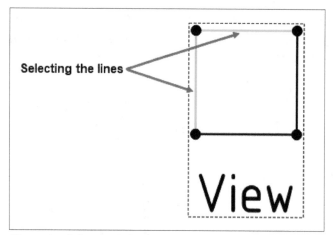

Figure-96. Selecting the lines to change the appearance

- Click on the **Change Appearance of selected Lines** tool from **Toolbar** or **TechDraw** menu; refer to Figure-97. The **Line Decoration and Restore Invisible Lines** dialogs will be displayed in the **Tasks** panel of **Combo View**; refer to Figure-98.

Figure-97. Change Appearance of selected Lines tool

Figure-98. Line Decoration and Restore Invisible Lines dialogs

- Select desired style for the lines from **Style** drop-down.
- Click on the **Color** button to specify desired color for the lines.
- Enter desired value in the **Weight** edit box to specify desired thickness for the lines.
- Select desired option from **Visible** drop-down to specify whether the line should be visible or not.
- After specifying desired parameters, click on the **OK** button from the dialog. The Appearance of the lines will be changed; refer to Figure-99.

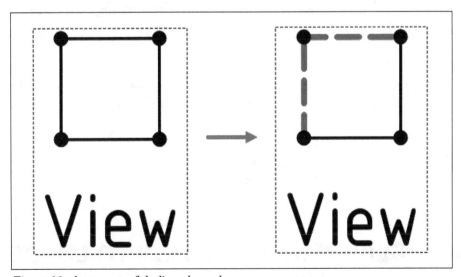

Figure-99. Appearance of the lines changed

- If you select the **False** option from **Visible** drop-down and click on the **OK** button then the lines will become invisible; refer to Figure-100.

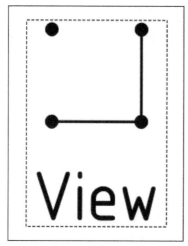

Figure-100. Lines become invisible

- Now, if you want to restore the lines then select the view from the **Model Tree** or from the 3D view area and click on the tool again. The **Restore Invisible Lines** dialog will be displayed.
- Click on desired button from the dialog to restore the lines. Click on the **All** button to restore all the invisible lines of the view. Click on the **Geometry** button to restore the invisible geometry lines. Click on the **Cosmetic** button to restore the invisible cosmetic lines. Click on the **CenterLine** button to restore the invisible centerlines.
- After specifying desired parameters, click on the **OK** button from the dialog. The invisible lines will be restored.

You can also restore the invisible lines using **Show/Hide Invisible Edges** tool.

Show/Hide Invisible Edges

The **Show/Hide Invisible Edges** tool is used to temporarily show and hide invisible lines in a View. Lines can be made invisible with the **TechDraw DecorateLine** tool. Note that "invisible" is a cosmetic state, not to be confused with hidden lines which are geometric constructs.

- Click on the **Show/Hide Invisible Edges** tool from the **Toolbar** or **TechDraw** menu; refer to Figure-101. All invisible lines in the view are shown/hidden based on tool selection.

Figure-101. Show hide invisible edges

ADDING WELDING INFORMATION TO LEADERLINE

The **Add Welding Information to Leaderline** tool adds welding specifications to an existing leader line. The procedure to use this tool is discussed next.

- Select desired leaderline from the 3D view area to which you want to add welding information; refer to Figure-102.

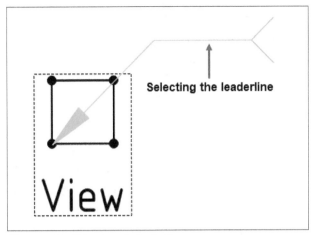

Figure-102. Selecting the leaderline

- Click on the **Add Welding Information to Leaderline** tool from the **Toolbar** or **TechDraw** menu; refer to Figure-103. The **Create Welding Symbol** dialog will be displayed in the **Tasks** panel of **Combo View**; refer to Figure-104.

Figure-103. Add Welding Information to Leaderline tool

Figure-104. Create Welding Symbol dialog

- Click on the **Pick arrow side symbol** button. The **Select a symbol** dialog box will be displayed; refer to Figure-105.

Figure-105. Select a symbol dialog

- Select desired symbol from the dialog box or from the symbol directory in your system and click on the **OK** button. The **Pick arrow side symbol** button will be replaced with the selected symbol.
- Specify desired text in the **Text above arrow side symbol**, **Text before arrow side symbol**, and **Text after arrow side symbol** edit boxes to place the text above, before, and after the arrow side symbol, respectively.
- In the same way, click on the **Pick other side symbol** button and select desired symbol from the **Select a symbol** dialog box.
- Specify desired text in the **Text below other side symbol**, **Text before other side symbol**, and **Text after other side symbol** edit boxes to place the text below, before, and after the other side symbol, respectively.
- Click on the **Delete** button to remove the other side symbol.
- Click on the **Flip Sides** button to flip the sides of the symbol.
- Select the **Field Weld** check box to add the Field Weld symbol (flag) at the kink in the leader line.
- Select the **All Around** check box to add the All Around symbol (circle) at the kink in the leader line.
- Select the **Alternating** check box to offset the lower symbol to indicate alternating welds.
- Specify desired text at the end of symbol in the **Tail Text** edit box.
- Click on the **Symbol Directory** button to specify directory to welding symbols to be used for the symbol selection.
- After specifying desired parameters, click on the **OK** button from the dialog. The weld information will be added to the leader line; refer to Figure-106.

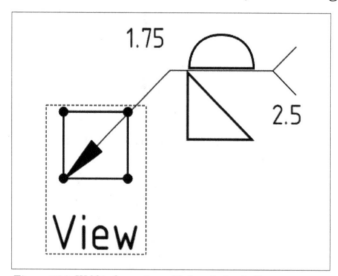

Figure-106. Weld information added to the leaderline

Create Surface Finish Symbol

The **Create a surface finish symbol** tool is used to add a surface finish symbol in drawing page.

- Click on the **Create a Surface finish symbol** tool from the **Toolbar** or **TechDraw** menu; refer to Figure-107. The **Surface Finish Symbols** dialog box will be displayed; refer to Figure-108.
- Select desired symbol from the dialog to create surface finish symbol.
- Specify desired rotation angle for the symbol in the **Symbol angle** edit box.

Figure-107. Create a surface finish symbol

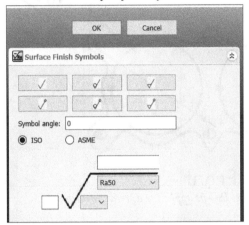

Figure-108. Surface Finish Symbols dialog

- Select desired radio button to select either **ISO** or **ASME** standard.
- Specify the other parameter as discussed earlier and click on the **OK** button from the dialog.

Add Hole or Shaft Fit

The **Add Hole or Shaft Fit** tool is used to add hole or shaft tolerance using ISO 286 to a linear dimension or a diameter dimension.

- Select the linear/diameter dimension from drawing and click on the **Add Hole or Shaft Fit** tool from the **Toolbar** or **TechDraw** menu; refer to Figure-109. The **Hole/Shaft Fit ISO 286** dialog will be displayed; refer to Figure-110.

Figure-109. Add hole or shaft tool

Figure-110. Add hole or shaft fit dialog box

- Select the **shaft fit** or **hole fit** radio button as applicable in the dialog and select fit option from the drop-down in the dialog.
- Click on the **OK** button from the dialog to apply fit tolerance; refer to Figure-111.

Figure-111. ISO fit applied to diameter dimension

WORKING WITH CLIP GROUP

The **Insert Clip group** tool is used to insert a clipping group in the drawing. You can add or remove objects from the clip group by dragging them inside or out of the group. The procedure to use this tool is given next.

- Click on the **Insert Clip group** tool from the **Toolbar** or **TechDraw** menu; refer to Figure-112. A clip group will be inserted in the drawing; refer to Figure-113.

Figure-112. Insert Clip Group tool

Figure-113. Clip group created

- You can drag the clip group to desired location as discussed for drawing views.
- Select the clip group recently created, select the object/view to be added in the clip group while holding the **CTRL** key and then click on the **Add a View to Clip group** tool from the **Toolbar** or **TechDraw** menu; refer to Figure-114. Selected view will be clipped and added in the clip group.

Figure-114. Adding view to clip group

- If you want to remove a view from clip group then select it from the clip group and click on the **Remove a View from Clip group** tool from the **Toolbar** or **TechDraw** menu. The view will be removed from clip group; refer to Figure-115.

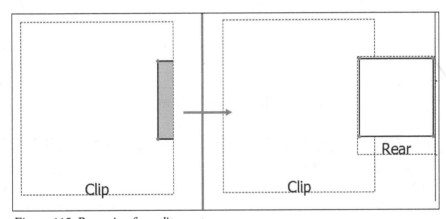

Figure-115. Removing from clip group

Note that you can use **CTRL+Z** and **CTRL+Y** shortcut keys to undo or redo previous operations.

INSERTING LENGTH DIMENSION

The **Insert a new length dimension** tool is used to apply a length dimension based on selected two vertices, an edge, or two edges. The procedure to use this tool is given next.

- After selecting two vertices, two edges (between which length dimension is to be created), or an edge to be dimensioned, click on the **Insert a new length dimension** tool from the **Toolbar** or **TechDraw** menu; refer to Figure-116. The dimension will be generated; refer to Figure-117 and Figure-118.

Figure-116. Insert Length Dimension tool

Figure-117. Dimension generated between two vertices

Figure-118. Dimension between two edges

Similarly, you can use other dimensioning tools to create horizontal distance dimension, vertical distance dimension, radius dimension, diameter dimension, angle dimension, 3 point angle dimensions, horizontal extent dimension, vertical extent dimension, balloon dimension, and landmark dimension.

LINKING DIMENSION TO 3D GEOMETRY

The **Link dimension to 3D geometry** tool is used to link selected dimension of **TechDraw** workbench to an edge or vertices of the 3D model. On doing so, the dimension in TechDraw changes as per the actual dimension in 3D model for same edge/vertices. The procedure to use this tool is given next.

• Create the dimension of edge/vertices that you want to be linked in **TechDraw** workbench page; refer to Figure-119.

Figure-119. Dimension created in page

- Select the model tab at the bottom of drawing area and select the edge that you want to be linked to available dimensions along with page from **Model Tree**; refer to Figure-120.

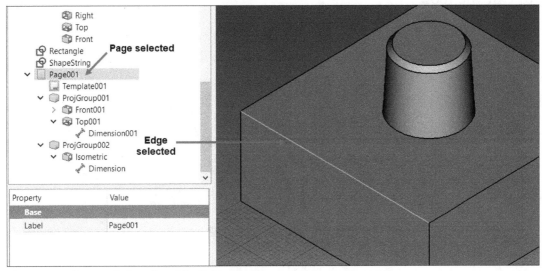

Figure-120. Page and edge selected for linking

- After selecting the edge and page, click on the **Link dimension to 3D geometry** tool from the **Toolbar** or **TechDraw** menu; refer to Figure-121. The **Link Dimension** dialog will be displayed; refer to Figure-122.

Figure-121. Link dimension to 3D geometry tool

Figure-122. Link Dimension dialog

- Select the dimension to be linked from the **Available** area in the dialog and click on the **Add** button ![add]. The dimension will be linked to selected edge.
- After adding desired dimension for linking, click on the **OK** button from the dialog. The linked dimension will reflect actual dimension of the edge in 3D model.

Insert Balloon Annotation

The **Insert balloon annotation** tool is used to add balloons with leader line in a drawing.

- Select desired object from the drawing to which balloon will be attached and click on the **Insert balloon annotation** tool from the **Toolbar** or **TechDraw** menu; refer to Figure-123. The balloon will be applied to the object; refer to Figure-124. Press **ESC** to exit the tool.

Figure-123. Insert Balloon Annotation

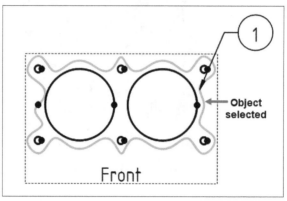

Figure-124. Balloon attached to object

- To move the bubble of a balloon, press and hold the left mouse button on its center and drag the mouse.
- To change the properties of balloon, double-click it in the drawing. The **Balloon** dialog will be displayed; refer to Figure-125. Set the parameters as desired and click on the **OK** button from the dialog.

Figure-125. Balloon dialog

Note that position of a balloon is relative to its Source View and uses the same scale factor.

Axonometric Length Dimension

The **Axonometric Length Dimension** tool is used to add a length dimension to selected entity in axonometric view. The dimension may be the length of an edge or the distance between two points. The procedure to use this tool is given next.

- Select the line/edge to be dimensioned and another line/edge to define direction of extension lines. Click on the **Axonometric Length Dimension** tool from the **Toolbar** or **TechDraw** menu; refer to Figure-126. The dimension will be generated; refer to Figure-127.

Figure-126. Axonometric length dimension

Figure-127. Axonometric length dimension created

Repair Dimension Reference

The **Repair Dimension Reference** tool is used to adjust the 2D or 3D geometry references of a dimension after the model has been modified.

- Select the dimension whose reference is to be repaired and click on the **Repair Dimension Reference** tool from the **Toolbar** or **TechDraw** menu; refer to Figure-128. The **Dimension Repair** dialog will be displayed; refer to Figure-129.

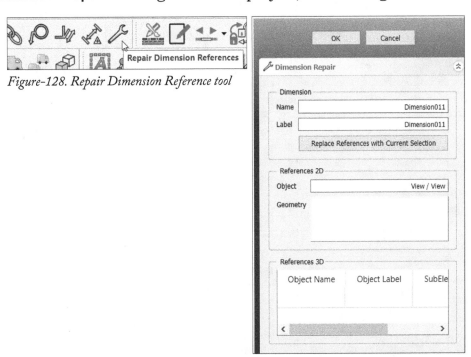

Figure-128. Repair Dimension Reference tool

Figure-129. Dimension Repair dialog

- Select new geometry references to be dimensioned and click on the **Replace References with Current Selection** button.
- Press the **OK** button to update the dimension; refer to Figure-130.

Figure-130. Repairing dimension

EXPORTING PAGE TO SVG FILE

The **Export a page to an SVG file** tool is used to export selected technical drawing page as an SVG image file. The procedure to use this tool is given next.

- Click on the **Export a page to an SVG file** tool from the **Toolbar** or **TechDraw** menu; refer to Figure-131. The **Export page as SVG** dialog box will be displayed; refer to Figure-132.

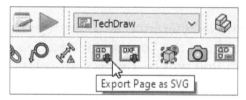

Figure-131. Export page as SVG file tool

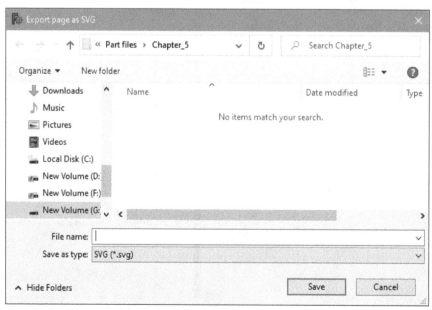

Figure-132. Export page as SVG dialog box

- Specify desired name of file to be exported in the **File name** edit box and click on the **Save** button to save the file. You can open this svg file in your default web browser.

EXPORTING PAGE AS DXF FILE

The **Export a page to a DXF file** tool is used to export current drawing page as DXF file generally used by CAD programs for generating engineering drawings. The procedure to use this tool is given next.

- Click on the **Export a page to a DXF file** tool from the **Toolbar** or **TechDraw** menu; refer to Figure-133. The **Save Dxf File** dialog box will be displayed; refer to Figure-134.

Figure-133. Export page as DXF tool

Figure-134. Save Dxf File dialog box

- Specify desired name of file in the **File name** edit box and click on the **Save** button from the dialog box. The file will be created. You can open this file using any Engineering drawing CAD software.

APPLYING HATCH PATTERN

The **Hatch a Face using image file** tool is used to apply hatch pattern in selected faces. The procedure to use this tool is given next.

- Select the face on which you want to apply hatch pattern and click on the **Hatch a Face using image file** tool from the **Toolbar** or **TechDraw** menu; refer to Figure-135. The **Apply Hatch to Face** dialog will be displayed in the **Tasks** panel of **Combo View** along with the preview of hatch pattern; refer to Figure-136.

Figure-135. Hatch face using image file tool

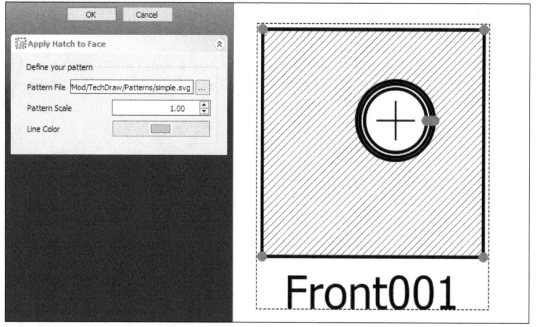

Figure-136. Apply Hatch to Face dialog along with preview of hatch pattern

- Click on the **Browse** button for **Pattern File** box from **Define your pattern** area of the dialog. The **Select a file** dialog box will be displayed; refer to Figure-137.

Figure-137. Select a file dialog box

- Select desired image file of hatch pattern from the dialog box and click on the **Open** button. The hatch pattern in drawing will be modified, accordingly.
- Enter desired value in **Pattern Scale** edit box to enlarges/shrinks the pattern.

- Click on the **Line Color** button to specify the color of pattern lines.
- After specifying desired parameters, click on the **OK** button from the dialog to create the pattern.

APPLYING GEOMETRIC HATCH TO A FACE

The **Apply geometric hatch to Face** tool is used to apply hatch based on specified geometric parameters. The procedure to use this tool is given next.

- Select a face and click on the **Apply geometric hatch to Face** tool from the **Toolbar** or **TechDraw** menu; refer to Figure-138. The **Apply Geometric Hatch to Face** dialog will be displayed; refer to Figure-139.

Figure-138. Apply geometric hatch to face tool

Figure-139. Apply Geometric Hatch to Face dialog

- Select desired option from the **Pattern Name** drop-down to define pattern to be created for hatching.
- Specify the scale, line weight, and line color for pattern in respective fields of the dialog.
- After setting desired parameters, click in the face on which you want to apply the pattern and click on the **OK** button from the dialog to create the pattern.

INSERTING SYMBOL FROM AN SVG FILE

The **Insert symbol from an svg file** tool is used to insert a symbol using a local svg file available. The procedure to use this tool is given next.

- Click on the **Insert symbol from an svg file** tool from the **Toolbar** or **TechDraw** menu; refer to Figure-140. The **Choose an SVG file to open** dialog box will be displayed; refer to Figure-141.

Figure-140. Insert symbol from SVG file tool

Figure-141. Choose an SVG file to open dialog box

- Select desired symbol file and click on the **Open** button. The symbol will be inserted at the center of the page; refer to Figure-142.

Figure-142. Symbol inserted in drawing

INSERTS A BITMAP FROM A FILE INTO A PAGE

The **Inserts bitmap image** tool is used to insert an image file in the page. The procedure to use this tool is given next.

- Click on the **Inserts bitmap image** tool from the **Toolbar** or **TechDraw** menu; refer to Figure-143. The **Select an Image File** dialog box will be displayed; refer to Figure-144.

Figure-143. Insert bitmap image tool

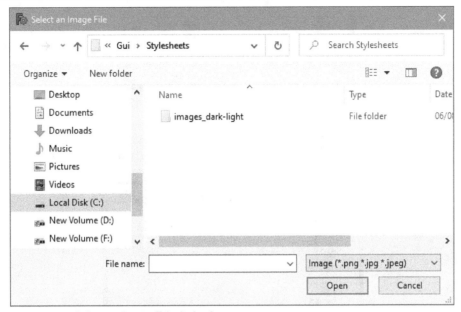

Figure-144. Select an Image File dialog box

- Select desired image file to be inserted in the page and click on the **Open** button. The selected image file will be inserted in the drawing. Note that only PNG, BMP, and JPG/JPEG formats are supported for image insertion.
- Select the **Image** feature from the **Model Tree** and specify desired parameters in the **Property Editor**.
- You can specify width and height of image in the **Width** and **Height** edit boxes of the **Image** section in **Property Editor**.

TURNING VIEW FRAMES ON/OFF

The **Turn View Frames On/Off** tool is used to turn ON or turn OFF frames around the views in drawing page. Click on the **Turn View Frames On/Off** tool from **Toolbar** or **TechDraw** menu to toggle display of frames around the views; refer to Figure-145.

Figure-145. Turn view frames on off tool

SELECTING LINE ATTRIBUTES, CASCADE SPACING, AND DELTA DISTANCE

This tool is used to select the attributes (style, width, and color) for new cosmetic lines and centerlines, and specifies the cascade spacing and delta distance. The procedure to use this tool is discussed next.

- Click on the **Select Line Attributes, Cascade Spacing, and Delta Distance** tool from the **TechDraw Attribute Toolbar** or **Techdraw** menu; refer to Figure-146. The **Select line** attributes dialog will be displayed; refer to Figure-147.

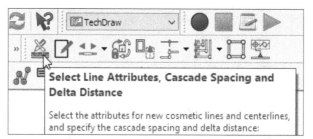

Figure-146. Select Line Attributes Cascade Spacing and Delta Distance tool

Figure-147. Select line attributes dialog

- Specify desired line style, line width, and line color by selecting desired radio button from **Line style**, **Line width**, and **Line color** sections of the dialog, respectively.
- Specify desired values in the **Cascade spacing** and **Delta distance** edit boxes of the dialog.
- After specifying desired parameters, click on the **OK** button from the dialog.

CHANGING LINE ATTRIBUTES

The **Change Line Attributes** tool is used to change the attributes (style, width, and color) of cosmetic lines and centerlines. The procedure to use this tool is discussed next.

- Select desired lines or centerlines of which you want to change the attributes; refer to Figure-148.

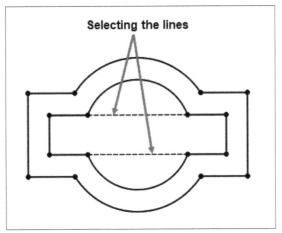

Figure-148. Selecting lines to change attributes

- Click on the **Change Line Attributes** tool from **Toolbar** or **Techdraw** menu; refer to Figure-149. The line attributes will be changed to specified attributes created by using the **Select Line Attributes, Cascade Spacing, and Delta Distance** tool; refer to Figure-150.

Figure-149. Change Line Attributes tool

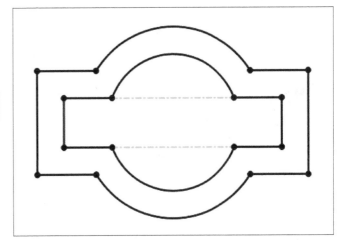

Figure-150. Line attributes changed

EXTENDING LINE

The **Extend Line** tool extends a cosmetic line or centerline at both ends. The procedure to use this tool is discussed next.

- Select desired line which you want to extend; refer to Figure-151.

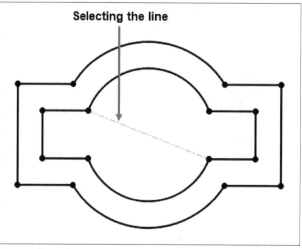

Figure-151. Selecting the line to extend

- Click on the **Extend Line** tool from drop-down in the **Toolbar** or **Techdraw** menu; refer to Figure-152. The line will be extended from both the ends; refer to Figure-153.

Figure-152. Extend Line tool

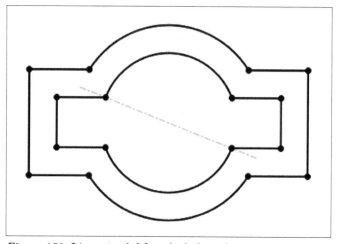

Figure-153. Line extended from both the ends

SHORTENING THE LINE

The **Shorten Line** tool is used to shorten a cosmetic line or centerline at both ends. The procedure to use this tool is same as discussed for **Extend Line** tool; refer to Figure-154.

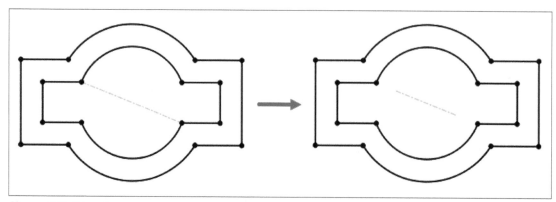

Figure-154. Line shorten

LOCKING/UNLOCKING THE VIEW

The **Lock/Unlock View** tool is used to lock or unlock the position of a view. A locked view has a fixed position relative to the page or the group it belongs to.

POSITIONING SECTION VIEW

The **Position Section View** tool is used to align a section view orthogonally with its source view. The procedure to use this tool is discussed next.

- Select the section view which you want to align orthogonally; refer to Figure-155.

Figure-155. Selecting the section view to align orthogonally

- Click on the **Position Section View** tool from **Toolbar** or **Techdraw** menu; refer to Figure-156. The section view will be aligned orthogonally with the source view; refer to Figure-157.

Figure-156. Position Section View tool

Figure-157. Section view aligned

POSITIONING HORIZONTAL CHAIN DIMENSIONS

The **Position Horizontal Chain Dimensions** tool is used to align horizontal dimensions to create a chain dimension. The procedure to use this tool is discussed next.

- Select two or more horizontal dimensions which you want to align to create a chain dimensions; refer to Figure-158.

Figure-158. Selecting the horizontal dimensions to align

- Click on the **Position Horizontal Chain Dimensions** tool from drop-down in the **Toolbar** or **Techdraw** menu; refer to Figure-159. The dimensions will be aligned to create chain dimensions; refer to Figure-160.

Figure-159. Position Horizontal Chain Dimensions tool

Figure-160. Dimensions aligned

POSITIONING VERTICAL CHAIN DIMENSIONS

The **Position Vertical Chain Dimensions** tool is used to align vertical dimensions to create a chain dimension. The procedure to use this tool is same as discussed for previous tool; refer to Figure-161.

Figure-161. Vertical dimensions aligned

POSITIONING OBLIQUE CHAIN DIMENSIONS

The **Position Oblique Chain Dimensions** tool is used to align oblique dimensions to create a chain dimension. The procedure to use this tool is same as discussed for previous tool; refer to Figure-162.

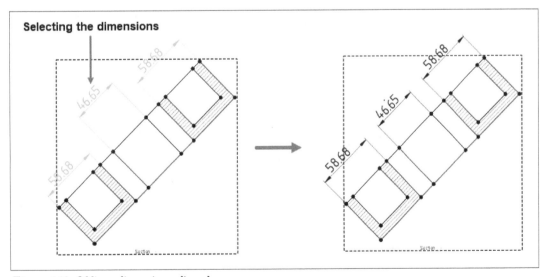

Figure-162. Oblique dimensions aligned

CASCADING HORIZONTAL DIMENSIONS

The **Cascade Horizontal Dimensions** tool is used to evenly space the horizontal dimensions. The procedure to use this tool is discussed next.

- Select two or more horizontal dimensions which you want to space evenly; refer to Figure-163.

Figure-163. Selecting horizontal dimensions to space evenly

- Click on the **Cascade Horizontal Dimensions** tool from drop-down in the **Toolbar** or **Techdraw** menu; refer to Figure-164. The dimensions will be evenly spaced; refer to Figure-165.

Figure-164. Cascade Horizontal Dimensions tool

Figure-165. Dimensions evenly spaced

CASCADING VERTICAL DIMENSIONS

The **Cascade Vertical Dimensions** tool is used to evenly space the vertical dimensions. The procedure to use this tool is same as discussed for previous tool; refer to Figure-166.

Figure-166. Vertical dimensions evenly spaced

CASCADING OBLIQUE DIMENSIONS

The **Cascade Oblique Dimensions** tool is used to evenly space the oblique dimensions. The procedure to use this tool is same as discussed for previous tool; refer to Figure-167.

Figure-167. Oblique dimensions evenly spaced

CALCULATING THE AREA OF SELECTED FACES

The **Area Annotation** tool is used to calculate the area of selected faces and inserts an area annotation. The tool can calculate the area of faces with only straight edges. The procedure to use this tool is discussed next.

- Select one or more faces with only straight edges; refer to Figure-168.

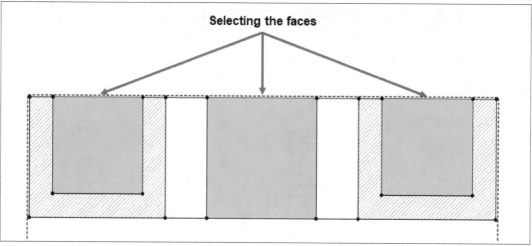

Figure-168. Selecting the faces with straight edges

- Click on the **Area Annotation** tool from **Toolbar** or **Techdraw** menu; refer to Figure-169. The area of selected faces will be calculated; refer to Figure-170.

Figure-169. Area Annotation tool

Figure-170. Area calculated of selected faces

CUSTOMIZING FORMAT LABEL

The **Customize Format Label** tool is used to customize the formatting of a balloon text or dimension text. The procedure to use this tool is discussed next.

- Select a balloon object or a dimension object from **Model Tree** or from 3D view area; refer to Figure-171.

Figure-171. Selecting the dimension to customize

- Click on the **Customize Format Label** tool from **Toolbar** or **Techdraw** menu; refer to Figure-172. The **Customize Format** dialog will be displayed; refer to Figure-173.

Figure-172. Customize Format Label tool

Figure-173. Customize Format dialog

- Select desired options from **GD&T**, **Modifiers**, **Radius & Diameter**, **Angles**, **Other** and **Greek Letters** sections of the dialog to customize the formatting of selected dimension. The selected options will be displayed in the **Format** edit box of the dialog.
- The **Preview** section of the dialog displays the result of the current formatting.
- After specifying desired parameters, click on the **OK** button from the dialog. The formatting of dimension will be customized; refer to Figure-174.

Figure-174. Formatting of dimension customized

ADDING CIRCLE CENTERLINES

The **Add Circle Centerlines** tool is used to add centerlines to circles and arcs. The procedure to use this tool is discussed next.

- Select one or more circles or arcs to which you want to add centerlines; refer to Figure-175.

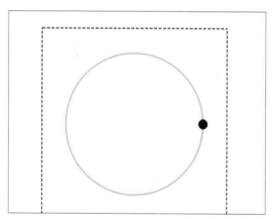

Figure-175. Selecting the circle to add centerlines

- Click on the **Add Circle Centerlines** tool from drop-down in the **Toolbar** or **Techdraw** menu; refer to Figure-176. The centerlines will be added to the circle; refer to Figure-177.

Figure-176. Add Circle Centerlines tool

Figure-177. Centerlines on circle added

ADDING BOLT CIRCLE CENTERLINES

The **Add Bolt Circle Centerlines** tool is used to add centerlines to a circular pattern of circles. The procedure to use this tool is same as discussed for **Add Circle Centerlines** tool; refer to Figure-145.

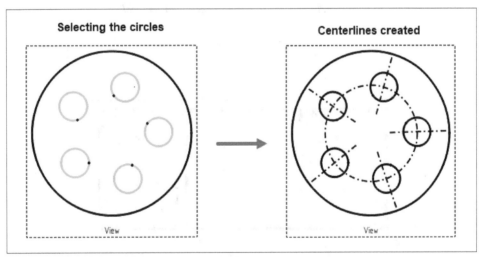

Figure-178. Centerlines on circular pattern created

ADDING COSMETIC THREAD HOLE SIDE VIEW

The **Add Cosmetic Thread Hole Side View** tool is used to add a cosmetic thread to the side view of a hole. The procedure to use this tool is discussed next.

• Select two parallel lines of the side view of a hole; refer to Figure-179.

Figure-179. Selecting two parallel lines to add cosmetic thread

- Click on the **Add Cosmetic Thread Hole Side View** tool from drop-down in the **Toolbar** or **Techdraw** menu; refer to Figure-180. The cosmetic threads will be added to the side view of a hole; refer to Figure-181.

Figure-180. Add Cosmetic Thread Hole Side View tool

Figure-181. Cosmetic threads added

ADDING COSMETIC THREAD HOLE BOTTOM VIEW

The **Add Cosmetic Thread Hole Bottom View** tool is used to add a cosmetic thread to the top or bottom view of holes. The procedure to use this tool is same as discussed for previous tool; refer to Figure-182.

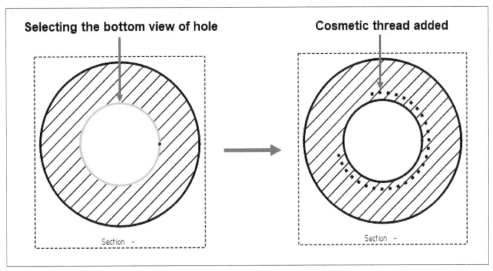

Figure-182. Cosmetic thread added to the bottom view of hole

ADDING COSMETIC THREAD BOLT SIDE VIEW

The **Add Cosmetic Thread Bolt Side View** tool is used to add cosmetic thread to the side view of a bolt/screw/rod. The procedure to use this tool is same as discussed for previous tool; refer to Figure-183.

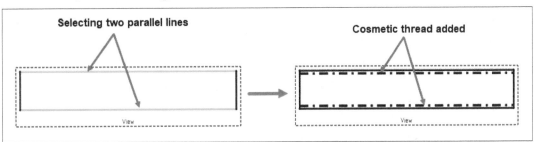

Figure-183. Cosmetic thread added to the side view of rod

ADDING COSMETIC THREAD BOLT BOTTOM VIEW

The **Add Cosmetic Thread Bolt Bottom View** tool is used to add a cosmetic thread to the top or bottom view of bolts/screws/rods. The procedure to use this tool is same as discussed for previous tool; refer to Figure-151.

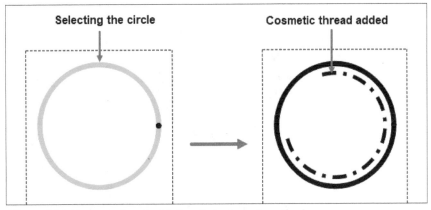

Figure-184. Cosmetic thread added to the bottom view of rod

ADDING COSMETIC INTERSECTION VERTEX(ES)

The **Add Cosmetic Intersection Vertex(es)** tool is used to add cosmetic vertex(es) at the intersection of selected edges. The procedure to use this tool is discussed next.

- Select two edges and a circle to which you want to add vertices; refer to Figure-185.

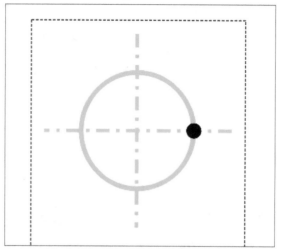

Figure-185. Selecting edges and a circle

- Click on the **Add Cosmetic Intersection Vertex(es)** tool from **Toolbar** or **Techdraw** menu; refer to Figure-186. The vertices will be added at the intersection of the edges; refer to Figure-187.

Figure-186. Add Cosmetic Intersection Vertex(es) tool

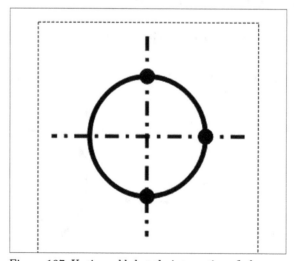

Figure-187. Vertices added at the intersection of edges

ADDING COSMETIC CIRCLE

The **Add Cosmetic Circle** tool is used to add a cosmetic circle based on two vertexes. The procedure to use this tool is discussed next.

- Select the first vertex to specify the center point of the circle and select the second vertex to specify the radius of the circle; refer to Figure-188.

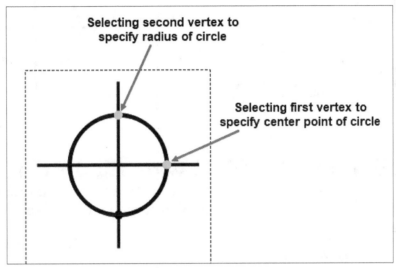

Figure-188. Selecting the vertices to add cosmetic circle

- Click on the **Add Cosmetic Circle** tool from drop-down in the **Toolbar** or **Techdraw** menu; refer to Figure-189. The cosmetic circle will be added; refer to Figure-190.

Figure-189. Add Cosmetic Circle tool

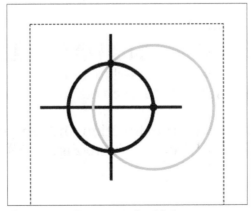

Figure-190. Cosmetic circle added

ADDING COSMETIC ARC

The **Add Cosmetic Arc** tool is used to add a cosmetic counter clockwise arc based on three vertexes. The procedure to use this tool is same as discussed for previous tool; refer to Figure-191.

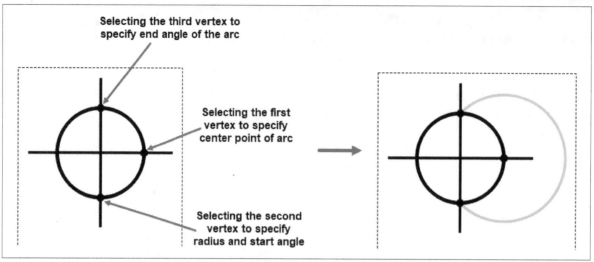

Figure-191. Cosmetic arc added

ADDING COSMETIC CIRCLE 3 POINTS

The **Add Cosmetic Circle 3 Points** tool is used to add a cosmetic circle based on three vertexes. The procedure to use this tool is same as discussed for previous tool; refer to Figure-192.

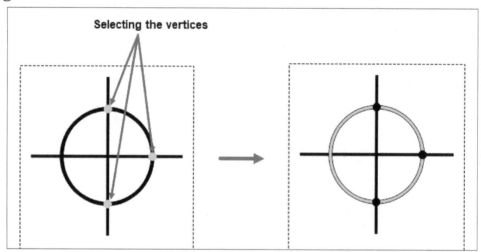

Figure-192. Cosmetic circle from 3 points added

ADDING COSMETIC PARALLEL LINE

The **Add Cosmetic Parallel Line** tool is used to add a cosmetic line parallel to another line through a vertex. The procedure to use this tool is discussed next.

• Select a straight edge to specify the direction and length of line and then select a vertex to specify the position of line; refer to Figure-193.

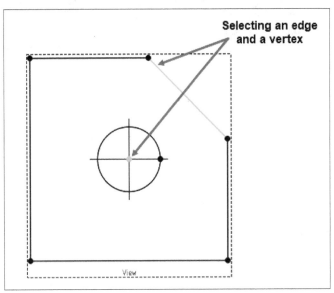

Figure-193. Selecting an edge and a vertex to add a cosmetic parallel line

- Click on the **Add Cosmetic Parallel Line** tool from drop-down in the **Toolbar** and **Techdraw** menu; refer to Figure-194. The cosmetic parallel line will be added; refer to Figure-195.

Figure-194. Add Cosmetic Parallel Line tool

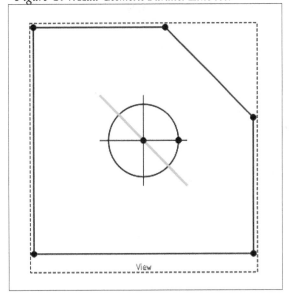

Figure-195. Cosmetic parallel line added

ADDING COSMETIC PERPENDICULAR LINE

The **Add Cosmetic Perpendicular Line** tool is used to add a cosmetic line perpendicular to another line through a vertex. The procedure to use this tool is same as discussed for previous tool; refer to Figure-196.

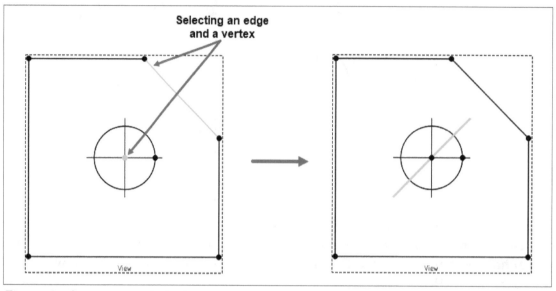

Figure-196. Cosmetic perpendicular line added

CREATING HORIZONTAL CHAIN DIMENSIONS

The **Create Horizontal Chain Dimensions** tool is used to create horizontal chain dimensions: a sequence of aligned dimensions. The procedure to use this tool is discussed next.

*   Select three or more vertices to create horizontal chain dimensions; refer to Figure-197. The two leftmost vertices specify the position of chain dimension.

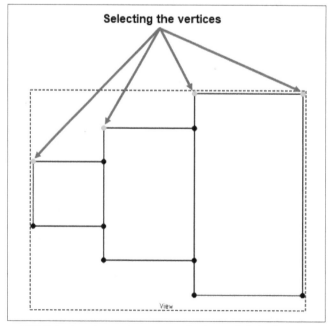

Figure-197. Selecting the vertices to create horizontal chain dimensions

*   Click on the **Create Horizontal Chain Dimensions** tool from drop-down in the **Toolbar** or **Techdraw** menu; refer to Figure-198. The horizontal chain dimensions will be created; refer to Figure-199.

Figure-198. Create Horizontal Chain Dimensions tool

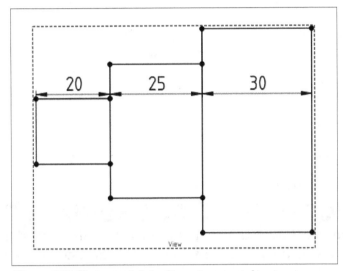

Figure-199. Horizontal chain dimensions created

CREATING VERTICAL CHAIN DIMENSIONS

The **Create Vertical Chain Dimensions** tool is used to create vertical chain dimensions: a sequence of aligned dimensions. The procedure to use this tool is same as discussed for previous tool; refer to Figure-200.

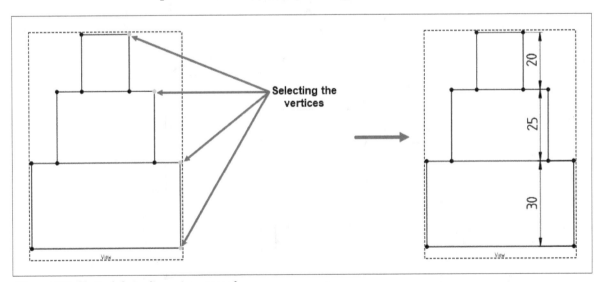

Figure-200. Vertical chain dimensions created

CREATING OBLIQUE CHAIN DIMENSIONS

The **Create Oblique Chain Dimensions** tool is used to create oblique chain dimensions; a sequence of aligned dimensions. The procedure to use this tool is same as discussed for previous tool; refer to Figure-201.

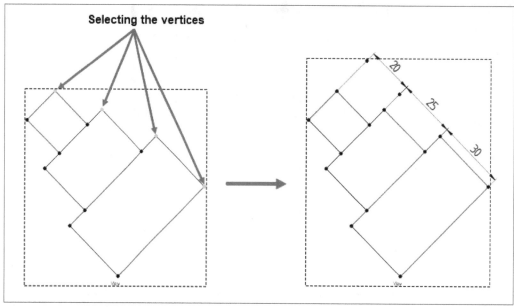

Figure-201. Oblique chain dimensions created

CREATING HORIZONTAL COORDINATE DIMENSIONS

The **Create Horizontal Coordinate Dimensions** tool is used to create horizontal coordinate dimensions: multiple evenly spaced dimensions starting from the same baseline. The procedure to use this tool is discussed next.

- Select three or more vertices to create horizontal coordinate dimensions; refer to Figure-202. The selection order of first two vertices specifies the position of the baseline.

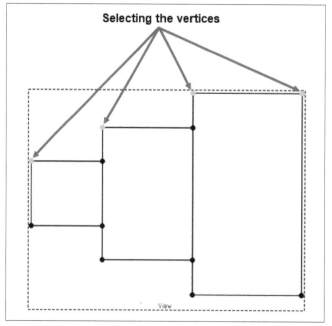

Figure-202. Selecting the vertices to create horizontal coordinate dimensions

- Click on the **Create Horizontal Coordinate Dimensions** tool from drop-down in the **Toolbar** or **Techdraw** menu; refer to Figure-203. The horizontal coordinate dimensions will be created; refer to Figure-204.

Figure-203. Create Horizontal Coordinate Dimensions tool

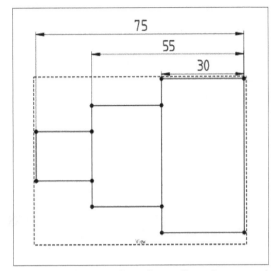

Figure-204. Horizontal coordinate dimensions created

CREATING VERTICAL COORDINATE DIMENSIONS

The **Create Vertical Coordinate Dimensions** tool is used to create vertical coordinate dimensions: multiple evenly spaced dimensions starting from the same baseline. The procedure to use this tool is same as discussed for previous tool; refer to Figure-205.

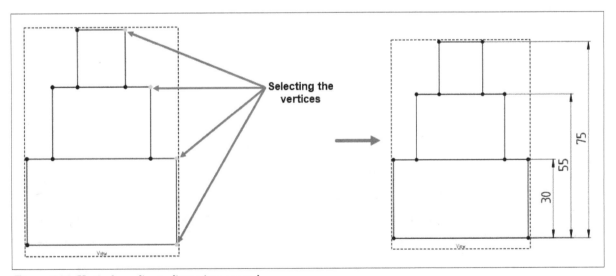

Figure-205. Vertical coordinate dimensions created

CREATING OBLIQUE COORDINATE DIMENSIONS

The **Create Oblique Coordinate Dimensions** tool is used to create oblique coordinate dimensions: multiple evenly spaced dimensions starting from the same baseline. The procedure to use this tool is same as discussed for previous tool; refer to Figure-206.

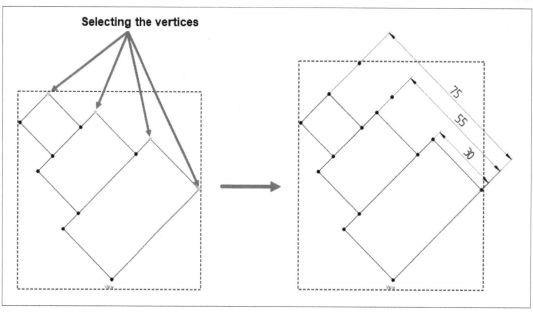

Figure-206. Oblique coordinate dimensions created

CREATING HORIZONTAL CHAMFER DIMENSION

The **Create Horizontal Chamfer Dimension** tool is used to create a horizontal size and angle dimension for a chamfer. The procedure to use this tool is discussed next.

*   Select two vertices to create a horizontal dimension for a chamfer; refer to Figure-207.

Figure-207. Selecting the vertices to create horizontal chamfer dimension

*   Click on the **Create Horizontal Chamfer Dimension** tool from drop-down in the **Toolbar** or **Techdraw** menu; refer to Figure-208. The horizontal chamfer dimension will be created; refer to Figure-209.

Figure-208. Create Horizontal Chamfer Dimension tool

Figure-209. Horizontal chamfer dimension created

CREATING VERTICAL CHAMFER DIMENSION

The **Create Vertical Chamfer Dimension** tool is used to create a vertical size and angle dimension for a chamfer. The procedure to use this tool is same as discussed for previous tool; refer to Figure-210.

Figure-210. Vertical chamfer dimension created

CREATING ARC LENGTH DIMENSION

The **Create Arc Length Dimension** tool is used to create an arc length dimension. The procedure to use this tool is discussed next.

• Select a single circular arc to create an arc length dimension; refer to Figure-211.

Figure-211. Selecting an arc to create dimension

- Click on the **Create Arc Length Dimension** tool from **Toolbar** or **Techdraw** menu; refer to Figure-212. The arc length dimension will be created; refer to Figure-213.

Figure-212. Create Arc Length Dimension tool

Figure-213. Arc length dimension created

INSERTING '⌀' PREFIX

The **Insert** '⌀' **Prefix** tool is used to insert a '⌀' symbol at the beginning of the dimension text. The procedure to use this tool is discussed next.

- Select one or more dimension to insert symbol at the beginning of that dimension; refer to Figure-214.

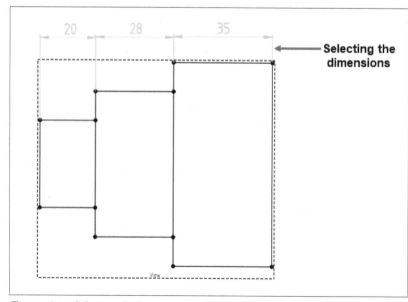

Figure-214. Selecting the dimensions to insert the symbol

- Click on the **Insert** '∅' **Prefix** tool from drop-down in the **Toolbar** or **Techdraw** menu; refer to Figure-215. The symbol will be inserted at the beginning of the dimension; refer to Figure-216.

Figure-215. Insert '∅' Prefix tool

Figure-216. Symbol inserted to the dimension

INSERTING '□' PREFIX

The **Insert** ' □ ' **Prefix** tool is used to insert a ' □ ' symbol at the beginning of the dimension text. The procedure to use this tool is same as discussed for previous tool; refer to Figure-217.

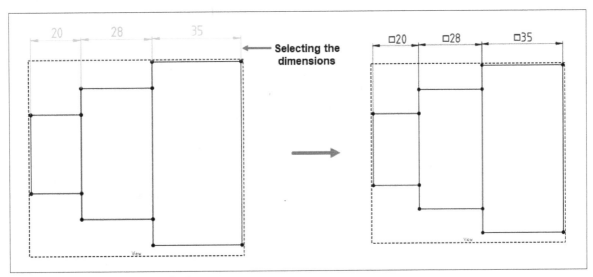

Figure-217. Symbol inserted to the dimension

REMOVING PREFIX

The **Remove Prefix** tool is used to remove all prefix symbols from the dimension text. The procedure to use this tool is same as discussed for previous tool; refer to Figure-218.

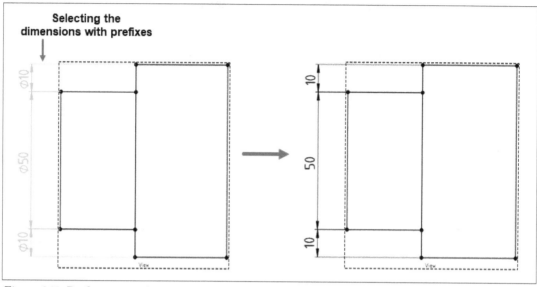

Figure-218. Prefixes removed

INCREASING DECIMAL PLACES

The **Increase Decimal Places** tool is used to increase the number of decimal places of the dimension text. The procedure to use this tool is discussed next.

*   Select one or more dimensions to increase the number of decimal places; refer to Figure-219.

Figure-219. Selecting the dimensions to increase number of decimal places

- Click on the **Increase Decimal Places** tool from drop-down in the **Toolbar** or **Techdraw** menu; refer to Figure-220. The number of decimal places of the dimensions will be increased by one; refer to Figure-221.

Figure-220. Increase Decimal Places tool

Figure-221. The number of decimal places of dimensions increased

DECREASING DECIMAL PLACES

The **Decrease Decimal Places** tool is used to decrease the number of decimal places of the dimension text. The procedure to use this tool is same as discussed for previous tool; refer to Figure-222.

Figure-222. The number of decimal places of dimensions decreased

PRACTICAL

Create the drawing views of the given model; refer to Figure-223.

Figure-223. Practical

Starting a New File

• Start FreeCAD if not started yet and open the model file to be used for creating drawing views
• Select the **TechDraw** option from the **Switch between workbenches** drop-down in the **Toolbar**. The drawing environment will be displayed.

Inserting a View

• Click on the **Insert Default Page** tool from the **Toolbar**. A new drawing page will be created.
• Select the tab of **Model** from the application menu and orient the model to desired view to be placed in the drawing; refer to Figure-224.
• Now, select the **Page** tab from the application window, select the model whose views are to be placed from **Model Tree**; refer to Figure-225.

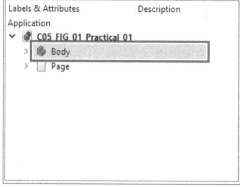

Figure-225. Selecting the model

Figure-224. Orientation of model

• Click on the **Insert Projection Group** tool from the **Toolbar**. The **Projection Group** dialog will be displayed.

- Specify the parameters in the dialog as shown in Figure-226, and click on the **OK** button. The views of the model will be created; refer to Figure-227.

Figure-226. Specifying parameters

Figure-227. Placing views of the model

Inserting the Dimensions

- Select the entities of the model one by one and place respective dimensions. The drawing view of the model with dimensions will be created; refer to Figure-228.

Figure-228. Drawing view of the model created

Practice

Create the model and drawing as shown in Figure-229.

Figure-229. GD&T

Chapter 11

Miscellaneous Workbench

Topics Covered

The major topics covered in this chapter are:

- *Introduction to Surface Workbench*
- *Creating Surfaces using Boundary Curves*
- *Introduction to Mesh Workbench*
- *Introduction to OpenSCAD Workbench*
- *Introduction to Plot Workbench*
- *Introduction to Points Workbench*
- *Introduction to Sheet Metal Workbench*

INTRODUCTION TO SURFACE WORKBENCH

The tools in **Surface** workbench are used to create surfaces using closed loop curve sections and boundary curves. Select the **Surface** option from **Switch between workbenches** drop-down in the **Toolbar**. The tools to create surfaces will be displayed in the **Toolbar**; refer to Figure-1.

Figure-1. Tools in Surface workbench

Creating Fill Surface using Curves and Vertices

The **Filling** tool is used to create surface using selected curves and vertices. The procedure to use this tool is given next.

- Click on the **Filling** tool from the **Surface** menu or **Toolbar** in **Surface** workbench; refer to Figure-2. The **Boundaries** dialog will be displayed; refer to Figure-3.

Figure-2. Filling tool

Figure-3. Boundaries dialog

- Click on the **Add Edge** button from dialog and select all the edges/curves forming a closed region for fill surface. Preview of the surface will be displayed; refer to Figure-4.

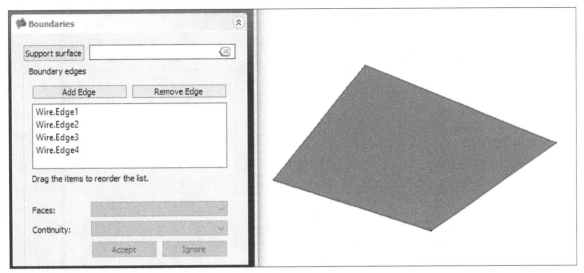

Figure-4. Preview of fill surface

- Expand the **Edge constraints** dialog in the **Tasks** panel and click on the **Add Edge** button from this dialog to define the curves to be used as guide for creating surface.
- Select the curves from the drawing area. Note that the curves can be from sketch or draft workbenches. The curves may or may not be intersecting with the boundary curves. The preview of surface will be displayed; refer to Figure-5.

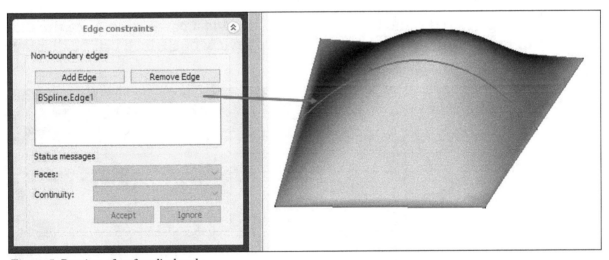

Figure-5. Preview of surface displayed

- Similarly, you can use vertexes to refine shape of surface.
- After setting desired parameters, click on the **OK** button to create the surface.

Creating Surface using Boundary Curves

The **Fill Boundary Curves** tool is used to create surface using 2, 3, or 4 boundary edges. The procedure to use this tool is given next.

- Click on the **Fill boundary curves** tool from the **Surface** menu or **Toolbar** in **Surface** workbench; refer to Figure-6. The **Surface** dialog will be displayed; refer to Figure-7.

Figure-6. Fill boundary curves tool

Figure-7. Surface dialog

- Select the **Stretch** radio button if you want to create surface as combination of flat patches.
- Select the **Coons** radio button if you want to surface with better finish compared to stretch but lesser finish as compared to curved.
- Select the **Curved** radio button if you want to create surfaces having finish following curvature of curves.
- After selecting desired radio button, click on the **Add Edge** button from the dialog and select two or more curve in succession. Preview of surface will be displayed; refer to Figure-8.

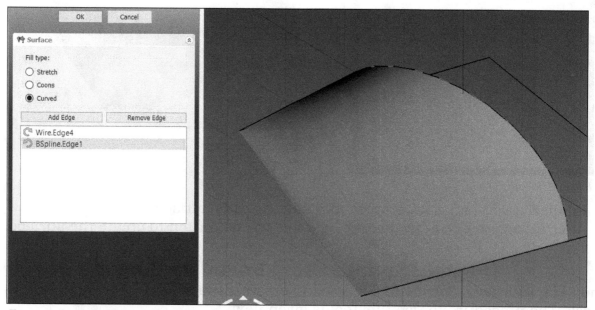

Figure-8. Preview of surface

- Click on the **OK** button from the dialog to create the surface.

Creating Surface using Edges

The **Sections** tool is used to create a surface from edges that represent transversal sections of a surface. The procedure to use this tool is discussed next.

- Click on the **Sections** tool from **Surface** menu or **Toolbar** in **Surface** workbench; refer to Figure-9. The **Sectional edges** dialog will be displayed in the **Tasks** panel of **Combo View**; refer to Figure-10.

Figure-9. Sections tool

Figure-10. Sectional edges dialog

- Click on the **Add Edge** button from the dialog and select two or more desired edges or curves in the 3D view area. The preview of surface will be displayed; refer to Figure-11.

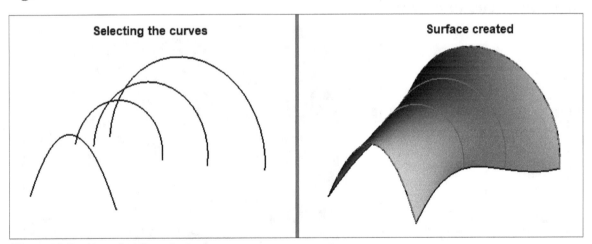

Figure-11. Preview of surface using curves

- After getting desired shape, click on the **OK** button from the dialog to create the surface.

Creating Surface by Extending a Face

The **Extend face** tool is used to create a surface by extending selected model face. The procedure to use this tool is given next.

- Select the face/surface you want to use for created extended surface and click on the **Extend face** tool from the **Surface** menu; refer to Figure-12. Preview of extended surface will be displayed; refer to Figure-13.

Figure-12. Extend face tool

Figure-13. Extend face surface created

- Select the newly created surface from drawing area or **Model Tree** and specify desired parameters in the **Property Editor** to specify extension length in various directions; refer to Figure-14.

Figure-14. Properties of extended face surface

Curve on mesh

The **Curve on mesh** tool is used to create approximated spline segments on top of a selected mesh. Note that Curve on mesh tool is active only if mesh object is available. The procedure to use this tool is discussed next.

- Click on the **Curve on mesh** tool from the **Surface** menu or **Toolbar** in **Surface** workbench; refer to Figure-15. The **Curve on mesh** dialog will be displayed; refer to Figure-16.

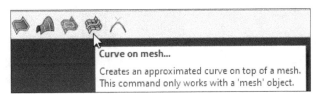

Figure-15. Curve on mesh tool

Figure-16. Curve on mesh dialog

- Click on the **Start** button from the dialog. The curve symbol will be attached to the cursor and you will be asked to create the curve on the mesh object.
- Click on the mesh object at desired location to specify first point of curve. Move the cursor away and click at desired location(s) to specify second point or multiple points of curve.
- After specifying desired number of points, right-click. A list of options will be displayed; refer to Figure-17.

Figure-17. Options on clicking RMB

- Click on the **Create** option to create the curve. The curve on the mesh object will be created; refer to Figure-18. Click on the **Close wire** option to join the last point to the first point if at least three points have been specified. Click on the **Clear** option to erase the specified points and allow to specify new points. Click on the **Cancel** option to erase the specified points and stop specifying new points.

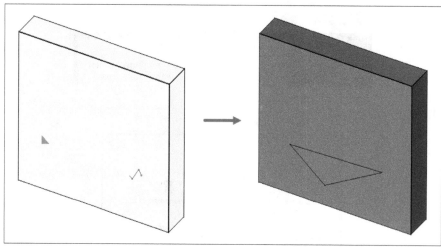

Figure-18. Curve on the mesh object created

- Specify desired minimum distance value between one point and another point in the **Snap tolerances to vertices** edit box from **Wire** area of the dialog.
- Specify desired angular deviation from one point in the mesh to another point in the **Split threshold** edit box from **Wire** area of the dialog.
- Select **Spline Approximation** check box to create spline objects.
- The **Tolerance to mesh** edit box display the imperfections of the mesh. The smaller the number, the mesh will be very fine.
- Select desired option from **Continuity** drop-down to determine the continuity of the spline.
- Select desired option from **Maximum curve degree** drop-down to determine the maximum degree of the spline to approximate the surface.
- Click on the **Close** button to close the dialog.

Blend Curve

The **Blend Curve** tool is used to create a bezier curve between two edges, with desired continuity. The procedure to use this tool is given next.

- Select two edges from the 3D view area and click on the **Blend Curve** tool from the **Surface** menu or **Toolbar** in **Surface** workbench; refer to Figure-19. The blend curve will be created; refer to Figure-20.

Figure-19. Blend Curve tool

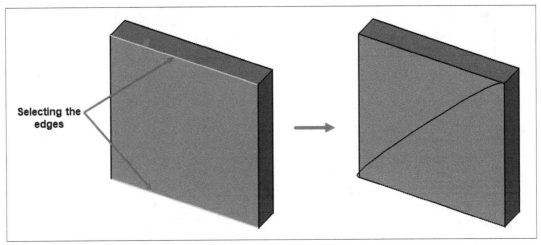

Figure-20. Blend curve created

INTRODUCTION TO MESH WORKBENCH

The **Mesh** workbench of FreeCAD is used to import, create, and modify mesh objects generated in FreeCAD as well as other CAD software. Select the **Mesh Design** option from the **Switch between workbenches** drop-down to activate the **Mesh** workbench. The tools will be displayed as shown in Figure-21. Various tools in this workbench are discussed next.

Figure-21. Mesh workbench tools

Importing Mesh Objects

The **Import mesh** tool is used to import mesh objects created in other CAD software like Alias, Inventor, and so on. The procedure to use this tool is given next.

- Click on the **Import mesh** tool from the **Meshes** menu or **Toolbar**. The **Import mesh** dialog box will be displayed; refer to Figure-22.

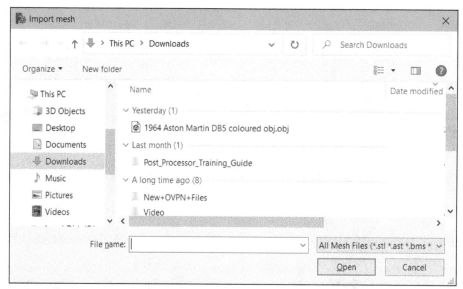

Figure-22. Import mesh dialog box

- Select desired mesh model file and click on the **Open** button. The mesh model will be displayed in the drawing area; refer to Figure-23.

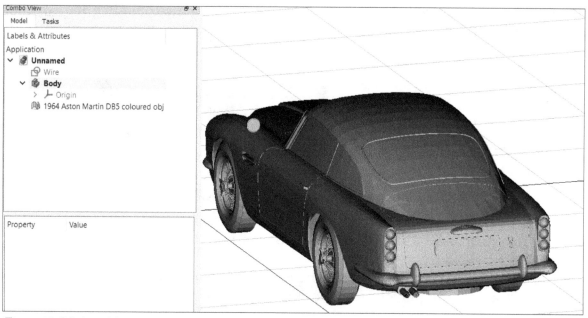

Figure-23. Mesh model imported

Exporting Mesh Model

The **Export mesh** tool is used to export mesh models created in FreeCAD for other CAD software. The procedure to use this tool is given next.

- Select the mesh model created in FreeCAD from **Model Tree** and click on the **Export mesh** tool from **Meshes** menu. The **Export mesh** dialog box will be displayed; refer to Figure-24.

Figure-24. Export mesh dialog box

- Select desired format in which you want to export the file in **Save as type** drop-down.
- Specify desired name of file in the **File name** edit box and click on the **Save** button to export the file.

Creating Mesh from Shape

The **Create mesh from shape** tool is used to create mesh from selected solid body created using **Part** or **Part Design** workbench. The procedure to use this tool is given next.

- Click on the **Create mesh from shape** tool from **Meshes** menu after selecting solid body to be converted. The **Tessellation** dialog will be displayed; refer to Figure-25.

Figure-25. Tessellation dialog

- Select desired tab from the **Meshing options** area to specify method to be used for generating mesh.

- Select the **Netgen** tab to choose fineness of mesh from predefined mesh qualities; refer to Figure-25. You can select **Very coarse**, **Coarse**, **Moderate**, **Fine**, and **Very fine** to use presets of mesh quality. If you want to manually define mesh quality then select the **User defined** option and specify related parameters in the edit boxes below the drop-down.

- Select the **Standard** tab if you want to specify surface deviation in the mesh to define quality of mesh; refer to Figure-26. Note that the smaller value of surface deviation gives finer mesh. The minimum value that can be specified for surface deviation is 0.001.

Figure-26. Standard mesh refinement

- Select the **Mefisto** tab if you want to specify maximum edge length in mesh to define quality of mesh; refer to Figure-27. Note that the smaller value of edge length gives finer mesh.

- Select the **gmsh** tab if you want to use gmsh mesher to define the quality of mesh; refer to Figure-28. Note that for Linux user, the external gmsh module is required.

Figure-27. Mefisto mesh refinement

Figure-28. Gmsh mesh refinement

- Specify the other parameters as desired in the dialog and click on the **OK** button. The mesh model will be generated.

Refinement

The **Refinement** tool remeshes a mesh object using the Gmsh mesher. The new mesh can be finer or coarser. The procedure to use this tool is discussed next.

- Select desired mesh object from the **Model Tree** or from the 3D view area which you want to remesh; refer to Figure-29.

Figure-29. Object selected to remesh

- Click on the **Refinement** tool from **Meshes** menu or **Toolbar** in the **Mesh Design** workbench. The **Remesh by gmsh** dialog will be displayed in the **Tasks** panel of **Combo View**; refer to Figure-30.

Figure-30. Remesh by gmsh dialog

- Select desired type of mesh from **Meshing** drop-down in the dialog.
- Enter desired value in the **Max element size** and **Min element size** edit boxes to specify the maximum and minimum size of the element, respectively.
- Specify desired angle of the element in the **Angle** edit box.
- Click on the **Path** button and specify the path for gmsh application.
- After specifying desired parameters, click on the **Apply** button. The gmsh operation will be running and the details of remeshing will be displaying in the box of the dialog; refer to Figure-31.

Figure-31. Running gmsh operation

• After the procedure is finished, click on the **Close** button to exit the tool.

Mesh Analysis Tools

The tools in **Analyze** cascading menu of **Meshes** menu are used to analyze and repair various aspects of mesh model imported/created in FreeCAD; refer to Figure-32. Various tools in this menu are discussed next.

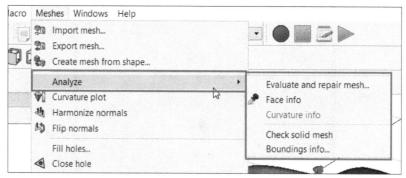

Figure-32. Analyze cascading menu

Evaluating and Repairing Mesh Model

The **Evaluate and repair mesh** tool is used to evaluate various aspects of mesh and repair them. The procedure to use this tool is given next.

• Click on the **Evaluate and repair mesh** tool from **Analyze** cascading menu of the **Meshes** menu. The **Evaluate & Repair Mesh** panel will be displayed; refer to Figure-33.

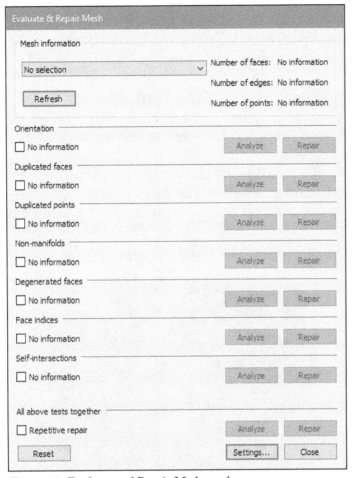

Figure-33. Evaluate and Repair Mesh panel

- There are various sections of panel to analyze various aspects of mesh model like orientation, duplicate faces, degenerated faces, and so on. Click on the **Analyze** button from the panel to check related results. If you want to analyze model based on all the analyses then click on the **Analyze** button from the **All above tests together** area of panel. The result of analyses will be displayed in respective sections of panel.
- Click on the **Repair** button from the panel to perform repairs in the model.
- Click on the **Settings** button from the panel to modify evaluation settings. The **Evaluation settings** dialog box will be displayed; refer to Figure-34.

Figure-34. Evaluation settings dialog box

- Select the **Check for non-manifold points** check box to check mesh model for non-manifold points. Non-manifold points are locations of model where you cannot define precision of vertices.
- Select the **Enable check for folds on surface** check box to check for folds in the surface of mesh.

- Select the **Only consider zero area faces as degenerated** check box to define which faces are to be considered as degenerated.
- After setting desired parameters, click on the **OK** button.
- Click on the **Close** button from the panel to exit the panel.

Face Info

The **Face info** tool is used to check information of a face in mesh. Using this tool, you can index of selected face. The procedure to use this tool is given next.

- Click on the **Face info** tool from the **Analyze** cascading menu of **Meshes** menu. The cursor will change to pick face cursor.
- Click on desired faces of mesh model. Indexes of the faces will be displayed; refer to Figure-35.

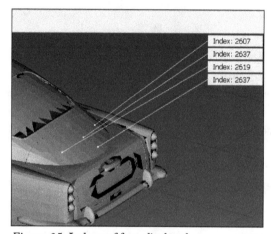

Figure-35. Indexes of faces displayed

- Right-click in the empty area of the drawing and select **Leave info mode** option from the shortcut menu.

Similarly, you can use the **Curvature info** tool of **Analyze** cascading menu in **Meshes** menu to check curvature information of mesh model. Note that this tool is active only when curvature plot has been generated.

Checking Solid Mesh

The **Check solid mesh** tool is used to check whether mesh is a solid mesh body or not. On clicking this tool, the **Solid Mesh** dialog box will be displayed showing information of mesh; refer to Figure-36.

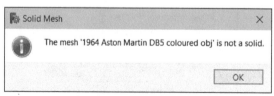

Figure-36. Solid Mesh dialog box

Checking Boundings Info

The **Boundings info** tool is used to check bounding parameters of the mesh body to define minimum and maximum point of the mesh; refer to Figure-37.

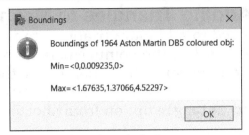

Figure-37. Boundings dialog box

Generating Curvature Plot

The **Curvature plot** tool is used to generate curvature plot for selected mesh model. Click on this tool after selecting mesh body, the curvature plot will be generated and displayed in the drawing area.

Harmonizing Normals

The **Harmonize normals** tool is used to make normals of faces in mesh share the same pattern. Click on the **Harmonize normals** tool from the **Meshes** menu after selecting the mesh body. All the normals of faces in mesh will point in same direction.

Flipping Normals of Mesh Faces

The **Flip normals** tool in **Meshes** menu is used to flip normal direction of mesh faces. After selecting mesh model, click on the **Flip normals** tool from the menu.

Filling Holes in Mesh

The **Fill holes** tool in **Meshes** menu is used to fill holes in the mesh based on specified edges. On clicking this tool, the **Fill holes** dialog box will be displayed; refer to Figure-38. Specify minimum number of edges to be considered for filling holes and click on the **OK** button. All the holes in mesh will be filled.

Figure-38. Fill holes dialog box

Closing Holes

The **Close hole** tool is used to close holes in selected faces of mesh. The procedure to use this tool is given next.

- Click on the **Close hole** tool from the **Meshes** menu. The cursor will change to face selection cursor.
- Click on the faces with empty holes/regions. The holes in selected faces will be filled.
- Right-click in the drawing area and select **Leave hole-filling mode** option to exit the hole closing mode.

Adding Triangles to Mesh

The **Add triangle** tool is used to add more triangles to the mesh body. On clicking this tool, you will be asked to select the points where more triangles for tessellation are to be added. One by one specify three points on the faces of mesh body to create a triangle. On selecting the points, a shortcut menu will be displayed; refer to Figure-39. Select the **Add triangle** option from shortcut menu. A triangle will be created in mesh.

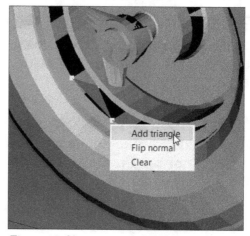

Figure-39. Shortcut menu for adding triangles

- Right-click in the drawing area and select **Finish** button to exit triangle creation mode.

Removing Components of Mesh

The **Remove components** tool is used to remove desired sections of the mesh model. The procedure to use this tool is given next.

- Click on the **Remove components** tool from the **Meshes** menu. The **Remove components** dialog will be displayed; refer to Figure-40.

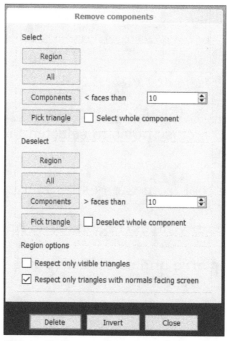

Figure-40. Remove components dialog

- Select the **Region** button from **Select** section and click & drag the cursor over faces of the mesh body to select them; refer to Figure-41. The faces in created region will get selected; refer to Figure-42.

Figure-42. Region selected

Figure-41. Selecting region on mesh body

- If you want to select full body of mesh then select the **All** button. If you want to select specified number of faces from center of body then specify number of faces to be selected in the **< faces than** edit box and click on the **Components** button. If you want to individually select triangles from the mesh body then click on the **Pick triangle** button and select desired faces.
- If you want to deselect faces then you can deselect the faces in the same way from the **Deselect** section of dialog.
- Select the **Respect only visible triangles** check box to select or deselect visible triangles from the model.
- Select the **Respect only triangles with normals facing screen** check box to only select/deselect the triangles which have normals pointing towards the screen.
- After selecting desired faces, click on the **Delete** button to delete faces.

Removing Components by Hand

The **Remove components by hand** tool is used to remove selected faces of the mesh model. The procedure to use this tool is given next.

- Click on the **Remove components by hand** tool from the **Meshes** menu. The cursor will change to a hand sign for selection.
- Click on desired faces of mesh model to select them and right-click in the drawing area. A shortcut menu will be displayed; refer to Figure-43.

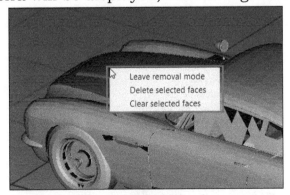

Figure-43. Shortcut menu for deleting faces

- Select the **Delete selected faces** option from the shortcut menu to delete selected faces. Select the **Clear selected faces** option to remove all faces from selection and start a new selection set. Select the **Leave removal mode** option to exit the face removal mode.

Creating Mesh Segments

The **Create mesh segments** tool is used to generate segments from the mesh model. The procedure to use this tool is given next.

- After selecting the mesh model, click on the **Create mesh segments** tool from the **Meshes** menu. The **Mesh segmentation** dialog will be displayed; refer to Figure-44.

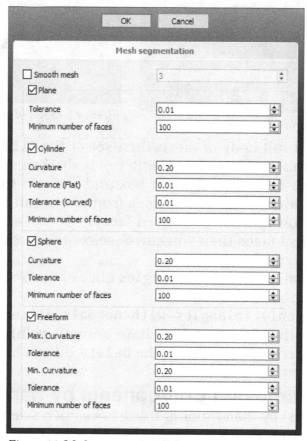

Figure-44. Mesh segmentation dialog

- Select check boxes for object types to be created in segmenting the mesh and specify related tolerances.
- After setting desired parameters, click on the **OK** button from the dialog. The segments will be created; refer to Figure-45.

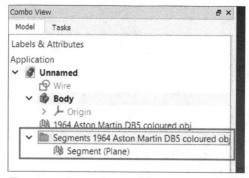

Figure-45. Segment created from mesh model

Similarly, you can use the **Create mesh segments from best-fit surfaces** tool to generate segments of mesh body.

Smoothening Mesh Model

The **Smooth** tool is used to smoothen selected mesh surface by specified degrees. The procedure to use this tool is given next.

- Select the mesh model from **Model Tree** or drawing area and then select the **Smooth** tool from the **Meshes** menu. The **Smoothing** dialog will be displayed; refer to Figure-46.

Figure-46. Smoothing dialog

- Select the **Taubin** radio button to use taubin smoothing and select the **Laplace** radio button to use Laplacian smoothing. Figure-47 shows difference between two smoothing methods.

Figure-47. Smoothing methods

- Specify desired parameters in the dialog and click on the **OK** button to smoothen mesh model.

Decimating Mesh Model

The **Decimation** tool reduces the number of faces in mesh objects. The procedure to use this tool is discussed next.

- After selecting mesh model, click on the **Decimation** tool from the **Meshes** menu. The **Decimating** dialog will be displayed; refer to Figure-48.

Figure-48. Decimating dialog

- Move the slider towards **None** or **Full** as desired to specify the reduction of faces.
- Select the **Absolute number** check box to specify the reduction of absolute number of faces by entering desired value in the edit box of **Reduction** area in the dialog.
- Specify desired tolerance value in the edit box of **Tolerance** area in the dialog.
- After specifying desired parameters, click on the **OK** button to complete the procedure.

Scaling Mesh Model

The **Scale** tool is used to scale up/down the mesh model. The procedure to use this tool is given next.

- After selecting mesh model, click on the **Scale** tool from the **Meshes** menu. The **Scaling** dialog box will be displayed; refer to Figure-49.

Figure-49. Scaling dialog box

- Specify desired value of scale factor in the edit box and click on the **OK** button. A value more than **1** will increase the size of mesh model and a value less than **1** but greater than **0** will decrease the size of mesh model.

Creating Regular Mesh Solids

The **Regular solid** tool is used to create regular mesh solids. The procedure to use this tool is given next.

- After selecting the mesh body, click on the **Regular solid** tool from **Meshes** menu. The **Regular Solid** dialog box will be displayed; refer to Figure-50.

Figure-50. Regular Solid dialog box

- Select desired option from the **Solid** drop-down to define shape of solid and specify related parameters in the dialog box.
- After setting parameters, click on the **Create** button to create solid object at the origin.
- After creating solid, click on the **Close** button from the dialog box to exit.

Performing Boolean Operations on Meshes

The tools in **Boolean** cascading menu of **Meshes** menu are used to perform boolean operations like performing union, intersection, and difference of two mesh objects. Note that you need to install OpenSCAD using web link http://www.openscad.org/ and specify path for exe file of OpenSCAD in **Preferences** dialog box. The procedure to perform boolean operations is given next.

- Select two meshes from the **Model Tree** on which you want to perform boolean operations. The tools in **Boolean** cascading menu of **Meshes** menu will become active.
- Click on the **Union** tool from the cascading menu if you want to join two mesh bodies; refer to Figure-51.

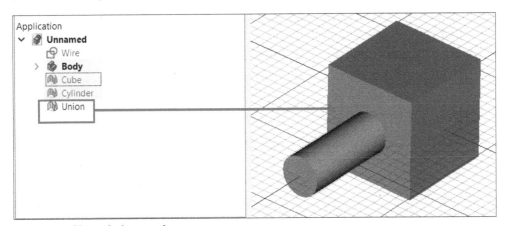

Figure-51. Union body created

- Click on the **Difference** tool to subtract 2nd selected object from 1st selected object. The **Difference** feature will be displayed; refer to Figure-52.

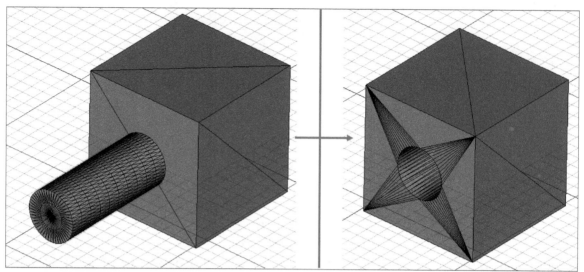

Figure-52. Difference feature

- Click on the **Intersection** tool from the **Boolean** cascading menu after selecting two mesh objects, the region common in both objects will be created as intersection object; refer to Figure-53.

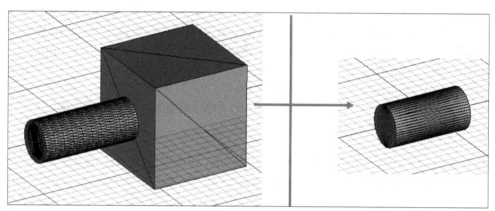

Figure-53. Intersection feature created

Mesh Cutting Operations

The tools in **Cutting** cascading menu of **Meshes** menu are used to cut mesh object using plane, section, or polygon selection; refer to Figure-54. Various tools in this cascading menu are discussed next.

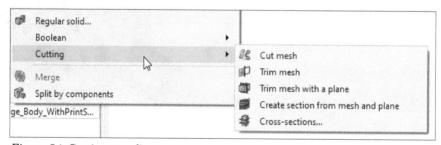

Figure-54. Cutting cascading menu

Cut Mesh Tool

The **Cut Mesh** tool is used to cut a mesh by drawing a polygon selection box. The procedure to use this tool is given next.

- Click on the **Cut Mesh** tool from the **Cutting** cascading menu of the **Meshes** menu after selecting the mesh body to be cut. The cursor will change to a scissor icon and you will be asked to draw a polygonal selection box.
- Draw a polygon on the body by specifying end points and right-click in the drawing area. A shortcut menu will be displayed; refer to Figure-55.

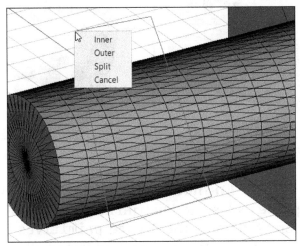

Figure-55. Shortcut menu for cutting

- Select the **Inner** option from shortcut menu to delete the mesh elements falling inside drawn polygon; refer to Figure-56. Select the **Outer** option from shortcut menu to delete mesh elements falling outside the selection polygon. Select the **Split** option from shortcut menu to delete the area falling inside selection polygon and create a new body. Select the **Cancel** option to abort the operation.

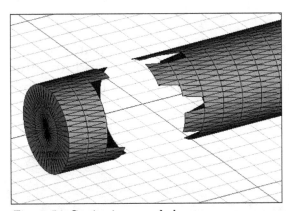

Figure-56. Cutting inner mesh elements

Trim Mesh Tool

The **Trim mesh** tool is used to trim a portion of mesh body using drawn selection polygon. Note that when you use **Cut mesh** tool then full elements falling inside selection polygon will be removed but when you use **Trim mesh** tool then these elements will also be trimmed at the boundary of selection polygon. The procedure to use **Trim mesh** tool is given next.

- Click on the **Trim mesh** tool from **Cutting** cascading menu of the **Meshes** menu after selecting the mesh body to be trimmed. You will be asked to draw a selection polygon.
- Create the polygon by specifying end points and right-click. A shortcut menu will be displayed.

- Select desired option from the shortcut menu to trim the mesh body; refer to Figure-57. The options in shortcut menu are same as discussed earlier.

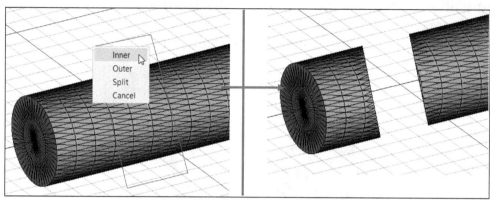

Figure-57. Trimming mesh

Trim mesh with a plane

The **Trim mesh with a plane** tool is used to trim mesh body using selected plane. Note that the plane must be created using the **Create primitives** tool of **Part** workbench; refer to Figure-58. The procedure to use this tool is given next.

Figure-58. Create primitives tool

- Click on the **Trim mesh with a plane** tool from the **Cutting** cascading menu of the **Meshes** menu after selecting mesh body and an intersecting plane. The **Trim by plane** dialog box will be displayed; refer to Figure-59.

Figure-59. Trim by plane dialog box

- Select the **Below** button from the dialog box to remove mesh body on the front side of plane and keep back side of mesh body.
- Select the **Above** button from the dialog box to remove mesh body on the back side of plane and keep front side of mesh body.
- Select the **Split** button from the dialog box to split the mesh body into two pieces.

Create section from mesh and plane

The **Create section from mesh and plane** tool is used to generate section curves at the intersection of mesh body and plane. The procedure to use this tool is given next.

- After selecting plane and mesh body, click on the **Create section from mesh and plane** tool from the **Cutting** cascading menu of the **Meshes** menu. The section will be created; refer to Figure-60.

Figure-60. Section created

Creating Cross Sections

The **Cross section** tool creates multiple cross-sections across mesh objects. The cross-sections are taken parallel to one of the main global planes (XY, XZ, or YZ). The procedure to use this tool is discussed next.

- After selecting one or more mesh objects, click on the **Cross-sections** tool from the **Cutting** cascading menu of the **Meshes** menu. The **Cross sections** dialog will be displayed; refer to Figure-61.

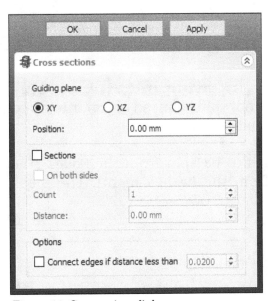

Figure-61. Cross sections dialog

- Select desired guiding plane from **XY, XZ,** or **YZ** radio button to be used to create the cross-sections.

- Enter desired value in the **Position** edit box to specify the position of guiding plane from the origin. The default position is based on the center of the bounding box of the selected mesh objects.
- Select the **Sections** check box to create multiple cross-sections.
- Select **On both sides** check box to create the cross sections on both sides of the guiding plane.
- Specify desired number of cross sections in the **Count** edit box.
- Specify desired distance between the cross sections in the **Distance** edit box.
- Select **Connect edges if distance less than** check box and specify desired value in the activated edit box.
- After specifying desired parameters, click on the **OK** button from the dialog to complete the procedure.

Merging Mesh

The **Merge** tool is used to combine two or more mesh bodies to form a single mesh. The procedure to use this tool is given next.

- Select the mesh bodies to be merged from the **Model Tree** and click on the **Merge** tool from the **Meshes** menu. The merged mesh body will be created; refer to Figure-62.

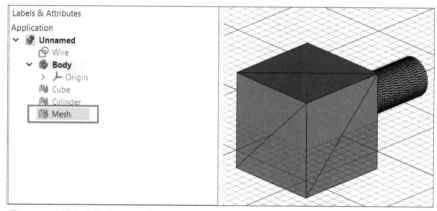

Figure-62. Mesh body created

Splitting Mesh Object

The **Split by components** tool splits a mesh object into its components. The procedure to use this tool is discussed next.

- Select the mesh bodies to be split from the **Model Tree** and click on the **Split by components** tool from the **Meshes** menu. The mesh body will be split to its components; refer to Figure-63.

Figure-63. Mesh body splitted into components

Unwrap Mesh

The **Unwrap Mesh** tool creates a flat representation of a mesh object by unwrapping or unfolding it. The procedure to use this tool is discussed next.

* Select desired mesh body from the **Model Tree** to unwrap and click on the **Unwrap Mesh** tool from the **Meshes** menu. The meshed body will be unwrapped as a Part Feature; refer to Figure-64.

Figure-64. Meshed object unwrapped

Unwrap Face

The **Unwrap Face** tool is used to create a flat representation of a face of a shape object. The procedure to use this tool is same as discussed for **Unwrap Mesh** tool.

ROBOT WORKBENCH

The tools in **Robot** workbench are used to simulate path of robot movements. A robot is generally used for performing repeating tasks like welding an edge, tightening some screws, and so on which have predefined paths. The tools in **Robot** workbench are shown in Figure-65. Various tools of this workbench are discussed next.

Figure-65. Robot workbench tools

Inserting Robot

The tools in **Insert Robots** cascading menu of **Robot** menu are used to insert different type of robots for path simulation; refer to Figure-66. Select desired robot from the menu for which you want to simulate the toolpath. The robot will be displayed in drawing area; refer to Figure-67.

Figure-66. Insert Robots cascading menu

Figure-67. Robot placed in drawing

Creating Robot Trajectory

The **Create trajectory** tool is used to create an empty trajectory path for robot. Click on the **Create trajectory** tool from the **Robot** menu. An empty trajectory feature will be added in the **Model Tree**; refer to Figure-68.

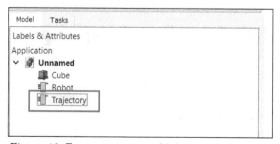

Figure-68. Empty trajectory added

Adding Edges to Trajectory

The **Edge to Trajectory** tool is used to add selected edges to the trajectory. The procedure to use this tool is given next.

- Click on the **Edge to Trajectory** tool from the **Robot** menu. The dialogs to add edges to trajectory will be displayed; refer to Figure-69.

Figure-69. Dialogs for adding edges to trajectory

- Select the edges to be added in the trajectory while holding the **CTRL** key; refer to Figure-70.

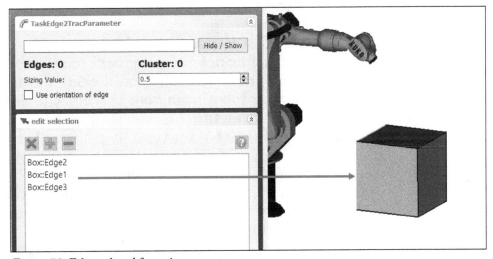

Figure-70. Edges selected for trajectory

- Click on the **Apply** button from the dialog after selecting edges and click on the **OK** button to add edges to trajectory.

Setting Default Orientation of Robot

The **Set default orientation** tool is used to set initial placement position and orientation of the selected robot. The procedure to use this tool is given next.

- Click on the **Set default orientation** tool from the **Robot** menu. The **Placement** dialog box will be displayed; refer to Figure-71.

Figure-71. Placement dialog box

- Set desired parameters in the **Translation** area to move robot to desired location.
- Set desired values in **X**, **Y**, and **Z** edit boxes of **Center** area to define location of center of Robot for translation.
- In the **Rotation** area, you can rotate robot about selected axis by specifying desired values in the **Axis** edit boxes and define rotation angle in the **Angle** edit box.
- After setting desired changes, click on the **Apply** button to apply orientation and click on the **OK** button to exit the dialog box.

Setting Default Values for Robot

The **Set default values** tool is used to define default movement parameters for robot like velocity of movement, acceleration, and so on. Click on the **Set default values** tool from the **Robot** menu after selecting robot from **Model Tree**. The **Set default speed** dialog box will be displayed; refer to Figure-72. Specify desired value of speed and click on the **OK** button. The **Set default continuity** dialog box will be displayed; refer to Figure-73. Select the **False** option if you want robot movements to be in fragments. Select the **True** option if you want robot movements to be continuous. After selecting desired option, click on the **OK** button. The **Set default acceleration** dialog box will be displayed; refer to Figure-74. Specify desired value in the edit box for acceleration and click on the **OK** button.

Figure-72. Set default speed dialog box   *Figure-73. Set default continuity dialog box*

Figure-74. Set default acceleration dialog box

Adding Tool to Robot

To add a tool to Robot arm, first create the model of tool using **Part Design** workbench. Now, select the **Robot** and part model created as tool from **Model Tree**, switch to **Tasks** panel in the **Combo View** and click on the **Add tool** option from **Robot tools** dialog; refer to Figure-75. The selected body will be added as tool to robot arm; refer to Figure-76.

Figure-75. Add tool option

Figure-76. Tool added to robot arm

Simulating Robot Trajectory

The **Simulate a trajectory** tool is used to simulate movements of robot on the trajectory earlier created. Select the Robot and trajectory features from the **Model Tree** and click on the **Simulate a trajectory** tool from the **Robot** menu. Dialogs will be displayed in the **Tasks** panel to simulate robot trajectory; refer to Figure-77. You can move the slider in **Trajectory** dialog to check movement of robot arm or use the buttons to run simulation.

Figure-77. Dialogs for simulating robot

3D PRINTING TOOLS WORKBENCH

The tools in **3D Printing Tools** workbench are used to scale and modify mesh models. To start **3D Printing Tools** workbench, select the **3D Printing Tools** option from the **Switch between workbenches** drop-down in the **Toolbar**; refer to Figure-78. Make sure to active the **3D Printing Tools** addon from the **Addon Manager**. Various tools in this workbench are discussed next.

Figure-78. 3D Printing Tools workbench

Converting Imperial Mesh to Metric and vice-versa

Select the mesh model and click on the **Converts Imperial Mesh to Metric** button from the **Toolbar**. The imperial mesh model will be converted to metric mesh model.

Similarly, click on the **Converts Metric Mesh to Imperial** button from the **Toolbar**. The metric mesh model will be converted to imperial mesh model.

Reducing and scaling mesh by 50%

Select the mesh model and click on the **Reduces Mesh by 50%** button from the **Toolbar**. The mesh model will be scaled down by 50%.

Similarly, click on the **Scale Mesh by 50%** button from the **Toolbar**. The mesh model will be scaled up by 50%.

Scaling mesh by variable scale factor

Select the mesh model and click on the **Scale Geometry by variable scale factor** button from the **Toolbar**. The mesh model will be scaled by variable scale factor.

Reducing mesh by 95%

Select the mesh model and click on the **Reduces Mesh by 95% to fit 150mm Printer Bed** button from the **Toolbar**. The mesh model will be scaled down by 95% to fit 150mm Printer Bed.

Creating mesh box

Select the mesh model and click on the **Create mesh box size of Printer** button from the **Toolbar**. The mesh box by the size of the Printer Bed will be generated.

Converting mesh to refined solids

Select the mesh model and click on the **Module to convert Mesh to Refined Solids** button from the **Toolbar**. The mesh model will be converted to refined solid model.

Changing objects transparency

Select the mesh model and click on the **Changes objects 50% Transparency** [icon] button from the **Toolbar**. The transparency of mesh model will be changed to 50%.

Similarly, click on the **Changes objects 0% Transparency** [icon] button from the **Toolbar**. The transparency of mesh model will be changed to 0%.

Hiding or showing the selected objects

Select the mesh model and click on the **Hides Selected Objects** [icon] button from the **Toolbar**. The selected mesh model will be hidden.

Similarly, click on the **Unhides Selected Objects** [icon] button from the **Toolbar**. The selected mesh model will be shown.

Modifying objects to random color

Select the mesh model and click on the **Modifies objects to Random Colour** [icon] button from the **Toolbar**. The color of model will be changed to 11 different colors.

Changing color of objects

Select the mesh model and click on desired color buttons, viz. **White, Yellow, Orange, Red, Pink, Purple, Blue, Cyan, Green, Brown,** and **Black** from the **Toolbar** to change the respective colors of the model.

Changing the line width

Select the mesh model and click on the **Changes Line Width to 2.0** [icon] button from the **Toolbar**. The line width of the selected model will be changed to 2.00.

Turning on or off the Bounding box size

Create and select the bounding box of the mesh model and then click on the **Turn on and off the Bounding box size** [icon] button from the **Toolbar**. The bounding box size of the model will be turned on or off.

Defining Printer bed size

Click on the **Change default Bed Length Size** [icon] button from the **Toolbar**. The **Printer Bed Length** dialog box will be displayed; refer to Figure-79. Specify desired Printer Bed Size in the edit box of the dialog box and click on the **OK** button. The value of size will be changed.

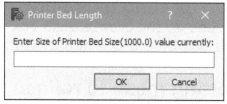

Figure-79. Printer Bed Length dialog box

Defining Die length size

Click on the **Change default Die Length Size** button from the **Toolbar**. The **Die Size** dialog box will be displayed; refer to Figure-80. Specify desired size of die to scale down in the edit box of the dialog box and click on the **OK** button. The value of size will be changed.

Figure-80. Die Size dialog box

Defining the Scale factor

Click on the **Change default Variable scale factor** 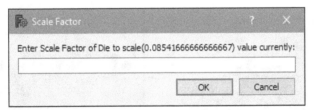 button from the **Toolbar**. The **Scale Factor** dialog box will be displayed; refer to Figure-81. Specify desired value of the scale factor of die to scale in the edit box of the dialog box. The value of scale factor of die will be changed.

Figure-81. Scale Factor dialog box

SHEET METAL WORKBENCH

The Sheetmetal workbench is external workbench to be added in FreeCAD using the Addon manager. After adding workbench, select the **Sheet Metal** option from the **Switch between workbenches** drop-down of the toolbar to activate sheet metal design tools; refer to Figure-82. Various tools of this workbench are discussed next.

Figure-82. Sheet Metal toolbar

Making Base Wall

The **Make Base Wall** tool is used to create base flange for sheetmetal design. Note that you must have a closed sketch for using this tool. The procedure to use this tool is given next.

- Make sure there is a closed loop sketch in the graphics area created using Sketcher workbench. Select the sketch from **Model** tab in **Combo View** and click on the **Make Base Wall** tool from **Toolbar** or **Sheet Metal** menu. The base wall/flange will be created; refer to Figure-83.

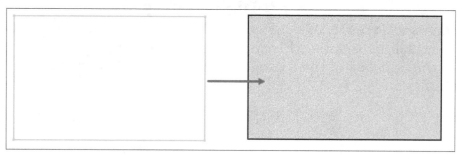

Figure-83. Base wall created

Creating Bend Wall

The **Make Wall** tool is used to create a bend wall using selected edge(s). The procedure to use this tool is given next.

- Select the edges of base flange to be used for creating bend walls and click on the **Make Wall** tool from the **Toolbar** or **Sheet Metal** menu. The bend walls will be created; refer to Figure-84.

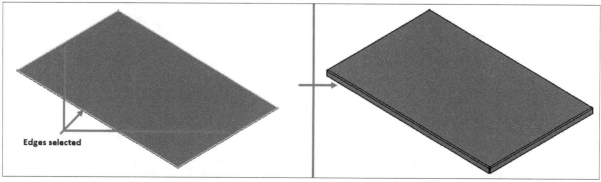

Figure-84. Creating bend wall

- To increase the length of bend walls, select the **Bend** from **Model** tab in **Combo View**. The Property form will be displayed with parameters related to bend walls; refer to Figure-85.

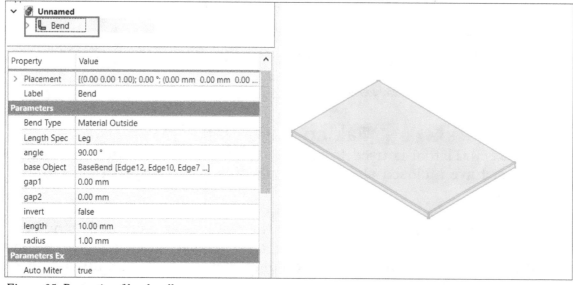

Figure-85. Properties of bend wall

- Click in the **length** edit box of **Parameters** section in **Property** form and enter desired length for bend walls; refer to Figure-86.

Figure-86. Changing length of bend wall

- Specify bend radius for flanges in the **radius** edit box of **Property** form. Similarly, you can specify other parameters in the form. Click in the empty area to exit the editing mode.

Extending Sheet Metal Flange Face

The **Extend Face** tool is used to extend selected edge of flange by specified value. The procedure to use this tool is given next.

- Select the edge of flange wall to be extended and click on the **Extend Face** tool from the **Toolbar** or **Sheet Metal** menu. The edge will be extended; refer to Figure-87.
- Select the newly created extend feature from the **Combo View** and specify desired parameters in the **Property** form.

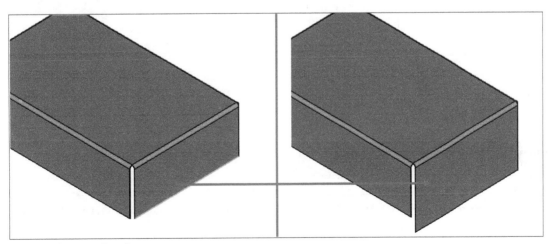

Figure-87. Extending flange

Applying Sketch Bend

The **Fold a Wall** tool is used to bend selected flange using sketch line. The procedure to use this tool is given next.

- Select face of flange and sketch line from graphics area. The **Fold a Wall** tool will become active. Click on the tool from the **Toolbar** or **Sheet Metal** menu. The sketch bend will be created; refer to Figure-88.

Figure-88. Applying sketch bend

Unfolding Sheet Metal Bends

The **Unfold** tool is used to unfold all the bends of the model while keeping a selected face fixed. The procedure to use this tool is given next.

- Select the face of sheet metal part that you want to be fixed and click on the **Unfold** tool from the **Toolbar** or **Sheet Metal** menu. The **Unfold sheet metal object** dialog will be displayed in the **Tasks** panel of **Combo View**; refer to Figure-89.

Figure-89. Face selected

- Select the **Generate projection sketch** check box to create projection sketch of unfolded sheet metal part.
- Select the **Separate projection layers** check box to create a new layer for projection.
- Click on color button from the dialog to change it.
- Select the **Manual K-factor** check box to manually specify K-factor for sheet metal design.
- Select the **Use Material Definition Sheet** check box if you want to use sheet metal data specified in material properties.

- After setting desired parameters, click on the **OK** button from the dialog. The unfolded model will be displayed; refer to Figure-90.

Figure-90. Unfolded sheet metal model

Adding Corner Relief

The **Add Corner Relief** tool is used to apply cut at the intersection for easy folding/ unfolding of bends at intersection. The procedure to use this tool is given next.

- Select two edges of bends intersecting at a corner point; refer to Figure-91. The **Add Corner Relief** tool will become active. Click on the tool from **Toolbar** or **Sheet Metal** menu. The **CornerRelief** feature will be added in the **Model** tab of **Combo View**. The **Property** form will be displayed with parameters related to corner relief; refer to.

Figure-91. Edges selected for corner relief

Figure-92. Property form

- Select desired option from the **Relief Sketch** field and set related parameters in the **Property** form. After setting parameters, click on the **Refresh** tool from **Toolbar**.
- Click in the empty area to exit editing mode.

Applying Solid Corner Relief

The **Make Relief** tool is used to apply corner relief at solid corner of model. The procedure to use this tool is given next.

- Select corner point of a flange/wall. The **Make Relief** tool will become active. Click on the tool from the **Toolbar** or **Sheet Metal** menu. The relief will be created; refer to Figure-93.

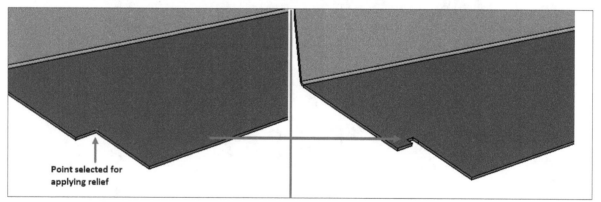

Figure-93. Relief created

- Select the newly created **Relief** feature from the **Model** panel of **Combo View** and set related parameters in the **Property** form.
- Click in the empty area to exit editing mode.

Applying Rip Feature

The rip feature is applied to cut gap at the edges of a solid body when you are converting a solid into sheet metal part. The **Make Junction** tool is used to create rip feature. The procedure to use this tool is given next.

- Select the edge of solid body which you want to convert into a sheetmetal part. The **Make Junction** tool will become active; refer to Figure-94. Select the tool from **Toolbar** or **Sheet Metal** menu. The rip feature will be created; refer to Figure-95.

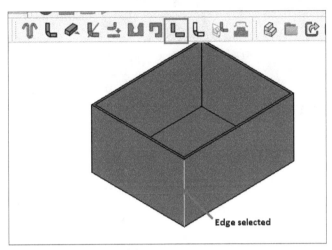

Figure-94. Edge selected for rip

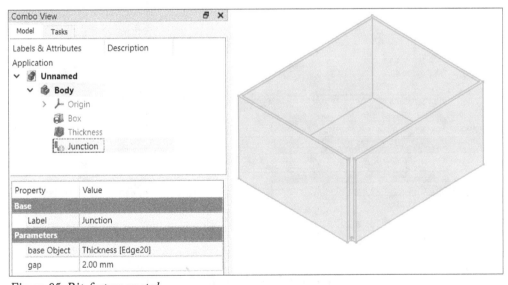

Figure-95. Rip feature created

- Select the recently created **Junction** feature from the **Combo View** and modify the gap as desired from the **Property** form.
- Click in the empty area to exit editing mode.

Creating Bend

The **Make Bend** tool is used to create bend at the intersection of two faces of a solid body. This tool is also useful in converting a solid into sheet metal part. The procedure to use this tool is given next.

- Select an edge of a solid body to be converted into sheet metal part. The **Make Bend** tool will become active; refer to Figure-96.

Figure-96. Make Bend tool

- Select the tool from **Toolbar** or **Sheet Metal** menu. The selected edge will be converted into a bend.
- Select the newly created **SolidBend** feature from **Combo View** and modify the parameters in **Property** form as desired; refer to Figure-97.

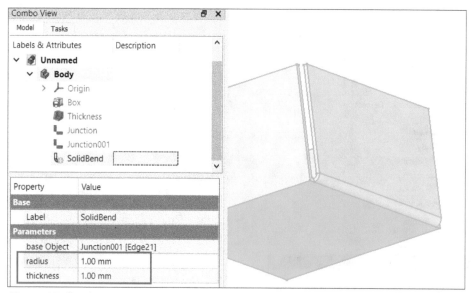

Figure-97. Bend parameters

Making Extruded Cut

The **Sketch On Sheet metal** tool is used to create an extruded cut from the sheet metal part. The procedure to use this tool is given next.

- Select a closed loop sketch created on sheet metal part face and sheet metal part from the graphics area. The **Sketch On Sheet metal** tool will become active; refer to Figure-98.
- Click on the tool from **Toolbar** or **Sheet Metal** menu. The extruded cut feature will be displayed; refer to Figure-99.

Figure-98. Selection for extruded cut

Figure-99. Extruded cut created

Applying Forming Tool

A forming tool in sheet metal design is used to indent its shape of sheet metal part surface. The procedure to use this tool is given next.

- Select the face of sheet metal part and face of solid body to be used as forming tool. The **Make Forming in Wall** tool will become active; refer to Figure-100.

Figure-100. Make Forming in Wall tool

- Click on the tool from the **Toolbar** or **Sheet Metal** menu. The form of solid body face will be applied; refer to Figure-101.

Figure-101. Forming

There are many other workbenches that can be used with FreeCAD to perform various tasks. You can download them and add them to the software using FreeCAD website links.

Chapter 12

Miscellaneous Tool

Topics Covered

The major topics covered in this chapter are:

- *Edit menu*
- *View menu*
- *Tools menu*
- *Macro menu*

INTRODUCTION

In previous chapters, you have learned about various workbenches of FreeCAD. There are some tools in FreeCAD which are common for all the workbenches and are used to modify basic parameters of software like changing orientation of model, moving object in 3D space, and so on. These tools are available in various menus of FreeCAD which are discussed next.

EDIT MENU

The tools in **Edit** menu are used to perform common tasks like Undo, Redo, Copy, Paste, and so on; refer to Figure-1. These tools are discussed next.

Figure-1. Edit menu

Undo Operation

The **Undo** tool in **Edit** menu is used to reverse the effect of any operation performed recently in the software. For example, you have drawn a line in FreeCAD sketcher and dragged it to a different location but now you want it to be back at its original location then you can use the **Undo** tool to do so. You can also use the shortcut key **CTRL+Z** to perform undo operation.

Redo Operation

The **Redo** tool in **Edit** menu is used to reverse the Undo operation. You can also use the shortcut key **CTRL+Y** to perform redo operation.

Cut, Copy, and Paste Operations

The **Cut** tool in **Edit** menu is used to copy selected objects in temporary memory of software and delete them from original location. You can also use **CTRL+X** shortcut keys for cut operation. The **Copy** tool in **Edit** menu is used to copy selected objects in temporary memory of software while keeping the original objects intact. You can also use **CTRL+C** shortcut keys for copy operation. After selecting **Cut** or **Copy** tool, click on the **Paste** tool from **Edit** menu to place copy of the objects in temporary memory. You can also use the **CTRL+V** shortcut keys for paste operation.

Duplicating Selected Objects

The **Duplicate selected objects** tool is used to generate a duplicate copy of selected objects. This tool is useful when creating assemblies using same components.

Refresh Operation

The **Refresh** tool is used to recompute any changes made in the model. Sometimes, changes made in the model do not reflect automatically in drawing area. For refreshing operation, you need to first mark an object for refreshing by selecting the object, right-clicking on it, and then selecting the **Mark to recompute** option from shortcut menu; refer to Figure-2. After marking an object to recompute, select the **Refresh** tool from **Edit** menu or press **F5** key from keyboard to update changes in the model.

Figure-2. Mark to recompute option

Box Selection

The **Box Selection** tool is used to select objects which fall inside the boundaries of box selected created. To use box selection, select the **Box selection** tool from **Edit** menu and draw a box by dragging cursor; refer to Figure-3.

Figure-3. Box selection created

Select All

The **Select All** tool is used to select all the objects in drawing area.

Deleting Objects

The **Delete** tool in **Edit** menu is used to delete selected objects. You can select the objects from drawing area as well as **Model Tree**. You can also use the shortcut key **DELETE** from keyboard to delete selected objects.

Sending to Python Console

The **Send to Python Console** tool creates a variable in the Python console referencing a selected object. If a subshape of the object is selected, two additional variables are created, one referencing the shape of the object and the other referencing the subshape itself. The variables and the code involved can be used to develop python code. You can also use the shortcut key **CTRL+SHIFT+P** from the keyboard.

Modifying Placement of Objects

The **Placement** tool in **Edit** menu is used to modify placement location and orientation of selected object. The procedure to do so is given next.

• Select the object whose placement is to be modified and click on the **Placement** tool from **Edit** menu. The **Placement** dialog will be displayed in **Tasks** panel of **Combo View**; refer to Figure-4.

Figure-4. Placement dialog

The options in this dialog have been discussed earlier. Note that you may need to refresh the model after performing placement operation.

Alignment Operation

The **Alignment** tool is used to align two selected objects about selected points. The procedure to use this tool is given next.

*   Select the two objects that you want to align and click on the **Alignment** tool from the **Edit** menu. The two selected objects will be displayed in different viewports of drawing area; refer to Figure-5.

Figure-5. Objects displayed for alignment

*   Select equal number of points on both the objects (movable as well as fixed object) and right-click in the drawing area. A shortcut menu will be displayed; refer to Figure-6.

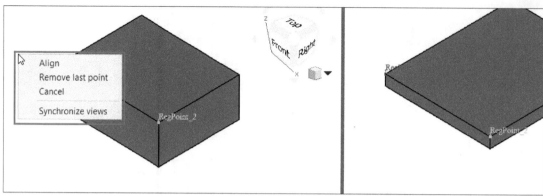

Figure-6. Shortcut menu for alignment

- Select the **Align** option from the shortcut menu to perform alignment operation. The two objects will be aligned as per the points selected on the bodies.

Activating Edit Mode

The **Toggle Edit mode** tool is used to modify a feature created in FreeCAD using its related dialog. The procedure to use this tool is given next.

- Select the feature that you want to be modified and click on the **Toggle Edit mode** tool from the **Edit** menu. The related dialog for modifying feature will be displayed. If you have selected an object/body which do not have any dialog associated with it then **Translation** and **Rotation** options will be displayed in the dialog; refer to Figure-7.

Figure-7. Increments dialog

- Set the parameters as desired and click on the **OK** button from the dialog.

Edit Mode options

The options in the **Edit Mode** cascading menu define the edit mode to be used when an object is double-clicked in the **Model Tree**; refer to Figure-8.

Figure-8. Edit mode cascading menu

- If you select the **Default** option then **Primitive parameters** dialog will be displayed on double-clicking the object in the **Model Tree**.
- If you select the **Transform** option then **Increments** dialog will be displayed on double-clicking the object in the **Model Tree**.
- If you select the **Color** option then **Set color per face** dialog will be displayed on double-clicking the object in the **Model Tree**.

The **Cutting** option in **Edit mode** cascading menu is still in experiment mode. All the dialogs have been discussed earlier in this book.

Preferences

The **Preferences** tool is used to modify underlying parameters of FreeCAD application. The procedure to use this tool is given next.

- Click on the **Preferences** tool from the **Edit** menu. The **Preferences** dialog box will be displayed; refer to Figure-9.

Figure-9. Preferences dialog box

The options in the dialog box are categorized in various pages like **General** page, **Display** page, **Python** page, and so on. To modify options related to a category, select the respective page option from the left area of the dialog box. Some of the most common options of this dialog box are discussed next.

General Page

The options in **General** page are used to define general parameters for interface of application, documents, code editor, and so on. These options are discussed next.

General tab

- Select desired option from the **Language** drop-down to specify language used by application interface.
- Select desired option from the **Unit system** drop-down to define unit system used for creating model.
- Specify desired value in the **Number of decimals** edit box to define precision up to which you want to specify values in FreeCAD.
- If you are using **Building US (ft-in/sqft/cft)** unit system then you need to specify minimum fractional value allowed in the **Minimum fractional inch** drop-down.
- Select desired option from **Number format** drop-down to specify the format of number. Select **Operating system** option to define the decimal separator by the selected operating system. Select **Selected language** option to define the decimal separator by the selected FreeCAD interface language. Select **C/POSIX** option to use the point as the decimal separator.
- Select **Substitute decimal separator** check box to substitute the numerical keypad decimal separator with the local separator.
- Select desired option from the **Style sheet** drop-down to define color and style of interface of application. For example, select the **Dark-blue** option to make background of application dark and selected tools/options highlighted in blue color.
- Select desired option from the **Size of toolbar icons** drop-down to define size of buttons in the toolbar.
- Select desired option from **Tree view mode** drop-down to customize how tree view is shown in the panel.
- Set desired value in the **Size of recent file list** edit box to specify number of recently worked files being displayed in **Recent files** cascading menu of **File** menu.
- Select the **Enable tiled background** check box to display background as tiled image.
- Select **Enable cursor blinking** check box to enable the text cursor to be blink.
- Select the **Enable splash screen at start up** check box to display splash screen showing modules being loaded in software when start the FreeCAD software.

Document tab

The options in **Document** tab are used to specify settings related to documents in FreeCAD; refer to Figure-10.

Figure-10. Document tab

- Select the **Create new document at start up** check box to automatically create a new document when you start the software.
- Set desired value in the **Document save compression level** edit box to define the compression level being used when saving the file. A higher value for compression will make saved files occupy less memory but it will also take longer to open the files in application. The minimum compression value is **0** and maximum compression value is **9**.
- Select the **Using Undo/Redo on documents** check box to allow undo/redo operations on the model. You can specify maximum number of undo/redo steps allowed in the **Maximum Undo/Redo steps** edit box.
- Select **Allow aborting recomputation** check box to allow the user to abort document recomputation by pressing **ESC**. This feature may slightly increase recomputation time.
- Select the **Run AutoRecovery at startup** check box to check for auto recovered documents at the startup.
- Select the **Save AutoRecovery information every** check box to automatically save file at regular interval specified in the edit box next to check box.
- Select the **Save thumbnail into project file when saving document** check box to add a thumbnail of model with project file while saving it.
- Select the **Add the program logo to the generated thumbnail** check box to also add logo of application with the thumbnail of model.
- Select the **Maximum number of backup files to keep when resaving document** check box to define the number of documents to be created as backup for a model.
- On selecting the **Use date and FCBak extension** check box, the backup files will get extension **'.FCBak'** and file names get date suffix according to the specified format in the **Date format** edit box.
- Select the **Allow duplicate object labels in one document** check box to allow duplicate labels of features in the **Model Tree** of a document.

- On selecting the **Disable partial loading of external linked objects** check box, the only referenced objects and their dependencies will be loaded when a linked document is auto-opened together with the main document.
- Specify desired values in **Author name** and **Company** edit boxes to define name of model author and designing company.
- If you select the **Set on save** check box then you will be asked to specify author name and company information while saving model.
- Similarly, you can specify license information in the **Default license** and **License URL** edit boxes.

Selection tab

The options in the **Selection** tab are used to specify the settings related to selection of documents in FreeCAD; refer to Figure-11.

Figure-11. Selection tab

- Specify desired value in the **Pick radius** edit box to define the circle radius within which objects will be selected when clicked at the cursor location.
- Select the **Auto switch to the 3D view containing the selected item** check box to enable the **Tree view SyncView** mode. If this mode is on, selecting an object from a different document in the Tree view automatically activates that document's last used docked 3D view.
- Select the **Auto expand tree item when the corresponding object is selected in 3D view** check box to enable the **Tree view SyncSelection** mode. If this mode is on, selecting an object in a 3D view will automatically expand the Tree view to show that object.
- Select the **Preselect the object in 3D view when mouse over the tree item** check box to enable the **Tree view PreSelection** mode. If this mode is on, an object that the mouse is over in the Tree view will be preselected in all 3D views belonging to that object's document.
- Select the **Record selection in tree view in order to go back/forward using navigation button** check box to enable the **Tree view RecordSelection** mode. If this mode is on, each Tree view selection is stored for future use.
- On selecting the **Add checkboxes for selection in document tree** check box, each document tree item will get a check box. This is useful for selecting multiple items on a touchscreen.

Cache tab

The options in **Cache** tab are related to the cache directory where FreeCAD stores temporary files; refer to Figure-12. Specify desired location for the **Location** edit box to store the temporary files. Select desired option from **Check periodically at program start** drop-down. Select desired option from **Cache size limit** drop-down so that the user will be notified if the cache size exceeds the specified limit. Click on the **Check now** button to check the current cache size stored.

Figure-12. Cache tab

Notification Area tab

The options in **Notification Area** tab are used to control the notifications in FreeCAD; refer to Figure-13.

Figure-13. Notification Area tab

- Select **Enable Notification Area** check box from **Settings** area to display Notification Area in the status bar.
- Select **Enable non-intrusive notifications** check box from **Settings** area to display non-intrusive notifications next to the Notification Area in the status bar.
- Select **Debug errors** check box from **Additional data sources** area to display errors in the Notification Area.
- Select **Debug warnings** check box from **Additional data sources** area to display warnings in the Notification Area.

- Specify desired maximum or minimum duration during which notifications are shown in the **Maximum Duration** or **Minimum Duration** edit boxes from the **Non Intrusive Notifications** area, respectively.
- Specify desired maximum number of notifications to be shown simultaneously in the **Maximum Number of Notifications** edit box.
- Specify desired width of the notification area in pixels in the **Notification width** edit box.
- Select **Hide when other window is activated** check box to disappear open notifications when another window is activated.
- Select **Do not show when inactive** check box to disappear the notifications if the FreeCAD window is not the active window.
- Specify desired value in the **Maximum Messages (0 = no limit)** edit box from **Message List** area to keep maximum number of messages in the list.
- Select **Auto-remove User Notifications** check box to remove the notifications from the message list.

Report view tab

The options in **Report view** tab are used to define how information will be displayed in the **Report view** panel; refer to Figure-14.

Figure-14. Report view tab

- Select the **Record normal messages** check box to record the normal messages in a file.
- Select the **Record log messages** check box to record the log messages displayed in **Report view** panel as a log file.
- Select the **Record warnings** check box to record warning messages in a file.
- Select the **Record error messages** check box to record error messages in a file.
- Select the **Show report view on error** check box to display the **Report View** panel on-screen showing the error, when an error has occurred.
- Select the **Show report view on warning** check box to display the **Report View** panel on-screen showing the warning, when a warning has occurred.
- Select the **Show report view on normal message** check box to display the **Report View** panel on-screen showing the message, when a normal message has occurred.
- Select the **Show report view on log message** check box to display the **Report View** panel on-screen showing the log message, when the log message has occurred.
- Select **Include a timecode for each entry** check box to include a timecode for each report.
- Set desired colors for Normal messages, Log messages, Warnings, and Errors in the **Colors** area of the dialog box.
- Select the check boxes from **Python interpreter** area of the tab to redirect outputs and errors of Python in **Report view** panel.

Display Page

The options in the **Display** page of the dialog box are used to set parameters for display of various objects in the interface; refer to Figure-15.

Figure-15. Display page

3D View tab

- Select the **Show coordinate system in the corner** check box to display Coordinate system at the bottom right corner of the drawing area. Specify desired value in the **Relative size** edit box from **General** area in the tab.
- Select the **Show axis cross by default** check box to show the axis cross by default at file opening or creation.
- Select the **Show counter of frames per second** check box to display a counter showing number of frames running in the software per second.
- Select **Use software OpenGL** check box for troubleshooting graphics card and driver problems.
- Select **Use OpenGL VBO (Vertex Buffer Object)** check box to use OpenGL buffer of your system for rendering objects in the drawing area.
- Select desired option from **Render cache** drop-down. Select **Auto** option from the drop-down to let the Coin3D decide where to cache. Select **Distributed** option to manually turn on cache for all view provider root nodes. Select **Centralized** option to manually turn off cache in all nodes of all view providers, and only cache at the scene graph root node.
- Select desired option from the **Anti-Aliasing** drop-down to smoothen jagged edges of the model.
- Select desired option from **Transparent objects** drop-down to specify render type of transparent objects.
- Select desired option from the **Marker size** drop-down to define size of point/constraint marks.
- Specify desired value in **Eye to eye distance for stereo modes** edit box to define distance between two eye for enhancing 3D view of model. For a mechanical engineer, it hardly makes any difference but if you are creating different views of model for each eye then you need to specify the value carefully. Default value for this parameter is 5.
- Select the **Backlight color** check box to define the color in which backside of model will be highlighted and select desired color by using button next to the check box.
- Select desired radio button from the **Camera type** area to specify whether you want to render model in perspective view or orthographic view.

Navigation tab

The options in the **Navigation** tab are used for the settings related to navigation of elements; refer to Figure-16.

Figure-16. Navigation tab

- Select the **Navigation cube** check box to place 3D navigation cube for changing view. Enter desired value in **Steps by turn** edit box to specify number of steps by turn when using arrows. Select the **Rotate to nearest** check box to rotate to nearest possible state when clicking a cube face. Enter desired value in **Cube size** edit box to specify the size of navigation cube. Select desired option from the **Corner and Font name** drop-down to define location of Cube in drawing area.

- Select desired option from the **3D Navigation** drop-down in the **Navigation** area of the tab to specify what type of mouse gestures will be used in FreeCAD for rotating view, pan view, zoom view, or other orientation functions. You can find some common CAD application styles in this drop-down. Click on the **Mouse** button next to drop-down to check functions of different mouse buttons for selected style.

- Select desired option from the **Orbit style** drop-down to define how object will rotate in 3D orbit. Select the **Turntable** option from the drop-down to allow rotation of object as if it is placed on a turn table. Select the **Trackball** option to rotate the object freely in 3D orbit.

- Select desired option from **Rotation mode** drop-down to define the rotation center.

- Select desired option from the **Default Camera Orientation** drop-down to define which orientation will be applied to the model by default when you start a new document.

- Enter desired value in the **Camera zoom** edit box to set the camera zoom for new documents. The value is the diameter of the sphere to fit on the screen.

- Select the **Enable animation** check box to animate the change of views in drawing area.

- Select the **Zoom at cursor** check box to zoom the object at the location of cursor.

- Enter desired value in the **Zoom step** edit box to specify how much the object will be zoomed.

- Select the **Invert zoom** check box to reverse the style of zoom when you scroll up/down in drawing.

- Select the **Disable touchscreen tilt gesture** check box to disable tilting of screen when you perform pinch zooming.

Colors tab

The options in the **Colors** tab are used to modify colors for selection and background of drawing area; refer to Figure-17.

Figure-17. Colors tab

- Select the **Enable preselection highlighting** check box to highlight the object under cursor in drawing area. You can specify the color of highlighting by selecting the button next to the check box.
- Select the **Enable selection highlighting** check box to highlight the object selected from drawing area in color set next to the check box.
- Select the **Simple color** radio button from the **Background color** area to apply simple color in the background of drawing area. You can select desired color by using button below the radio button.
- Select the **Linear gradient** radio button to apply vertical color gradient in the background of drawing area by specifying colors in the **Top** and **Bottom** section.
- Select the **Radial gradient** radio button to apply radial color gradient in the background of drawing area by specifying colors in the **Central** and **End** section.
- If you want to create a three color gradient background then select the **Middle color** check box.
- Click on the **Object being edited** color button from **Tree view** area of the tab to define the background color for objects in the tree view that are currently edited.
- Click on the **Active container** color button to define the background color for active containers in the tree view.
- Similarly, you can set colors for objects in **Model Tree** by using options in the **Tree View** area of the dialog box.

The options of **Mesh view** tab can be modified in the same way. After setting desired parameters, click on the **OK** button from the dialog box to apply changes.

Python Page

The options in the **Python** page of the dialog box are discussed next; refer to Figure-18.

Figure-18. Python page

Macro tab

The options in the **Macro** tab are used to define path and parameters for running macros in FreeCAD; refer to Figure-18.

- Select the **Run macros in local environment** check box to run the macros internally in FreeCAD environment.
- Specify desired location in the **Macro path** area to define location where macro file will be saved and searched.
- Select check boxes from the **Gui commands** area to define whether graphic user interface commands and comments will be recorded in the macro.
- Select the **Show script commands in python console** check box to show scripts in python console when you activate a tool.
- Specify desired number of macros which should be listed in recent macros list in the **Size of recent macro list** edit box from **Recent macros menu** area of the dialog box.
- Specify desired number of recent macros which should have shortcuts in the **Keyboard shortcut count** edit box.
- Specify the default keys for the macro in the **Keyboard Modifiers** edit box.

Python console tab

The options in **Python console** tab are used to control FreeCAD processes, and create and modify objects and their properties; refer to Figure-19. Select **Enable word wrap** check box from **Settings** area of the tab to wrap the words when they exceed available horizontal space in Python console. If you select the **Enable block cursor** check box then the cursor shape will be a block. Select **Save history** check box to save Python history across sessions.

Figure–19. Python console tab

Editor tab

The options in **Editor** tab are used to specify font type and size for different types of texts in Python Console; refer to Figure-20. Select desired text type from **Display Items** list box and specify related parameters in the right area of dialog box. Select the **Enable line numbers** check box from the **Options** area of this tab to display line numbers in Python Console. Select **Enable block cursor** check box to allow the cursor to be a block shape. Similarly, you can specify the indent sizes in the **Indentation** area of the tab.

Figure–20. Editor tab

VIEW MENU

The tools in the **View** menu are used to create and manage views of drawing area. You can also display/hide various elements of user interface of software by using the tools in this menu; refer to Figure-21. Some of the important tools in this menu have been discussed earlier in Chapter 1. The other tools in this menu are discussed next.

Figure-21. View menu

Stereoscopic Views

The tools in the **Stereo** cascading menu are used to display the model stereoscopic color style; refer to Figure-22.

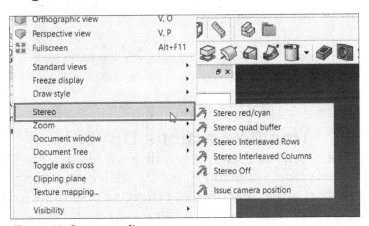

Figure-22. Stereo cascading menu

- Select the **Stereo red/cyan** option from the cascading menu to apply red and cyan color filters on the model, so that stereoscopic 3D effect can be achieved when viewing model using 3D glasses.
- Select the **Stereo quad buffer** option from the cascading menu to switch to Quad buffer stereo display mode if supported by your hardware. Quad-buffered stereo mode provides separate left and right eye frame buffers. It allows the window display in stereo while all other windows functions appear as normal.
- Select the **Stereo Interleaved Rows** or **Stereo Interleaved Columns** option to apply stereo color filters alternatively to every pixel row or column in the display, respectively.

- Select the **Stereo Off** option to stop stereoscopic display of model.
- Select the **Issue camera position** tool from the cascading menu to display current position of camera in 3D space. Note that the parameters will be displayed in **Report view** panel.

Zoom Options

The options in the **Zoom** cascading menu are used to zoom in/out and zoom specific objects; refer to Figure-23. Select the **Zoom In** option to enlarge the view of model and select the **Zoom Out** option to diminish the view of model. If you want to zoom on a specific object then select the **Box zoom** option and create a selection box around the object to be zoomed at.

Figure-23. Zoom cascading menu

Document Window Options

The options in the **Document window** cascading menu are used to dock, undock, and full screen the document windows where models are displayed; refer to Figure-24. Select the **Undocked** option from menu to make document window free floating in interface. Select the **Docked** option to make document window attached as tile in interface. Select the **Fullscreen** option to display document window on full screen. You can also use **F11** shortcut key to enter and exit full screen mode.

Figure-24. Document window cascading menu

TreeView actions Options

The options in the **TreeView actions** cascading menu are used to modify display parameters for document tree; refer to Figure-25. Select the **Single Document** option from cascading menu to display features of current document in **Model Tree**. Select the **Multi Document** option from the cascading menu to display features of all the documents currently open in **Model Tree**. Select the **Collapse/Expand** option from the cascading menu to expand or collapse all the feature nodes displayed in **Model Tree**. Select the **Initiate dragging** option to initiate a drag operation for selected objects in the Tree view. Select the **Go to selection** option to scroll the tree view to the first created object in a 3D view selection. The other options in the cascading menu have been discussed earlier in this chapter.

Figure-25. TreeView actions cascading menu

Toggle Axis Cross

The **Toggle Axis Cross** option in **View** menu is used to toggle global coordinate system On/Off. When the Axis Cross is On, it will be displayed at the center of drawing area as shown in Figure-26.

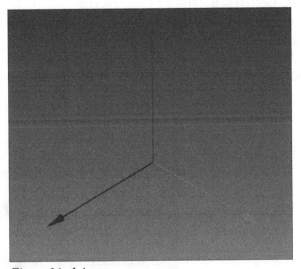

Figure-26. Axis cross

Bounding Box

The **Bounding box** tool in the **View** menu toggles the global bounding box highlighting mode. If this mode is switched on, selected objects are marked in a 3D view with a highlighted bounding box even if their selection style is set to 'Shape'.

Clipping Plane

The **Clipping plane** tool in **View** menu is used to clip the model for checking interior of model. The procedure to use this tool is given next.

• Click on the **Clipping plane** tool from the **View** menu. The **Clipping** dialog box will be displayed; refer to Figure-27.

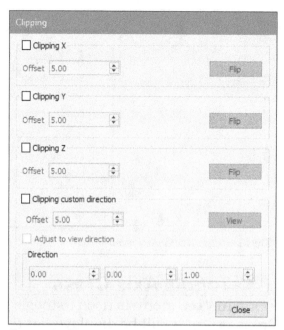

Figure-27. Clipping dialog box

- Select desired check box from the **Clipping** dialog to define in which direction you want to perform clipping of model. For example, select the **Clipping X** check box to clip model along X direction.
- After selecting check box, specify desired offset value to define offset distance of clipping plane from origin parallel to selected direction; refer to Figure-28.

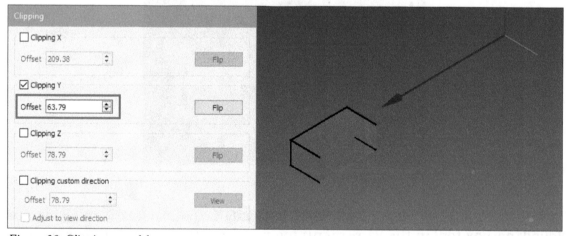

Figure-28. Clipping a model

- Using the **Flip** button, you can reverse the direction of clipping.
- Click on the **Close** button from the dialog to exit the tool.

Persistent section cut

The **Persistent section cut** tool is used to create a persistent cut of objects and assemblies. This tool only works for Part and PartDesign objects and assemblies of those. The procedure to use this tool is discussed next.

- Click on the **Persistent section cut** tool from the **View** menu. The **Section Cutting** dialog will be displayed; refer to Figure-29.

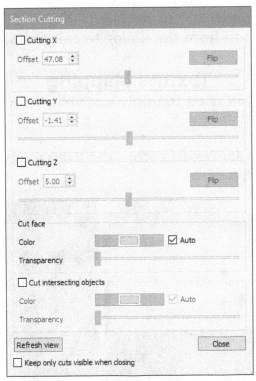

Figure-29. Section Cutting dialog

- Select desired check box from the dialog to define in which direction you want to perform section cutting of model. For example, select the **Cutting X** check box to cut model along **X** direction.
- After selecting check box, specify desired offset value to define offset distance of cutting plane from origin parallel to selected direction; refer to Figure-30.

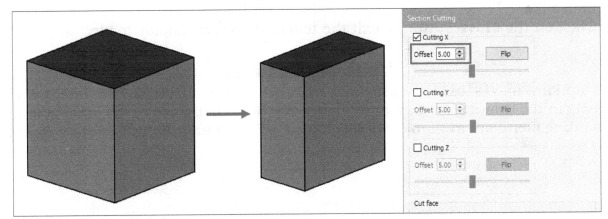

Figure-30. Section cutting a model

- Using the **Flip** button, you can reverse the direction of cutting.
- Click on the **Color** button ⬚⬚⬚ from **Cut face** section of the dialog to select the color for cut objects.
- Using the **Transparency** slider, adjust the transparency of cut face.
- If **Auto** check box is selected, the color and transparency of the cut face will be taken from the cut objects.
- The **Cut intersecting objects** check box allows to cut objects that intersects each other.
- Click on the **Refresh view** button to refresh the list of visible objects.

- Select the **Keep only cuts visible when closing** check box to visible only created cuts, when dialog is closed.
- Click on the **Close** button from the dialog to exit the tool.

Texture Mapping

The **Texture mapping** tool is used to apply texture on the model. The procedure to use this tool is given next.

- Click on the **Texture mapping** tool from the **View** menu. The **Texture** dialog will be displayed; refer to Figure-31.

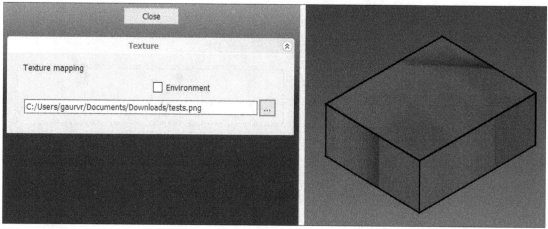

Figure-31. Texture dialog

- Select the **Environment** check box to apply selected image file on all the faces of model as environment.
- Click on the browse button next to edit box in the dialog to selected desired texture image file.
- Click on the **Close** button to exit the tool and texture display mode.

Visibility Options

The options in **Visibility** cascading menu are used to display or hide various objects in the drawing area; refer to Figure-32. Note that some of the options in this cascading menu are available when you have selected an object from drawing area.

Figure-32. Visibility cascading menu

- Select the **Toggle visibility** option from the **Visibility** cascading menu to hide selected component if it is visible and display the component if it is hidden in drawing area. You can also use the **SPACEBAR** key to toggle visibility. Note that although an object is hidden in drawing area, you can still select it from **Model Tree**. Hidden objects are displayed as washed out Black & White icons in the **Model Tree**.
- Select the **Show selection** option from the cascading menu to display selected object in drawing area.
- Select the **Hide selection** option from the cascading menu to hide selected object in drawing area.
- Select the **Select visible objects** option to select all the objects currently visible in drawing area.
- Select the **Toggle all objects** option from the **Visibility** cascading menu to hide currently visible objects and display currently hidden objects.
- Select the **Show all objects** option from the cascading menu to display all the objects of **Model Tree** in drawing area.
- Select the **Hide all objects** option from the cascading menu to hide all the objects of **Model Tree** from drawing area.
- Select the **Toggle selectability** option to make currently selected objects unselectable in drawing area. Note that you can still select those objects from **Model Tree**. If you have made an object unselectable then you can use this tool again to make it selectable.
- Similarly, you can use the **Toggle measurement** and **Clear measurement** options to display or hide measurement annotations in drawing area.

The **Toggle navigation/Edit mode** tool is used to switch the editing mode of a tool temporarily for rotation/reorientation of model.

The **Appearance** tool is used to apply different colors and appearances to the model. This tool has been discussed earlier.

Apply Random Colors to Model

The **Random color** tool is used to apply random colors to the faces of selected model. After selecting model, click on the **Random color** tool from the **View** menu.

Workbench Options

The options in **Workbench** cascading menu are used to switch between various workbenches. You can also use the **Switch between workbenches** drop-down of toolbar to select desired workbench.

Showing/Hiding Toolbars

The options in **Toolbars** cascading menu are used to display/hide various toolbars from the interface of software; refer to Figure-33.

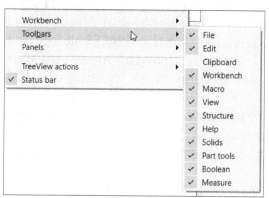

Figure-33. Toolbars cascading menu

Select desired option from the cascading menu to display respective toolbar. Note that toolbars for options which have tick mark before them are showing in interface.

Showing/Hiding Panels

The options in the **Panels** cascading menu are used to display/hide various panels of the interface like report view, panel view, selection view, and so on; refer to Figure-34.

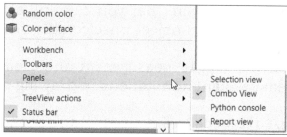

Figure-34. Panels cascading menu

Select desired option from **Panels** cascading menu to display respective panel in interface.

Status Bar

The **Status Bar** is used to display information like coordinates, mouse style, and so on. Select the **Status Bar** option from **View** menu to display **Status Bar** in interface.

MACRO MENU

The options in **Macro** menu are used to create and manage macros of FreeCAD. Macros are set of instructions for software to automate repetitive tasks; refer to Figure-35. For example, if you want to perform scale up of 10 sketches one by one then you can create a macro to do so after recording for just one sketch. Various tools of this menu are discussed next.

Figure-35. Macro menu

Recording Macro

The **Macro recording** tool in **Macro** menu is used to record sequence of operations performed in FreeCAD. The procedure to use this tool is given next.

- Click on the **Macro recording** tool from the **Model** menu. The **Macro recording** dialog box will be displayed; refer to Figure-36.

Figure-36. Macro recording dialog box

- Specify desired name for macro in the **Macro name** edit box.
- Click on the **Browse** button for **Macro path** edit box to specify location for saving macro and click on the **Record** button. The macro recording mode will start.
- Perform desired operations in the drawing area to be recorded and then click on the **Stop macro recording** tool from **Macro** menu. The macro will be created and saved at earlier specified location.

Executing and Managing Macros

The **Macros** tool in **Macro** menu is used to execute, edit, delete, or perform other tasks for macros. The procedure to use this tool is given next.

- Click on the **Macros** tool from the **Macro** menu. The **Execute macro** dialog box will be displayed; refer to Figure-37, with the list of macros available in location specified in **User macros location** edit box.

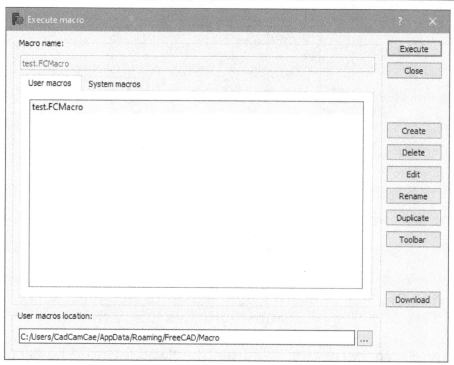

Figure-37. Execute macro dialog box

- Select desired macro from the list and click on the **Execute** button to run the macro.
- If you want to edit a macro then select it from list and click on the **Edit** button in the **Execute macro** dialog box. The **Macro Editor document** window will be displayed; refer to Figure-38. Use the Python coding to modify macro and save the file.

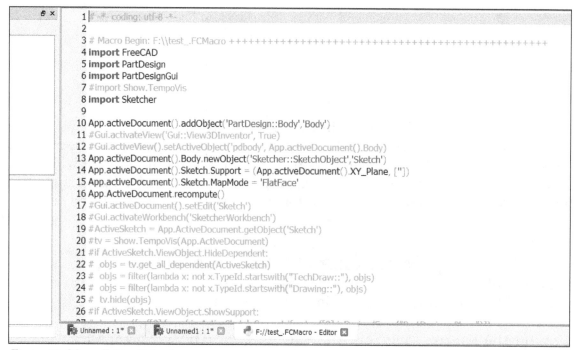

Figure-38. Macro Editor dialog box

- You can use the **Delete**, **Rename**, and **Duplicate** buttons in dialog box to delete, rename, or create duplicate copy of selected macro, respectively.
- Click on the **Toolbar** button to set up a macro in custom global toolbar.

- Click on the **Download** button to import macros available online. The **Addon Manager** dialog box will be displayed. Select **Macros** option from **Show addons containing** drop-down to check available macros; refer to Figure-39. Select desired macro from list and execute or Install/update as desired.

Figure-39. Macros option in Addon manager

- Click on the **Close** button to exit dialog boxes.

FOR STUDENT NOTES

Chapter 13

Assembly Workbench

Topics Covered

The major topics covered in this chapter are:

- *Introduction to Assembly 4 Workbench*
- *Starting New Assembly*
- *Adding New Parts, Bodies, and Groups*
- *Editing Part Information*
- *Inserting Parts and Fasteners*
- *Creating and Importing Datum Features*
- *Creating and Modifying Configurations*
- *Animating Assembly*

INTRODUCTION TO ASSEMBLY WORKBENCH

The **Assembly** workbench of FreeCAD is used to perform assembly of different bodies contained in a single file or in multiple documents. Use the Addon Manager to install Assembly 4 workbench as discussed earlier. To start creating assembly, select the **Assembly 4** option from the **Switch between workbenches** drop-down in the **Toolbar**; refer to Figure-1. On selecting the Assembly 4 workbench, tools will be displayed as shown in Figure-2.

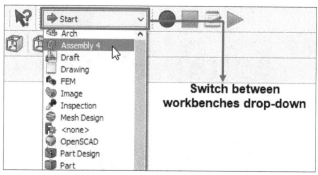

Figure-1. Assembly 4 workbench option

Figure-2. Assembly toolbar

Various tools of this workbench are discussed next.

NEW ASSEMBLY

The **New Assembly** tool is used to create a new empty model. Parts must be added to this model before using them in the assembly. The procedure to use this tool is discussed next.

- Click on the **New Assembly** tool from the **Toolbar** or **Assembly** menu; refer to Figure-3. A new model will be created; refer to Figure-4.

Figure-3. New Assembly tool

Figure-4. New assembly created

NEW PART

The **New Part** tool is used to create a new part. The procedure to use this tool is discussed next.

- Click on the **New Part** tool from **Toolbar** or **Assembly** menu; refer to Figure-5. The **Create a new Part** dialog box will be displayed; refer to Figure-6.

Figure-5. New Part tool

Figure-6. Create a new Part dialog box

- Specify desired part name in the **Enter new Part name** edit box and click on the **OK** button. A new part will be created; refer to Figure-7.

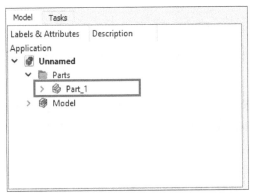

Figure-7. New part created

NEW BODY

The **New Body** tool is used to create a new body. The procedure to use this tool is discussed next.

- Click on the **New Body** tool from **Toolbar** or **Assembly** menu; refer to Figure-8. The **Create a new Body** dialog box will be displayed; refer to Figure-9.

Figure-8. New Body tool

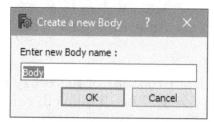

Figure-9. Create a new Body dialog box

- Specify desired body name in the **Enter new Body name** edit box and click on the **OK** button. A new body will be created; refer to Figure-10.

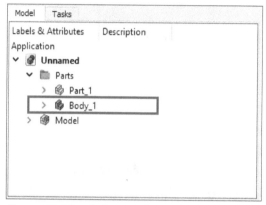

Figure-10. New body created

NEW GROUP

The **New Group** tool is used to create a new group. The procedure to use this tool is discussed next.

- Click on the **New Group** tool from **Toolbar** or **Assembly** menu; refer to Figure-11. The **Create a new Group** dialog box will be displayed; refer to Figure-12.

Figure-11. New Group tool

Figure-12. Create a new Group dialog box

- Specify desired group name in the **Enter new Group name** edit box and click on the **OK** button. A new group will be created; refer to Figure-13.

Figure-13. New group created

EDIT PART INFORMATION

The **Edit Part Information** tool is used to edit the default part information. The procedure to use this tool is discussed next.

- Select the model created from the **Model Tree** and click on the **Edit Part Information** tool from **Toolbar** or **Assembly** menu; refer to Figure-14. The **Edit Part Information** dialog will be displayed; refer to Figure-15.

Figure-14. Edit Part Information tool

Figure-15. Edit Part Information dialog

- Specify desired name in the **PartName**, **Reference**, **PartLength**, **PartWidth**, and **PartVolume** edit boxes of the dialog.
- Click on the **Configure fields** button from the dialog. The updated **Edit Part Information** dialog will be displayed; refer to Figure-16.

Figure-16. Updated Edit Part Information dialog

- All the check boxes are selected by default. Deselecting the check box will remove that data from the dialog.
- Click on the **Add** button to add new data to the dialog and click on the **Delete** button to delete the added data from the dialog.
- After specifying desired parameters, click on the **OK** button from the dialog.

INSERT PART

The **Insert Part** tool is used to insert a part into the assembly. This will create a dynamic link to the part, which can be in the same document or another open in the current session. The procedure to use this tool is discussed next.

- Click on the **Insert Part** tool from **Toolbar** or **Assembly** menu; refer to Figure-17. The **Insert a Part** dialog box will be displayed; Figure-18.

Figure-17. Insert Part tool

Figure-18. Insert a Part dialog box

- Select desired part from the **Select Part to be inserted** box of the dialog.
- Specify desired name of the part in **Name for the link** edit box of the dialog and click on the **Insert** button. The **Place linked Part** dialog will be displayed; refer to Figure-19.

Figure-19. Place linked Part dialog

- The **Selected Link** box show current selected part. Select desired part in which the attachment coordinate system is to be found from **Attach to** drop-down.
- Select desired coordinate system from the list in the **Select LCS in Part** and **Select LCS in Parent** boxes.
- Specify desired translation along X-axis, Y-axis, and along Z-axis in the **X Translation**, **Y Translation**, and **Z Translation** edit boxes of the dialog, respectively.
- After specifying desired parameters, click on the **OK** button from the dialog. The part will be inserted.

INSERT SCREW

The **Insert Screw** tool is used to insert a screw in the assembly. Note that you need to install **Fasteners** workbench from **Addon Manager** to use the tools of **Fasteners** drop-down. The procedure to use this tool is discussed next.

- Click on the **Insert Screw** tool from **Fasteners** drop-down in the **Toolbar** or **Assembly** menu; refer to Figure-20. The screw will be inserted; refer to Figure-21.

Figure-20. Insert Screw tool

Figure-21. Screw inserted

INSERT NUT

The **Insert Nut** tool is used to insert a nut in the assembly. The procedure to use this tool is discussed next.

- Click on the **Insert Nut** tool from **Fasteners** drop-down in the **Toolbar** or **Assembly** menu; refer to Figure-22. The nut will be inserted; refer to Figure-23.

Figure-22. Insert Nut tool

Figure-23. Nut inserted

INSERT WASHER

The **Insert Washer** tool is used to insert a washer in the assembly. The procedure to use this tool is discussed next.

- Click on the **Insert Washer** tool from **Fasteners** drop-down in the **Toolbar** or **Assembly** menu; refer to Figure-24. The washer will be inserted; refer to Figure-25.

Figure-24. Insert Washer tool

Figure-25. Washer inserted

CHANGE FASTENER PARAMETERS

The **Change Fastener Parameters** tool is used to change the parameters of fastener. The procedure to use this tool is discussed next.

- Click on the **Change Fastener parameters** tool from **Fasteners** drop-down in the **Toolbar** or **Assembly** menu; refer to Figure-26. The **Change fastener parameters** dialog will be displayed; refer to Figure-27.

Figure-26. Change Fastener parameters tool

Figure-27. Change fastener parameters dialog

- Select desired type of fastener from **Fastener type** drop-down.
- Select **Auto set diameter** check box to set the diameter automatically.
- Select desired value of diameter from **Diameter** drop-down.
- After specifying desired parameters, click on the **OK** button. The parameters of fastener will be changed; refer to Figure-28.

Figure-28. Parameters of fastener changed

NEW SKETCH

The **New Sketch** tool is used to create a new sketch in a part. The procedure to use this tool is discussed next.

- Click on the **New Sketch** tool from **Toolbar** or **Assembly** menu; refer to Figure-29. The **Create new Sketch** dialog box will be displayed; refer to Figure-30.

Figure-29. New Sketch tool

Figure-30. Create new Sketch dialog box

- Specify desired sketch name in the **Enter Sketch name** edit box and click on the **OK** button. The **Attachment** dialog will be displayed which have been discussed earlier in this book.
- After specifying desired parameters, click on the **OK** button from the dialog. A new sketch will be created; refer to Figure-31.

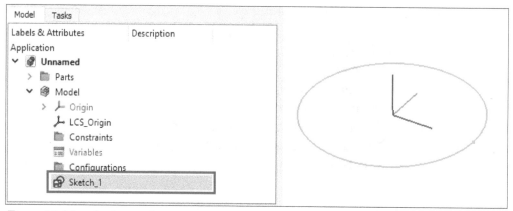

Figure-31. New sketch created

NEW COORDINATE SYSTEM

The **New Coordinate System** is used to create a new coordinate system for the part. The procedure to use this tool is discussed next.

- Click on the **New Coordinate System** tool from drop-down in the **Toolbar** or **Assembly** menu; refer to Figure-32. The **Create new LCS** dialog box will be displayed; refer to Figure-33.

Figure-32. New Coordinate System tool

Figure-33. Create new LCS dialog box

- Specify desired coordinate system name in the **Enter LCS** name edit box of the dialog and click on the **OK** button. The **Attachment** dialog will be displayed which have been discussed earlier in this book.
- After specifying desired parameters, click on the **OK** button from the dialog. The new coordinate system will be created; refer to Figure-34.

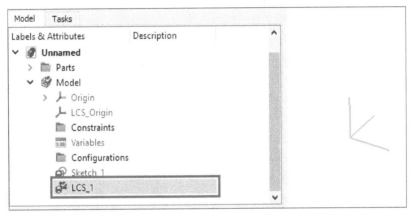

Figure-34. New coordinate system created

NEW PLANE

The **New Plane** tool is used to create a new datum plane in a part. The procedure to use this tool is discussed next.

- Click on the **New Plane** tool from drop-down in the **Toolbar** or **Assembly** menu; refer to Figure-35. The **Create new Plane** dialog box will be displayed; refer to Figure-36.

Figure-35. New Plane tool

Figure-36. Create new Plane dialog box

- Specify desired plane name in the **Enter Plane name** edit box of the dialog box and click on the **OK** button. The **Attachment** dialog will be displayed which have been discussed earlier in this book.

- After specifying desired parameters, click on the **OK** button from the dialog. A new datum plane will be created; refer to Figure-37.

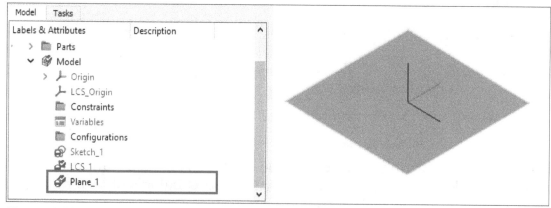

Figure-37. New datum plane created

NEW AXIS

The **New Axis** tool is used to create a new datum axis in a part. The procedure to use this tool is discussed next.

- Click on the **New Axis** tool from drop-down in the **Toolbar** or **Assembly** menu; refer to Figure-38. The **Create new Axis** dialog box will be displayed; refer to Figure-39.

Figure-38. New Axis tool

Figure-39. Create new Axis dialog box

- Specify desired axis name in the **Enter Axis name** edit box of the dialog box and click on the **OK** button. The **Attachment** dialog will be displayed which have been discussed earlier in this book.
- After specifying desired parameters, click on the **OK** button from the dialog. A new datum axis will be created; refer to Figure-40.

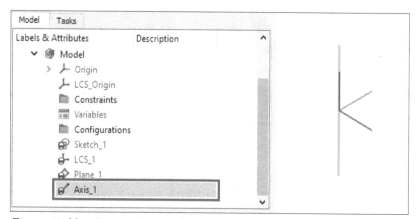

Figure-40. New datum axis created

NEW POINT

The **New Point** tool is used to create a new datum point in a part. The procedure to use this tool is discussed next.

- Click on the **New Point** tool from drop-down in the **Toolbar** or **Assembly** menu; refer to Figure-41. The **Create new Point** dialog box will be displayed; refer to Figure-42.

Figure-41. New Point tool

Figure-42. Create new Point dialog box

- Specify desired point name in the **Enter Point name** edit box of the dialog box and click on the **OK** button. The **Attachment** dialog will be displayed which have been discussed earlier in this book.
- After specifying desired parameters, click on the **OK** button from the dialog. A new datum point will be created; refer to Figure-43.

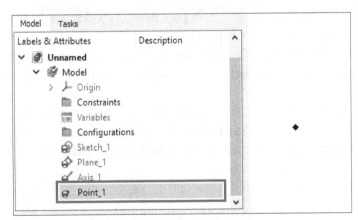

Figure-43. New datum point created

NEW HOLE AXIS

The **New Hole Axis** tool is used to create a datum axis attached to a hole. The procedure to use this tool is discussed next.

- Select the hole to which you want to create a datum axis; refer to Figure-44 and click on the **New Hole Axis** tool from drop-down in the **Toolbar** or **Assembly** menu; refer to Figure-45. The datum axis will be created attached to the hole; refer to Figure-46.

Figure-44. Selecting the hole

Figure-45. New Hole Axis tool

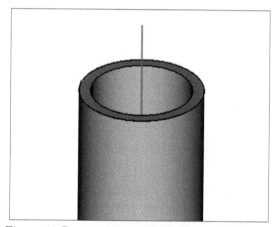

Figure-46. Datum axis created attached to the hole

IMPORT DATUM OBJECT

The **Import Datum Object** tool is used to import the selected datum object from the sub-part into the root assembly. This creates a new datum of the same type with the same global placement. The procedure to use this tool is discussed next.

- Select desired datum object from the **Model Tree** or from the 3D view area and click on the **Import Datum object** tool from **Toolbar** or **Assembly** menu; refer to Figure-47. The **Import Datum** dialog box will be displayed; refer to Figure-48.

Figure-47. Import Datum object tool

Figure-48. Import Datum dialog box

- Specify desired name for the datum in the **Enter name for this datum** edit box and click on the **OK** button. A new datum object will be imported; refer to Figure-49.

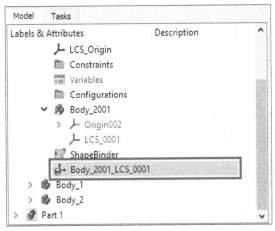

Figure-49. New datum object imported

CREATE A SHAPE BINDER

The **Create a shape binder** tool is used to create a reference to an external shape. This creates a sub-shape binder of the selected shapes (face, edge, point) in the root assembly. The procedure to use this tool has already been discussed earlier in this book.

EDIT PLACEMENT OF A PART

The **Edit Placement of a Part** tool is used to move or attach a part in the assembly. The procedure to use this tool is discussed next.

- Select desired part from the **Model Tree** or from the 3D view area which you want to move or attach in the assembly and click on the **Edit Placement of a Part** tool from **Toolbar** or **Assembly** menu; refer to Figure-50. The **Place a Part in the assembly** dialog will be displayed; refer to Figure-51.

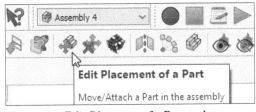

Figure-50. Edit Placement of a Part tool

Figure-51. Place a Part in the assembly dialog

- The **Selected Object** box display the object selected to move or attach in the assembly.
- Select desired option from **Attach to** drop-down to which you want to attach the selected part.
- Select desired attachment from **Select attachment LCS in parent Part** box.
- Specify desired translation along X-axis, Y-axis, and along Z-axis in the **X Translation**, **Y Translation**, and **Z Translation** edit boxes of the dialog, respectively.
- After specifying desired parameters, click on the **OK** button from the dialog. The part will be attached in the assembly.

RELEASE FROM ATTACHMENT

The **Release from Attachment** tool is used to release an object from all attachments to any geometry. The procedure to use this tool is discussed next.

- Select desired object from the **Model Tree** or from the 3D view area which you want to release from all attachments and click on the **Release from Attachment** tool from **Toolbar** or **Assembly** menu; refer to Figure-52. **A FreeCAD Warning** dialog box will be displayed asking you to release the attachments or not; refer to Figure-53.

Figure-52. Release from Attachment tool

Figure-53. FreeCAD Warning dialog box

- Click on the **OK** button from the dialog box to release the attachments.

SOLVE AND UPDATE ASSEMBLY

The **Solve and Update Assembly** tool is used to update assembly.

CREATE SYMMETRIC OF PART

The **Create Symmetric of part** tool is used to create a mirrored part of a part. Use only on individual parts, not assemblies. You must re-create the attachment datums in the resulting part. The procedure to use this tool is discussed next.

- Select desired model from the **Model Tree** or from the 3D view area which you want to mirror and click on the **Create symmetric of part** tool from **Toolbar** or **Assembly** menu; refer to Figure-54. The **Create mirrored Part** dialog box will be displayed; refer to Figure-55. The **Selected Part** box shows the part selected to be symmetric.

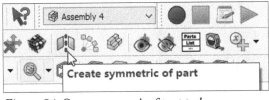

Figure-54. Create symmetric of part tool

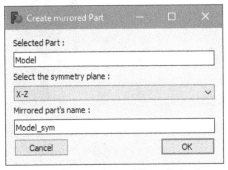

Figure-55. Create mirrored Part dialog box

- Select desired symmetry plane from **Select the symmetry plane** drop-down.
- Specify desired name for the mirrored part in the **Mirrored part's name** edit box.
- After specifying desired parameters, click on the **OK** button from the dialog box. The part will be mirrored.

CREATE A CIRCULAR ARRAY

The **Create a circular Array** tool is used to create an array of the selected object around an axis. The procedure to use this tool is discussed next.

- Select desired object of which you want to create circular array and then select an axis; refer to Figure-56. Note that the standard axis should be selected to create a circular array.

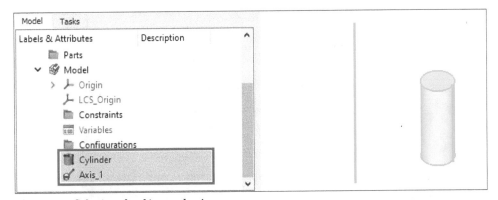

Figure-56. Selecting the object and axis

- Click on the **Create a circular Array** tool from **Toolbar** or **Assembly** menu; refer to Figure-57. The circular array of the object created; refer to Figure-58.

Figure-57. Create a circular Array tool

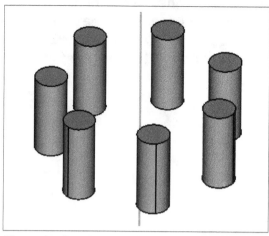

Figure-58. Circular array of the object created

CREATE A LINEAR ARRAY

The **Create a linear array** tool is used to create a linear array along an axis. The procedure to use this tool is similar to **Create a circular Array** tool.

CREATE AN EXPRESSION DRIVEN ARRAY

The **Create an expression driven Array** tool is used to create an array of the selected object where the placement of each element is calculated using expressions and an index property. The procedure to use this tool is discussed next.

- Select a source object to create an array and optionally select an axis; refer to Figure-59. Click on the **Create an expression driven Array** tool from **Toolbar** or **Assembly** menu; refer to Figure-60. The expression driven array will be created; refer to Figure-61.

Figure-59. Selecting an object

Figure-60. Create an expression driven Array tool

Figure-61. Expression driven array created

SHOW LCS

The **Show LCS** tool is used to show LCS (Local Coordinate System) and datums of selected part and its children. The procedure to use this tool is discussed next.

- Click on the **Show LCS** tool from **Toolbar** or **Assembly** menu; refer to Figure-62. The LCS will be shown; refer to Figure-63.

Figure-62. Show LCS tool

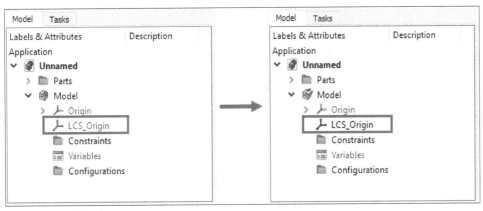

Figure-63. LCS shown

HIDE LCS

The **Hide LCS** tool is used to hide LCS and datums of selected part and its children. The procedure to use this tool is discussed next.

- Click on the **Hide LCS** tool from **Toolbar** or **Assembly** menu; refer to Figure-64. The LCS will be hidden; refer to Figure-65.

Figure-64. Hide LCS tool

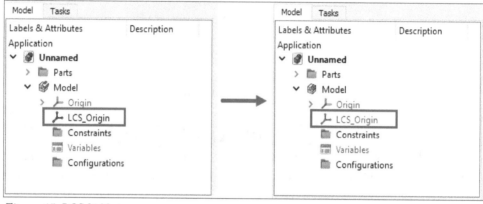

Figure-65. LCS hidden

TREE OF LINKED FILES

The **Tree of Linked Files** tool is used to show hierarchical tree structure of parts in the selected container. The procedure to use this tool is discussed next.

- Select desired entity from the **Model Tree** and click on the **Tree of Linked Files** tool from **Toolbar** or **Assembly** menu; refer to Figure-66. The **Tree structure of the selected object** dialog box will be displayed along with the hierarchical tree structure of selected object; refer to Figure-67.

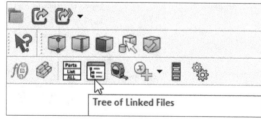

Figure-66. Tree of Linked Files tool

Figure-67. Tree structure of the selected object dialog box

MEASURE

The **Measure** tool is used to measure the features of parts. The procedure to use this tool is discussed next.

- Click on the **Measure** tool from **Toolbar** or **Assembly** menu; refer to Figure-68. The **Measure** dialog will be displayed; refer to Figure-69.

Figure-68. Measure tool

Figure-69. Measure dialog

- Select **Measure Radius of Arc or Circle Measure Length of Edge** radio button to measure the arc or circle radius and to measure the edge length.
- Select **Measure Distance** radio button to measure the distance.
- Select **Measure Angle** radio button to measure the angle.
- Select **Snap to EndPoint, MiddlePoint, Center** radio button to snap the endpoints, middle points and center.
- Select the **Select Shape** radio button to select the shapes.
- Select desired object to be measured from graphics area and it will be shown in **Selection 1** box; refer to Figure-70.

Figure-70. Selecting the edge for measurement

- Similarly, select the second object and it will be shown in **Selection 2** box.
- Select the **Show Label in 3D view** check box from **Results** area to enable the extra label.
- Select the **Show Components** check box to show all the dimension components.
- The **Results** box will display the measurement results; refer to Figure-71.

Figure-71. Measurement results

ADD VARIABLE

The **Add Variable** tool is used to add a variable into the 'Variables' placeholder in the document. This variable can be used in any formula using the Expression Engine of any compatible input field. The procedure to use this tool is discussed next.

- Click on the **Add Variable** tool from drop-down in the **Toolbar** or **Assembly** menu; refer to Figure-72. The **Add Variable** dialog box will be displayed; refer to Figure-73.

Figure-72. Add Variable tool

Figure-73. Add Variable dialog box

- Select desired type of variable from **Type** drop-down of the dialog box.
- Specify desired name for the variable in the **Name** edit box.
- Specify desired variable value in the **Value** edit box.
- Specify desired description for the variable in the **Description** box.
- After specifying desired parameters, click on the **OK** button from the dialog box. The variable will be added; refer to Figure-74.

Figure-74. Variable added

DELETE VARIABLE

The **Delete Variable** tool is used to delete a variable from the document. The procedure to use this tool is discussed next.

- Click on the **Delete Variable** tool from drop-down in the **Toolbar** or **Assembly** menu; refer to Figure-75. The **Delete Variable** dialog box will be displayed; refer to Figure-76.

Figure-75. Delete Variable tool

Figure-76. Delete Variable dialog box

- Select desired variable from **Variable** drop-down which you want to delete. The name and value of that variable will be displayed.
- Click on the **Delete** button from the dialog box. The variable will be deleted.

OPEN CONFIGURATIONS PANEL

The **Open configurations panel** tool is used to set visibilities and offsets of parts. The procedure to use this tool is discussed next.

- Click on the **Open configurations panel** tool from **Toolbar** or **Assembly** menu; refer to Figure-77. The **Assembly Configurations** dialog will be displayed; refer to Figure-78.

Figure-77. Open configurations panel tool

Figure-78. Assembly Configurations dialog

- Click on the **New** button from the dialog. The **Create a new assembly configuration** dialog box will be displayed; refer to Figure-79.

Figure-79. Create a new assembly configuration dialog box

- Specify desired configuration name in the **Enter configuration name** edit box.
- Specify desired description for the configuration in the **Description** edit box.
- After specifying desired parameters, click on the **OK** button from the dialog box. The new assembly configuration will be created; refer to Figure-80.

Figure-80. New assembly configuration created

ANIMATE ASSEMBLY

The **Animate Assembly** tool is used to animate the assembly. The procedure to use this tool.

- Click on the **Animate Assembly** tool from **Toolbar** or **Assembly** menu; refer to Figure-81. The **Animate Assembly** dialog box will be displayed; refer to Figure-82.

Figure-81. Animate Assembly tool

Figure-82. Animate Assembly dialog box

- Select desired document from **Document** drop-down.
- Select desired variable from **Variable** drop-down.
- Specify desired values in the **Range Begin**, **Range End**, **Step Size**, and **Sleep (s)** edit boxes of the dialog.
- Select **Force-render every step** check box to force GUI to update on every step.
- After specifying desired parameters, click on the **Close** button from the dialog box.

PRACTICAL 1

Create the Assembly as shown in Figure-83.

Figure-83. Assembly Practical

Creating Assembly Rod

- Create the assembly rod as shown in Figure-84.

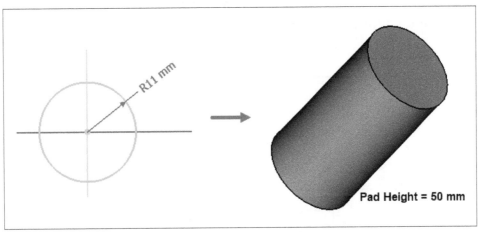

Figure-84. Rod created

Creating coordinate system on Rod

- Select the assembly rod created from the **Model Tree** and click on the **New Coordinate System** tool from the **Toolbar** or **Assembly** menu. The **Create new LCS** dialog box will be displayed.
- Specify desired name in the **Enter LCS name** edit box and click on the **OK** button. The **Attachment** dialog will be displayed.
- Select the edge of circle for **Reference 1** selection box; refer to Figure-85 and select the **Concentric** option from **Attachment mode** area of the dialog box.

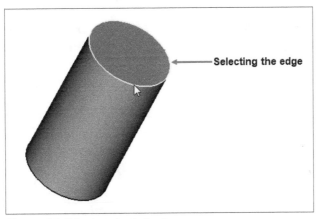

Figure-85. Selecting the circle edge

- Click on the **OK** button from the dialog. The LCS will be created; refer to Figure-86.

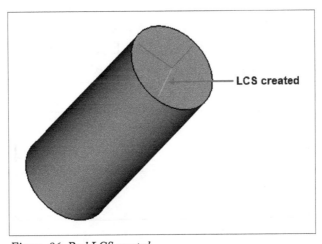

Figure-86. Rod LCS created

Creating Master Sketch

- Create the master sketch as shown in Figure-87.

Figure-87. Master sketch created

Creating coordinate system on Master Sketch

- Select the master sketch created from the **Model Tree** and click on the **New Coordinate System** tool from the **Toolbar** or **Assembly** menu. The **Create new LCS** dialog box will be displayed.
- Specify desired name in the **Enter LCS name** edit box and click on the **OK** button. The **Attachment** dialog will be displayed.
- Select the edge of circle for **Reference 1** selection box; refer to Figure-88 and select the **Concentric** option from **Attachment** mode area of the dialog box.

Figure-88. Selecting the edge of circle

- Click on the **OK** button from the dialog. The LCS will be created.
- Similarly, select another edge of circle and create LCS on it; refer to Figure-89.

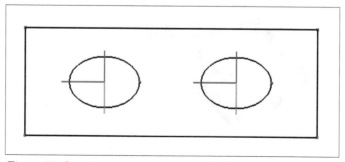

Figure-89. Coordinate system created on the sketch

Creating Main Model

- Create the assembly main model as shown in Figure-90.

Figure-90. Main model created

Creating coordinate system on main model

- Select the assembly main model created from the **Model Tree** and click on the **New Coordinate System** tool from the **Toolbar** or **Assembly** menu. The **Create new LCS** dialog box will be displayed.
- Specify desired name in the **Enter LCS name** edit box and click on the **OK** button. The **Attachment** dialog will be displayed.
- Select the edge of circle for **Reference 1** selection box; refer to Figure-91 and select the **Concentric** option from **Attachment** mode area of the dialog box.

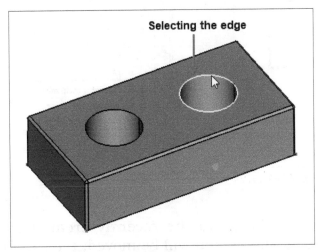

Figure-91. Selecting the circle edge

- Click on the **OK** button from the dialog. The LCS will be created; refer to Figure-92.

Figure-92. Main model LCS created

Save and close all the models and sketch recently created.

Creating New Assembly File

- Click on the **New Model** tool from **Toolbar** or **Assembly** menu. The new assembly file will be created; refer to Figure-93.
- Open the models and sketch recently created; refer to Figure-94.

Figure-93. New model file created

Figure-94. Models and sketch recently created

- Move the models and sketch to the recently created assembly file; refer to Figure-95. The models and sketch will be moved; refer to Figure-96.
- Select the **Assembly rod** from the **Model Tree** and specify the position of rod from **Placement** section of **Property editor**; refer to Figure-97.

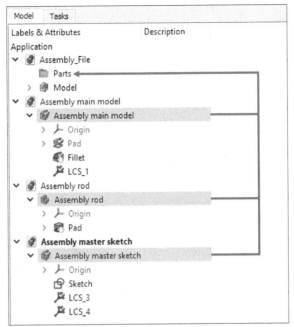

Figure-95. Moving models and sketch to the assembly file

Figure-96. Models and sketch moved

Figure-97. Placing the assembly rod

Importing the Parts

- Toggle the visibility of models and sketch as shown in Figure-98.

Figure-98. Toggled the visibility of parts

- Click on the **Insert Part** tool from **Toolbar** or **Assembly** menu. The **Insert a Part** dialog box will be displayed; refer to Figure-99.
- Select the assembly master sketch from the dialog box and click on the **Insert** button. The **Place linked Part** dialog will be displayed.
- Specify the parameters in the dialog as shown in Figure-100 and click on the **OK** button. The master sketch will be inserted; refer to Figure-101.

Figure-99. Insert a Part dialog box

Figure-100. Parameters specified for master sketch

Figure-101. Master sketch inserted

- Select the assembly rod from the dialog box and click on the **Insert** button. The **Place linked Part** dialog will be displayed.
- Specify the parameters in the dialog as shown in Figure-102 and click on the **OK** button. The rod will be inserted; refer to Figure-103.

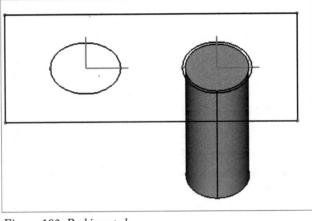

Figure-103. Rod inserted

Figure-102. Parameters specified for rod

- Select the assembly rod again from the dialog box and click on the **Insert** button. The **Place linked Part** dialog will be displayed.
- Specify the parameters in the dialog as shown in Figure-104 and click on the **OK** button. The rod will be inserted; refer to Figure-105.

Figure-105. Rod inserted

Figure-104. Parameters specified for rod

- Select the assembly main model from the dialog box and click on the **Insert** button. The **Place linked Part** dialog will be displayed.
- Specify the parameters in the dialog as shown in Figure-106 and click on the **OK** button. The main model will be inserted; refer to Figure-107.

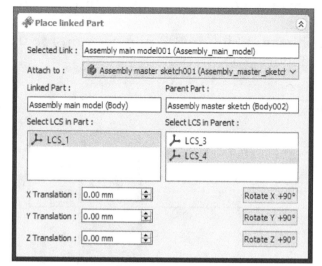

Figure-106. Parameters specified for main model

Figure-107. Main model inserted

Positioning the Rods

- Select the **Assembly rod001** from the **Model Tree** and specify the position of rod from **Attachment** area of **Property editor**; refer to Figure-108.

Figure-108. Specifying the position of rod

- Click on the **Add Variable** tool from **Toolbar** or **Assembly** menu. The **Add Variable** dialog box will be displayed.
- Specify the parameters in the dialog box as shown in Figure-109 and click on the **OK** button.
- Click on the button from **Attachment** section in the **Property editor**; refer to Figure-110. The **Formula editor** dialog box will be displayed.
- Specify the parameters in the dialog box as shown in Figure-111 and click on the **OK** button. The rod will be positioned.
- Select the **Assembly rod002** from the **Model Tree** ; refer to Figure-112 and specify the parameters in the **Formula editor** dialog box as shown in Figure-113. The rod will be positioned.

Figure-109. Specifying the variable

Figure-110. Clicking on the button

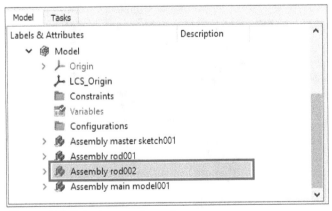

Figure-111. Parameters specified in Formula editor dialog box

Figure-112. Selecting another rod

Figure-113. Parameters specified in Formula editor dialog box

Animating the Assembly

- Click on the **Animate Assembly** tool from **Toolbar** or **Assembly** menu. The **Animate Assembly** dialog box will be displayed.
- Specify the parameters in the dialog box as shown in Figure-114 and click on the **Run** button. The assembly will animate; refer to Figure-115.

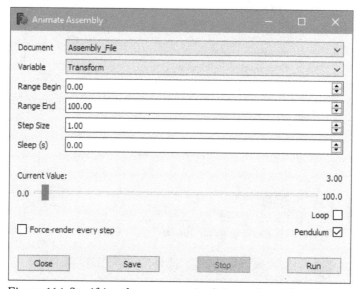

Figure-114. Specifying the parameters in Animate Assembly dialog box

Figure-115. Animating the assembly

Chapter 14

Introduction to Computational Fluid Dynamics

Topics Covered

The major topics covered in this chapter are:

- *Introduction to CFD*
- *Basic Properties of Fluids, Types of Fluids, and Thermodynamic Properties of Fluids*
- *Variations Of Navier-Strokes Equation*
- *Steps Of Computational Fluid Dynamics*
- *Introduction To Cfdof*
- *Activating Cfdof Workbench*
- *Installing Cfdof Related Software*
- *Performing Cfd On A Model*

INTRODUCTION

In this chapter, you will know various aspects of CFD for various practical problems. But, keep in mind that all computer software work on same concept of GIGO which means Garbage In - Garbage Out. So, if you have specified any wrong parameter while defining properties of analysis then you will not get the correct results. This problem demands a good knowledge of Fluid Mechanics so that you are well conversant with the terms of classical fluid mechanics and can relate the results to the theoretical concepts. In this chapter, we will discuss the basics of Fluid Mechanics and we will try to relate them with analysis wherever possible.

BASIC PROPERTIES OF FLUIDS

There are various basic properties required while performing analysis on fluid. These properties are collected by performing experiments in labs. Most of these properties are available in the form of tables in Steam Tables or Design Data books. These properties are explained next.

Mass Density, Weight Density, and Specific Gravity

- Density or Mass Density is the mass of fluid per unit volume. In SI units, mass is measured in kg and volume is measured in m³. So, mathematically we can say,

$$\text{Density (or Mass Density) } \rho = \frac{Mass\ of\ Fluid}{Volume\ Occupied\ by\ Fluid}\ kg/m^3$$

- If you are asked for weight density then multiply mass by gravity coefficient. Mathematically it can be expressed as:

$$\text{Weight Density } w = \frac{Mass\ of\ Fluid \times Gravity\ Coefficient}{Volume\ Occupied\ by\ Fluid}\ N/m^3$$

- Most of the time, fluid density is available as **Specific Gravity**. Specific gravity is the ratio of weight density of fluid to weight density of water in case of liquid. In case of gases, it is the ratio of weight density of fluid to weight density of air. Note that weight density of water is 1000 kg/m³ at 4 °C and weight density of air is 1.225 kg/m³ at 15 °C.

Note that as the temperature of liquid rises, its density is reduced and vice-versa. But as the temperature of gas rises, its density is increased and vice-versa.

Viscosity

Viscosity is the coefficient of friction between different layers of fluid. In other terms, it is the shear stress required to produce unit rate of shear strain in one layer of fluid. Mathematically it can be expressed as:

$$\mu = \frac{\tau}{\left(\frac{dx}{dy}\right)}\ N.s/m^2 \text{ or Pa.s}$$

Where, μ is viscosity, τ is shear stress (or force applied tangentially to the layer of fluid) and (dx/dy) is the shear strain.

As the density of fluid changes with temperature so does the viscosity. The formula for viscosity of fluid at different temperature is given next.

For Liquids, $\mu = \mu_0 \dfrac{1}{1 + \alpha t + \beta t^2}$

For Gases, $\mu = \mu_0 + \alpha t - \beta t^2$

here, μ_0 is viscosity at 0 °C

α and β are constants for fluid (for water α is 0.03368 and β is 0.000221)

(for air α is 5.6x10⁻⁸ and β is 1.189x10⁻¹⁰)

t is the temperature

PROBLEM ON VISCOSITY

Dynamic viscosity of lubricant oil used between shaft and sleeve is 8 poise. The shaft has a diameter of 0.4 m and rotates at 250 r.p.m. Find out the power lost due to viscosity of fluid if length of sleeve is 100 mm and thickness of oil film is 1.5 mm; refer to below figure.

Oil film 1.5mm

Shaft Dia 0.4 m

Sleeve length 100mm

Solution:

Viscosity μ = 8 poise = 8/10 N.s/m² =0.8 N.s/m²

Tangential velocity of shaft $u = \dfrac{\pi \times D \times N}{60} = \dfrac{\pi \times 0.4 \times 250}{60}$ = 5.236 m/s

Using the relation, $\tau = \mu \dfrac{dx}{dy}$

where dx is 5.236
and dy is 1.5x10⁻³

$\tau = 0.8 \times \dfrac{5.236}{1.5 \times 10^{-3}}$ = 2792.53 N/m²

Shear force F = τ x Area

$F = \tau \times \pi D \times L = 2792.53 \times \pi \times 0.4 \times 100 \times 10^{-3} = 350.92$ N

Torque (T) = Force x Radius = 350.92 x 0.2 = 70.184 N.m

Power = 2 π.N.T/60 = (2 π x 250 x 70.184)/60=1837.41 W Ans.

Now, you may ask how this problem relates with CFD. As discussed earlier, the viscosity changes with temperature and as fluid flows through pipe or comes in contact with rolling shaft, its temperature rises. In such cases, CFD gives the approximate viscosity and temperature of fluids in the system at different locations. This data later can be used to find solution for other engineering problems.

TYPES OF FLUIDS

There are mainly 5 types of fluids:

Ideal Fluids: These fluids are incompressible and have no viscosity which means they flow freely without any resistance. This category of fluid is imaginary and used in some cases of calculations.

Real Fluids: These are the fluids found in real world. These fluids have viscosity values as per their nature and can be compressible in some cases.

Newtonian Fluids: Newtonian fluids are those in which shear stress is directly proportional to shear strain. In a specific temperature range, water, gasoline, alcohol etc. can be Newtonian fluids.

Non-Newtonian Fluids: Those fluids in which shear stress is not directly proportional to shear strain. Most of the time Real Fluids fall in this category.

Ideal Plastic Fluids: Those fluids in which shear stress is more than yield value and so fluid deforms plastically. The shear stress in these fluids is directly proportional to shear strain.

THERMODYNAMIC PROPERTIES OF FLUID

Most of the liquids are not considered as compressible in general applications as their molecules are already bound closely to each other. But, Gas have large gap between their molecules and can be compressed easily relative to liquids. As we pick pressure to compress the gas, other thermodynamic properties also come into play. The relationship between Pressure, Temperature, and specific Volume is given by;

$$P.\forall = RT$$

P = Absolute pressure of a gas in N/m²
$\forall$ = Specific Volume = 1/ ρ
R = Gas Constant (for Air is 287 J/Kg-K
T = Absolute Temperature
ρ = Density of gas

If the density of gas changes with constant temperature then the process is called Isothermal process and if density changes with no heat transfer then the process is called Adiabatic process.

For Isothermal process,   p/ρ = Constant
For Adiabatic process,   p/ ρ^k = Constant

Here, k is Ratio of specific heat of a gas at constant pressure and constant volume (1.4 for air).

Universal Gas Constant

By Pressure, Temperature, volume equation,

$$p.\forall = nMRT$$

Here,

p = Absolute pressure of a gas in N/m²
$\forall$ = Specific Volume = 1/ ρ
n = Number of moles in a Volume of gas
M = Mass of gas molecules/ Mass of Hydrogen atom = n x m (m is mass gas in kg)
R = Gas Constant (for Air is 287 J/Kg-K
T = Absolute Temperature

MxR is called Universal Gas constant and is equal to 8314 J/kg-mole K for water.

Compressibility of Gases

Compressibility is reciprocal of bulk modulus of elasticity K, which is defined as ratio of Compressive stress to volumetric strain.

Bulk Modulus = Increase in pressure/ Volumetric strain

$K = -(dp/d\forall)x\forall$

Vapour Pressure and Cavitation

When a liquid converts into vapour due to high temperature in a vessel then vapours exert pressure on the walls of vessel. This pressure is called **Vapour pressure**.

When a liquid flows through pipe, sometimes bubbles are formed in the flow. When these bubbles collapse at the adjoining boundaries then they erode the surface of tube due to high pressure burst of bubble. This erosion is in the form of cavities at the surface of tube and the phenomena is called **Cavitation**.

PASCAL'S LAW

Pascal's Law states that pressure at a point in static fluid is same in all directions. In mathematical form $p_x=p_y=p_z$ in case of static fluids.

FLUID DYNAMICS

Up to this point, the rules stated in this chapter were for static fluid that is fluid at rest. Now, we will discuss the rules for flowing fluid.

Bernoulli's Incompressible Fluid Equation

Bernoulli's equation states that the total energy stored in fluid is always same in a closed system. In the language of mathematics,

$$p + \frac{1}{2}\rho V^2 + \rho gh = constant$$

Here, p is pressure

ρ is density

V is velocity of fluid

h is height of fluid

g is gravitational acceleration

Eulerian and Lagrangian Method of Analysis

There are two different points of view in analyzing problems in fluid mechanics. The first view, appropriate to fluid mechanics, is concerned with the field of flow and is called the eulerian method of description. In the eulerian method, we compute the pressure field p(x, y, z, t) of the flow pattern, not the pressure changes p(t) that a particle experiences as it moves through the field.

The second method, which follows an individual particle moving through the flow, is called the lagrangian description. The lagrangian approach, which is more appropriate to solid mechanics, will not be treated in this book. However, certain numerical analyses of sharply bounded fluid flows, such as the motion of isolated fluid droplets, are very conveniently computed in lagrangian coordinates.

Fluid dynamic measurements are also suited to the eulerian system. For example, when a pressure probe is introduced into a laboratory flow, it is fixed at a specific position (x, y, z). Its output thus contributes to the description of the eulerian pressure field p(x, y, z, t). To simulate a lagrangian measurement, the probe would have to move downstream at the fluid particle speeds; this is sometimes done in oceanographic measurements, where flow meters drift along with the prevailing currents.

Now, we know the two methods of analyzing fluid mechanics problem. But, there are further three approaches for these two methods by which problems are derived to solution. These approaches are:

1. Control Volume
2. Differential
3. Experimental

Control volume analysis, is accurate for any flow distribution but is often based on average or "one dimensional" property values at the boundaries. It always gives useful "engineering" estimates. In principle, the differential equation approach can be applied to any problem. Only a few problems, such as straight pipe flow, yield to exact analytical solutions. But the differential equations can be modeled numerically, and the flourishing field of computational fluid dynamics (CFD) can now be used to give good estimates for almost any geometry. Finally, the dimensional analysis applies to any problem, whether analytical, numerical, or experimental. It is particularly useful to reduce the cost of experimentation.

Since the Differential Equation approach is more concerned to CFD so we will discuss this approach a little deeper.

DIFFERENTIAL APPROACH OF FLUID FLOW ANALYSIS

As discussed earlier, in this approach, the fluid is divided in to very small finite number of elements via a computer process called meshing. Various equations for different properties of a fluid are given next.

Acceleration

Acceleration **a** can be given as:

$$\mathbf{a} = \frac{d\mathbf{V}}{dt} = \mathbf{i}\frac{du}{dt} + \mathbf{j}\frac{dv}{dt} + \mathbf{k}\frac{dw}{dt}$$

Various components of acceleration **a** are:

$$a_x = \frac{du}{dt} = \frac{\partial u}{\partial t} + u\frac{\partial u}{\partial x} + v\frac{\partial u}{\partial y} + w\frac{\partial u}{\partial z} = \frac{\partial u}{\partial t} + (\mathbf{V}\cdot\nabla)u$$

$$a_y = \frac{dv}{dt} = \frac{\partial v}{\partial t} + u\frac{\partial v}{\partial x} + v\frac{\partial v}{\partial y} + w\frac{\partial v}{\partial z} = \frac{\partial v}{\partial t} + (\mathbf{V}\cdot\nabla)v$$

$$a_z = \frac{dw}{dt} = \frac{\partial w}{\partial t} + u\frac{\partial w}{\partial x} + v\frac{\partial w}{\partial y} + w\frac{\partial w}{\partial z} = \frac{\partial w}{\partial t} + (\mathbf{V}\cdot\nabla)w$$

Summing these into a vector, we obtain the total acceleration as

$$a = \frac{dV}{dt} = \frac{\partial V}{\partial t} + \left(u\frac{\partial V}{\partial x} + v\frac{\partial V}{\partial y} + w\frac{\partial V}{\partial z}\right) = \frac{\partial V}{\partial t} + (V.\nabla)V$$

Similarly, you can divide other parameters as vector like Force, Pressure, Temperature, and so on.

INTRODUCTION TO CFD

The CFD stands for Computational Fluid Dynamics. Computational Fluid Dynamics constitutes a new "third approach" in the philosophical study and development of the whole discipline of fluid dynamics. In the seventeenth century, the foundations for experimental fluid dynamics were laid in France and England. The eighteenth and nineteenth centuries saw the gradual development of theoretical fluid dynamics, again primarily in Europe. As a result, throughout most of the twentieth century the study and practice of fluid dynamics involved the use of theory on the one hand and pure experiment on the other hand.

However, to keep things in context, CFD provides a new third approach-but nothing more than that. It nicely and synergistically complements the other two approaches of pure theory and pure experiment, but it will never replace either of these approaches. There will always be a need for theory and experiment. The future advancement of fluid dynamics will rest upon a balance of all three approaches, with computational fluid dynamics helping to interpret and understand the results of theory, experiment and vice-versa; refer to Figure-1.

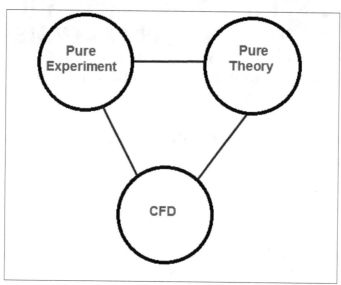

Figure-1. The three dimensions of fluid dynamics

Finally, we note that computational fluid dynamics is commonplace enough today that the composition CFD is universally accepted for the phrase "Computational Fluid Dynamics". We will use this composition throughout the book.

Computational fluid dynamics (CFD) is the use of applied math, physics, and computational software package to examine how a gas or liquid flows, also as how the gas or liquid affects objects as it flows past. Computational fluid dynamics is based on the Navier-Stokes equations. These equations describe how the velocity, pressure, temperature, and density of a moving fluid are related.

Navier-Stokes equation

The Navier-Stokes equations are the elemental partial differentials equations that describe the flow of incompressible fluids. Using the rate of stress and rate of strain, it can be shown that the parts of a viscous force F during a non rotating frame are given by-

$$
\begin{aligned}
\frac{F_i}{V} &= \frac{\partial}{\partial x_j}\left[\eta\left(\frac{\partial u_i}{\partial x_j} + \frac{\partial u_j}{\partial x_i}\right) + \lambda\delta_{ij}\nabla\cdot\mathbf{u}\right] \\
&= \frac{\partial}{\partial x_j}\left[\eta\left(\frac{\partial u_i}{\partial x_j} + \frac{\partial u_j}{\partial x_i} - \frac{2}{3}\delta_{ij}\nabla\cdot\mathbf{u}\right) + \mu_B\delta_{ij}\nabla\cdot\mathbf{u}\right],
\end{aligned}
$$

Where η is the dynamic viscosity, λ is the second viscosity coefficient, δ_{ij} is the Kronecker delta, $\nabla\cdot\mathbf{u}$ is the divergence, μ_B is the bulk viscosity and Einstein summation has been used to sum over j=1,2,and 3.

- ## Dynamic Viscosity

Dynamic viscosity is the force required by a fluid to overcome its own internal molecular friction so the fluid can flow. In other words, dynamic viscosity is defined as the tangential force per unit area required to move the fluid in one horizontal plane with reference to other plane with a unit velocity whereas the fluid's molecules maintain a unit distance apart.

A parameter η is defined as

$$[\text{Shear Stress}] = \eta \, [\text{Strain Rate}]$$

Written explicitly.

$$\sigma = \eta \dot{e} = \eta \frac{1}{l}\frac{dl}{dt} = \eta \frac{u}{l},$$

Where l is the length scale and u is the velocity scale. In cgs η has units of g cm^{-1} s^{-1}. Dynamic viscocity is related to kinematic viscocity ν by

$$\eta = \rho \nu$$

Where ρ is the density.

• **Second Viscosity Coefficient**

For a compressible fluid, i.e. one for which $\nabla . u$=0, where $\nabla . u$ is the divergence Σ of the velocity field, the stress tensor of the fluid can be written

$$S_{ij} = \eta \left(\frac{\partial u_i}{\partial x_j} + \frac{\partial u_j}{\partial x_i} \right) + \lambda \delta_{ij} \nabla \cdot \mathbf{u},$$

Where δ_{ij} is the Kronecker delta, η is the dynamic viscosity, and λ is the second coefficient of viscosity. λ is analogous to the first Lame constant. For an incompressible fluid, the term involving λ drops out from the equation, so λ can be ignored.

• **Kronecker Delta**

The simplest interpretation of the Kronecker delta is as the discrete version of the delta function defined by

$$\delta_{ij} = \begin{cases} 1 & i = j \\ 0 & i \neq j \end{cases}$$

The Kronecker delta is implemented in the Wolfram language as KroneckerDelta[i, j], as well as in a generalized form KroneckerDelta[i, j, ...] that returns 1 if all arguments are equal and 0 otherwise.
It has the contour integral representation

$$\delta_{mn} = \frac{1}{2\pi i} \oint_\gamma z^{m-n-1} \, dz,$$

Where γ is a contour corresponding to the unit circle and m and n are integers.

In three space, the Kronecker delta satisfies the identities

$$\delta_{ii} = 3$$
$$\delta_{ij}\,\epsilon_{ijk} = 0$$
$$\epsilon_{ipq}\,\epsilon_{jpq} = 2\,\delta_{ij}$$
$$\epsilon_{ijk}\,\epsilon_{pqk} = \delta_{ip}\,\delta_{jq} - \delta_{iq}\,\delta_{jp},$$

where Einstein summation is implicitly assumed. i,j=1,2,3,and ϵ_{ijk} is the permutation symbol.

Technically, the Kronecker delta is a tensor defined by the relationship

$$\delta_i^k \frac{\partial x_i'}{\partial x_k} \frac{\partial x_l}{\partial x_j'} = \frac{\partial x_i'}{\partial x_k} \frac{\partial x_k}{\partial x_j'} = \frac{\partial x_i'}{\partial x_j'}.$$

Since, by definition, the coordinates x_i and x_j are independent for i=j,

$$\frac{\partial x_i'}{\partial x_j'} = \delta''_j{}^i,$$

So

$$\delta''_j{}^i = \frac{\partial x_i'}{\partial x_k} \frac{\partial x_l}{\partial x_j'} \delta_i^k,$$

and $\delta''_j{}^i$ is really a mixed second-rank tensor. It satisfies

$$\delta_{ab}{}^{jk} = \epsilon_{abi}\,\epsilon^{jki}$$
$$= \delta_a^j\,\delta_b^k - \delta_a^k\,\delta_b^j$$
$$\delta_{abjk} = g_{aj}\,g_{bk} - g_{ak}\,g_{bj}$$
$$\epsilon_{aij}\,\epsilon^{bij} = \delta_{ai}{}^{bi}$$
$$= 2\,\delta_a^b.$$

- **Divergence**

The divergence of a vector field **F**, denoted div(**F**) or $\nabla.\mathbf{F}$ (the notation used in this work), is defined by a limit of the surface integral

$$\nabla \cdot \mathbf{F} \equiv \lim_{V \to 0} \frac{\oint_S \mathbf{F} \cdot d\mathbf{a}}{V}$$

Where the surface integral provides the value of **F** integrated over a closed small boundary surface $S=\partial V$ surrounding a volume component V that is taken to size zero using a limiting method. The divergence of vector field is thus a scalar field. If $\nabla.\mathbf{F}=0$, then the filed is alleged to be a divergence less field. The symbol ∇ is referred to as nabla or del.

- **Bulk Viscosity**

The Bulk viscosity μ_B of a fluid is defined as

$$\mu_B = \lambda + \tfrac{2}{3}\mu,$$

Where λ is the second viscosity coefficient and μ is the shear viscosity.

- **Reynolds Number**

The Reynolds number for a flow through a pipe is defined as

$$\mathrm{Re} \equiv \frac{\rho \bar{u} d}{h} = \frac{\bar{u} d}{\nu},$$

Where ρ is the density of the fluid, **u** is the velocity scale, d is the pipe diameter and v is the kinematic viscosity of the fluid. Poiseuille (laminar) flow is experimentally found to occur for **Re<30**. At larger
Reynolds numbers flow become turbulent.

- **Density**

A measure of a substance's mass per unit of volume. For a substance with mass **m** and volume **V**,

$$\rho \equiv \frac{m}{V}.$$

For a body weight w_a placed in a fluid of weight w_w,

$$\rho = G_s \rho_w = \frac{w_a}{w_a - w_w} \rho_w,$$

Where G_s is the specific gravity for an ideal gas,

$$\rho = \frac{mP}{kT}.$$

- **Kinematic Viscosity**

A coefficient which describes the diffusion of momentum. Let η be the dynamic viscosity, then

$$\nu \equiv \frac{\eta}{\rho}.$$

The unit of kinematic viscosity is Stoke, equal to 1 cm^2 s^{-1}
V_{water} = 1.0 X 10^{-6} m^2 s^{-1} = 0.010 cm^2 s^{-1}
V_{air} = 1.5 X 10^{-5} m^2 s^{-1} = 0.15 cm^2 s^{-1}

The story of CFD starts with Navier-Strokes equation. This equation is based on conservation laws of mass, momentum, and energy. These law's can be defined as:

CONSERVATION OF MASS

The conservation of mass law states that the mass in the control volume can be neither created nor destroyed in accordance with physical laws. The conservation of mass, also expressed as Continuity Equation, states that the mass flow difference throughout system between inlet- and outlet-section is zero. In equation terms, we can write as:

$$\frac{D_\rho}{D_t} + \rho \left(\nabla \cdot \vec{V} \right) = 0$$

Where, ρ is density, V is velocity and gradient operator ∇

$$\vec{\nabla} = \vec{i} \frac{\partial}{\partial x} + \vec{j} \frac{\partial}{\partial y} + \vec{k} \frac{\partial}{\partial z}$$

When density of fluid is constant, the flow is assumed as incompressible and then this equation represents a steady state process:

$$\frac{D_\rho}{D_t} = 0 \quad \text{so,} \quad \nabla \cdot \vec{V} = \frac{\partial u}{\partial x} + \frac{\partial v}{\partial y} + \frac{\partial w}{\partial z} = 0$$

here, u, v, and w are components of velocity at point (x,y,z) at time t.

Note that incompressible fluid is also called Newtonian fluid when stress/strain curve is linear.

CONSERVATION OF MOMENTUM

The momentum in a control volume is kept constant, which implies conservation of momentum that we call 'The Navier-Stokes Equations'. The description is set up in accordance with the expression of Newton's Second Law of Motion:

$$F=m.a$$

where, F is the net force applied to any particle, a is the acceleration, and m is the mass. In case the particle is a fluid, it is convenient to divide the equation to volume of particle to generate a derivation in terms of density as follows:

$$\rho \frac{DV}{D_t} = f = f_{body} + f_{surface}$$

in which f is the force exerted on the fluid particle per unit volume, and f_{body} is the applied force on the whole mass of fluid particles as below:

$$f_{body} = \rho \cdot g$$

Where, ρ is density, g is gravitational acceleration. External forces which are deployed through the surface of fluid particles, f$_{surface}$ is expressed by pressure and viscous forces as shown below:

$$f_{surface} = \nabla \cdot \tau_{ij} = \frac{\partial \tau_{ij}}{\partial x_i} = f_{pressure} + f_{viscous}$$

where τ$_{ij}$ is expressed as stress tensor. According to the general deformation law of Newtonian viscous fluid given by Stokes, τ$_{ij}$ is expressed as

$$\tau_{ij} = -p\delta_{ij} + \mu \left(\frac{\partial u_i}{\partial x_j} + \frac{\partial u_j}{\partial x_i} \right) + \delta_{ij}\lambda \nabla \cdot V$$

Hence, Newton's equation of motion can be specified in the form as follows:

$$\rho \frac{DV}{D_t} = \rho \cdot g + \nabla \cdot \tau_{ij}$$

Navier-Stokes equations of Newtonian viscous fluid in one equation gives:

$$\underbrace{\rho \frac{DV}{D_t}}_{I} = \underbrace{\rho \cdot g}_{II} - \underbrace{\nabla p}_{III} + \underbrace{\frac{\partial}{\partial x_i} \left[\mu \left(\frac{\partial v_i}{\partial x_j} + \frac{\partial v_j}{\partial x_i} \right) + \delta_{ij}\lambda \nabla \cdot V \right]}_{IV}$$

I : Momentum convection

II: Mass force

III: Surface force

IV: Viscous force

where, static pressure ρ and gravitational force ρ.g. The equation is convenient for fluid and flow fields both transient and compressible. D/D$_t$ indicates the substantial derivative as follows:

$$\frac{D()}{D_t} = \frac{\partial()}{\partial t} + u\frac{\partial()}{\partial x} + v\frac{\partial()}{\partial y} + w\frac{\partial()}{\partial z} = \frac{\partial()}{\partial t} + V \cdot \nabla()$$

If the density of fluid is accepted to be constant, the equations are greatly simplified in which the viscosity coefficient μ is assumed constant and ∇·V=0 in equation. Thus, the Navier-Stokes equations for an incompressible three-dimensional flow can be expressed as follows:

$$\rho \frac{DV}{Dt} = \rho g - \nabla p + \mu \nabla^2 V$$

For each dimension when the velocity is V(u,v,w):

$$\rho\left(\frac{\partial u}{\partial t}+u\frac{\partial u}{\partial x}+v\frac{\partial u}{\partial y}+w\frac{\partial u}{\partial z}\right)=\rho g_x-\frac{\partial p}{\partial x}+\mu\left(\frac{\partial^2 u}{\partial x^2}+\frac{\partial^2 u}{\partial y^2}+\frac{\partial^2 u}{\partial z^2}\right)$$

$$\rho\left(\frac{\partial v}{\partial t}+u\frac{\partial v}{\partial x}+v\frac{\partial v}{\partial y}+w\frac{\partial v}{\partial z}\right)=\rho g_y-\frac{\partial p}{\partial y}+\mu\left(\frac{\partial^2 v}{\partial x^2}+\frac{\partial^2 v}{\partial y^2}+\frac{\partial^2 v}{\partial z^2}\right)$$

$$\rho\left(\frac{\partial w}{\partial t}+u\frac{\partial w}{\partial x}+v\frac{\partial w}{\partial y}+w\frac{\partial w}{\partial z}\right)=\rho g_z-\frac{\partial p}{\partial z}+\mu\left(\frac{\partial^2 w}{\partial x^2}+\frac{\partial^2 w}{\partial y^2}+\frac{\partial^2 w}{\partial z^2}\right)$$

p , u, v, and w are unknowns where a solution is sought by application of both continuity equation and boundary conditions. Besides, the energy equation has to be considered if any thermal interaction is available in the problem.

CONSERVATION OF ENERGY

The Conservation of Energy is the first law of thermodynamics which states that the sum of the work and heat added to the system will result in the increase of energy of the system:

$$dE_t = dQ + dW$$

where dQ is the heat added to the system, dW is the work done on the system, and dE$_t$ is the increment in the total energy of the system. One of the common types of energy equation is :

$$\rho\left[\underbrace{\frac{\partial h}{\partial t}}_{I}+\underbrace{\nabla\cdot(hV)}_{II}\right]=\underbrace{-\frac{\partial p}{\partial t}}_{III}+\underbrace{\nabla\cdot(k\nabla T)}_{IV}+\underbrace{\phi}_{V}$$

I : Local change with time

II: Convective term

III: Pressure work

IV: Heat flux

V: Heat dissipation term

The Navier-Stokes equations have a non-linear structure and various complexities so it is hardly possible to conduct an exact solution of those equations. Thus, with regard to the physical domain, both approaches and assumptions are partially applied to simplify the equations. Some assumptions also need to be applied to provide a reliable model in which the equation is carried out to further step in terms of complexity such as turbulence.

VARIATIONS OF NAVIER-STROKES EQUATION

The solution of the Navier-Stokes equations can be realized with either analytical or numerical methods. The analytical method is the process that only compensates solutions in which non-linear and complex structures of the Navier-Stokes equations are ignored within several assumptions. It is only valid for simple / fundamental cases such as Couette flow, Poisellie flow, etc. Almost every case in fluid dynamics comprises non-linear and complex structures in the mathematical model which cannot be ignored to sustain reliability. Hence, the solution of the Navier-Stokes equations are carried out with several numerical methods. Various parameters based on which Navier-Strokes equation can vary are given next.

Time Domain

The analysis of fluid flow can be conducted in either steady (time-independent) or unsteady (time-dependent) condition depending on the physical incident. In case the fluid flow is steady, it means the motion of fluid and parameters do not rely on change in time, the term $\partial()/\partial\tau=0$ where the continuity and momentum equations are re-derived as follows:

Continuity equation:

$$\frac{\partial(\rho u)}{\partial x} + \frac{\partial(\rho v)}{\partial y} + \frac{\partial(\rho w)}{\partial z} = 0$$

The Navier-Stokes equation in x direction:

$$\rho\left(u\frac{\partial u}{\partial x} + v\frac{\partial u}{\partial y} + w\frac{\partial u}{\partial z}\right) = \rho g_x - \frac{\partial p}{\partial x} + \mu\left(\frac{\partial^2 u}{\partial x^2} + \frac{\partial^2 u}{\partial y^2} + \frac{\partial^2 u}{\partial z^2}\right)$$

While the steady flow assumption negates the effect of some non-linear terms and provides a convenient solution, variation of density is a hurdle that keeps the equation in a complex formation.

Compressibility

Due to the flexible structure of fluids, the compressibility of particles is a significant issue. Despite the fact that all types of fluid flow are compressible in a various range regarding molecular structure, most of them can be assumed to be incompressible in which the density changes are negligible. Thus, the term $\partial\rho/\partial t = 0$ is thrown away regardless of whether the flow is steady or not, as below:

Continuity equation:

$$\frac{\partial u}{\partial x} + \frac{\partial v}{\partial y} + \frac{\partial w}{\partial z} = 0$$

The Navier-Stokes equation in x direction:

$$\rho\left(\frac{\partial u}{\partial t} + u\frac{\partial u}{\partial x} + v\frac{\partial u}{\partial y} + w\frac{\partial u}{\partial z}\right) = \rho g_x - \frac{\partial p}{\partial x} + \mu\left(\frac{\partial^2 u}{\partial x^2} + \frac{\partial^2 u}{\partial y^2} + \frac{\partial^2 u}{\partial z^2}\right)$$

As incompressible flow assumption provides reasonable equations, the application of steady flow assumption concurrently enables us to ignore non-linear terms where $\partial()/\partial t=0$. Moreover, the density of fluid in high speed cannot be accepted as incompressible in which the density changes are important. "The Mach Number" is a dimensionless number that is convenient to investigate fluid flow, whether incompressible or compressible.

$$Ma = \frac{V}{a} \leq 0.3$$

where, Ma is the Mach number, V is the velocity of flow, and a is the speed of sound at 340.29 m/s at sea level.

As in above equation, when the Mach number is lower than 0.3, the assumption of incompressibility is acceptable. On the contrary, the change in density cannot be negligible in which density should be considered as a significant parameter. For instance, if the velocity of a car is higher than 100 m/s, the suitable approach to conduct credible numerical analysis is the compressible flow. Apart from velocity, the effect of thermal properties on the density changes has to be considered in geophysical flows.

Low and High Reynolds Numbers

The Reynolds number, the ratio of inertial and viscous effects, is also effective on Navier-Stokes equations to truncate the mathematical model. While Re -> ∞, the viscous effects are presumed negligible and viscous terms in Navier-Stokes equations are removed. The simplified form of Navier-Stokes equation, described as Euler equation, can be specified as follows.

The Navier-Stokes equation in x direction:

$$\rho\left(\frac{\partial u}{\partial t} + u\frac{\partial u}{\partial x} + v\frac{\partial u}{\partial y} + w\frac{\partial u}{\partial z}\right) = \rho g_x - \frac{\partial p}{\partial x}$$

Even though viscous effects are relatively important for fluids, the inviscid flow model partially provides a reliable mathematical model as to predict real process for some specific cases. For instance, high-speed external flow over bodies is a broadly used approximation where inviscid approach reasonably fits. While Re<<1, the inertial effects are assumed negligible where related terms in Navier-Stokes equations drop out. The simplified form of Navier-Stokes equations is called either creeping flow or Stokes flow.

The Navier-Stokes equation in x direction:

$$\rho g_x - \frac{\partial p}{\partial x} + \mu\left(\frac{\partial^2 u}{\partial x^2} + \frac{\partial^2 u}{\partial y^2} + \frac{\partial^2 u}{\partial z^2} +\right) = 0$$

Having tangible viscous effects, creeping flow is a suitable approach to investigate the flow of lava, swimming of microorganisms, flow of polymers, lubrication, etc.

Turbulence

The behavior of the fluid under dynamic conditions is a challenging issue that is compartmentalized as laminar and turbulent. The laminar flow is orderly at which motion of fluid can be predicted precisely. Except that, the turbulent flow has various hindrances, therefore it is hard to predict the fluid flow which shows a chaotic behavior. The Reynolds number, the ratio of inertial forces to viscous forces, predicts the behavior of fluid flow whether laminar or turbulent regarding several properties such as velocity, length, viscosity, and also type of flow. Whilst the flow is turbulent, a proper mathematical model is selected to carry out numerical solutions. Various turbulent models are available in literature and each of them has a slightly different structure to examine chaotic fluid flow.

Turbulent flow can be applied to the Navier-Stokes equations in order to conduct solutions to chaotic behavior of fluid flow. Apart from the laminar, transport quantities of the turbulent flow, it is driven by instantaneous values. Direct numerical simulation (DNS) is the approach to solving the Navier-Stokes equation with instantaneous values. Having district fluctuations varies in a broad range, DNS needs enormous effort and expensive computational facilities. To avoid those hurdles, the instantaneous quantities are reinstated by the sum of their mean and fluctuating parts as follows:

$$instantaneous\ value = \overline{mean\ value} + fluctuating\ value'$$

$$u = \overline{u} + u'$$

$$v = \overline{v} + v'$$

$$w = \overline{w} + w'$$

$$T = \overline{T} + T'$$

where u, v, and w are velocity components and T is temperature.

Instead of instantaneous values which cause non-linearity, carrying out a numerical solution with mean values provides an appropriate mathematical model which is named "The Reynolds-averaged Navier-Stokes (RANS) equation". The fluctuations can be negligible for most engineering cases which cause a complex mathematical model. Thus, RANS turbulence model is a procedure to close the system of mean flow equations. The general form of The Reynolds-averaged Navier-Stokes (RANS) equation can be specified as follows:

Continuity equation:

$$\frac{\partial \overline{u}}{\partial x} + \frac{\partial \overline{v}}{\partial y} + \frac{\partial \overline{w}}{\partial z} = 0$$

The Navier-Stokes equation in x direction:

$$\rho \left(\frac{\partial \overline{u}}{\partial t} + \overline{u}\frac{\partial \overline{u}}{\partial x} + \overline{v}\frac{\partial \overline{u}}{\partial y} + \overline{w}\frac{\partial \overline{u}}{\partial z} \right) = \rho g_x - \frac{\partial \overline{p}}{\partial x} + \mu \left(\frac{\partial^2 u}{\partial x^2} + \frac{\partial^2 u}{\partial y^2} + \frac{\partial^2 u}{\partial z^2} \right)$$

The turbulence model of RANS can also vary regarding methods such as k-omega, k-epsilon, k-omega-SST, and Spalart-Allmaras which have been used to seek a solution for different types of turbulent flow.

Likewise, large eddy simulation (LES) is another mathematical method for turbulent flow which is also comprehensively applied for several cases. Tough LES ensures more accurate results than RANS, it requires much more time and computer memory. As in DNS, LES considers to solve the instantaneous Navier-Stokes equations in time and three-dimensional space.

Favre Time Averaging

In Favre-averaging, the time averaged equations can be simplified significantly by using the density weighted averaging procedure suggested by Favre. If Reynolds time averaging is applied to the compressible form of the Navier-Stokes equations, some difficulties arise. In particular, the original form of the equations is significantly altered. To see this, consider Reynolds averaging applied to the continuity equation for compressible flow.

Favre time averaging can be defined as follows. The instantaneous solution variable, ϕ, is decomposed into a mean quantity, $\tilde{\phi}$, and fluctuating component, ϕ'', as follows:

$$\phi = \tilde{\phi} + \phi''$$

The Favre time-averaging is then

$$\overline{\rho\phi(x_i, t)} = \frac{1}{T} \int_{t-T/2}^{t+T/2} \rho(x_i, t')\phi(x_i, t')\,\mathrm{d}t' = \overline{\rho\tilde{\phi}} + \overline{\rho\phi''} = \bar{\rho}\tilde{\phi}$$

Where,

$$\tilde{\phi}(x_i, t) \equiv \frac{1}{\bar{\rho}T} \int_{t-T/2}^{t+T/2} \rho(x_i, t')\phi(x_i, t')\,\mathrm{d}t', \qquad \overline{\rho\phi''} \equiv 0$$

Favre-Averaged Navier-Stokes (FANS) Equations can be given as:

Continuity Equation:

$$\frac{\partial}{\partial t}(\bar{\rho}) + \frac{\partial}{\partial x_i}(\bar{\rho}\tilde{u}_i) = 0$$

Momentum Equation:

$$\frac{\partial}{\partial t}(\bar{\rho}\tilde{u}_i) + \frac{\partial}{\partial x_j}(\bar{\rho}\tilde{u}_i\tilde{u}_j + \bar{p}\delta_{ij}) = \frac{\partial}{\partial x_j}\left(\bar{\tau}_{ij} - \overline{\rho u_i'' u_j''}\right)$$

Favre-Averaged Reynolds Stress Tensor:

$$\lambda = -\overline{\rho u_i'' u_j''}$$

Turbulent Kinetic Energy:

$$\frac{1}{2}\overline{\rho u_i'' u_i''} = -\frac{1}{2}\lambda_{ii} = \bar{\rho}\tilde{k}$$

STEPS OF COMPUTATIONAL FLUID DYNAMICS

The process of solving Computational Fluid Dynamics can be defined in 5 stages which are: Creating Mathematical Model, Discretization of model, Analyzing with Numerical Scheme, Getting solution, and Post Processing (Visualization). These steps are discussed next.

Creating Mathematical Model

In this stage, various equations are defined based on physical properties, boundary conditions, and other real-world parameters of the problem. The mathematical model generally consists of partial differential equations, integral equations, and combinations of both. In this case, it will be Navier-Stokes equation with other mathematical equations to define boundary conditions and thermodynamic properties. Note that at this step there will be some assumptions which converting real problem into mathematical model like material might be assumed isotropic, heat conditions might be considered adiabatic and so on. These assumptions should generate permissible level of error in the solution and as a designer you should be aware of the quantum of those errors.

Discretization of Model

While studying the behavior of fluid, you will find that there are infinite number of particles with different physical properties in the fluid stream. These infinite particles will generate infinite equations to solve if mathematical model is to be solved directly which is not possible. So, we convert the mathematical model into finite number of elements and then we can solve equations for each of the element. In case of CFD, elements are called cells. The process of converting a mathematical model into finite element equations is called discretization. In simple words, the partial differential and integral equations are converted into algebraic equations of **A.x = B**.

There are many methods to perform discretization of mathematical model like Finite Difference Method, Finite Volume Method, and Finite Element Method which use mesh type structuring of the model. There are also methods which do not use mesh like Smooth Particle Hydrodynamics (SPH) and Finite Pointset Method (FPM). Note that there is again loss of information at the stage of discretization.

Analyzing with Numerical Schemes

Numerical scheme is the complete setup of equations with all the parameters and boundary conditions defined as needed. The scheme also includes numerical methods by which these equations will be solved and limiting points upto which the equations will be solved. When you start analyzing the problem with specified numerical scheme, there is no input required from your side in system; everything is automatic by computer. Every numerical scheme need to satisfy some basic requirements which are consistency, stability, convergence, and accuracy.

Solution

At this stage, we get the solution of equations as different basic variables like pressure, speed, volume, and so on. The resulting flow variables are obtained at each grid point/mesh point whether the scheme is time dependent (transient) or steady.

Visualization (Post-processing)

At this stage, everything is based on interpretation of designer. Various desired parameters are derived from basic calculated parameters at this stage. You also need to check whether results of CFD analysis are realistic or not.

Now, one question left here to understand in more detail is discretization. Discretization can be performed by various methods like FDM, FVM, FEM, and so on as discussed earlier. Out of these methods, Finite Difference Method and Finite Volume Methods are the most used methods for CFD. Here, we will discuss FDM in detail.

FINITE DIFFERENCE METHOD

At discretization stage, the model need to be converted into numerical grid with different cells defined by nodes; refer to Figure-2. For a 2D problem, i and j will be used for direction in horizontal and vertical directions.

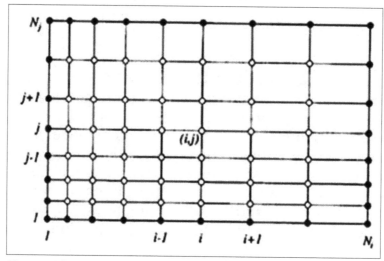

Figure-2. Grid with cartesian coordinates of nodes

If you recall the definition of derivative: $\left(\dfrac{\partial u}{\partial x}\right)_{x_i} = \lim\limits_{\Delta x \to 0} \dfrac{u(x_i + \Delta x) - u(x_i)}{\Delta x}$

The equation represents slope of tangent to a curve u(x) which can be geometrically represented as shown in Figure-3.

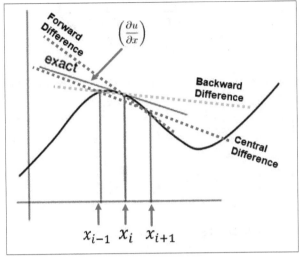

Figure-3. Geometrical interpretation of derivative

Now, assume that we have value of x_i, x_{i+1}, and x_{i-1} then there are three ways in which we can find the slope of curve.

Backward Difference which uses x_i and x_{i-1}.
Forward Difference which uses x_i and x_{i+1}.
Central Difference which uses x_{i+1} and x_{i-1}.

If we assume that $\boxed{\Delta x}$ is difference between two consecutive nodes then lower the value of $\boxed{\Delta x}$ more accurate we will get the value of slope.

For a uniform grid where $\boxed{\Delta x}$ is same for each node then based on

Backward Difference : $\boxed{\dfrac{du}{dx} \cong \dfrac{u_i - u_{i-1}}{\Delta x}}$

Forward Difference : $\boxed{\dfrac{du}{dx} \cong \dfrac{u_{i+1} - u_i}{\Delta x}}$

Central Difference : $\boxed{\dfrac{du}{dx} \cong \dfrac{u_{i+1} - u_{i-1}}{2\Delta x}}$

Now, if you expand the backward difference derivative using Taylor's series then

Consider a function u(x) and its derivative at point x,

$$\boxed{\left(\frac{\partial u}{\partial x}\right)_{x_i} = \lim_{\Delta x \to 0} \frac{u(x_i + \Delta x) - u(x_i)}{\Delta x}}$$

If u(x + $\boxed{\Delta x}$) is expanded in Taylor series about u(x), we obtain

$$\boxed{u(x + \Delta x) = u(x) + \Delta x \frac{\partial u(x)}{\partial x} + \frac{(\Delta x)^2}{2} \frac{\partial^2 u(x)}{\partial x^2} + \frac{(\Delta x)^3}{3!} \frac{\partial^3 u(x)}{\partial x^3} + \cdots}$$

If we substitute this value in derivative equation then

$$\frac{\partial u(x)}{\partial x} = \lim_{\Delta x \to 0} \left(\frac{\partial u(x)}{\partial x} + \frac{\Delta x}{2} \frac{\partial^2 u(x)}{\partial x^2} + \cdots \right)$$

Based on this equation, we can write u in Taylor Series at i+1 and i-1 as:

$$u_{i+1} = u_i + \Delta x \left(\frac{\partial u}{\partial x} \right)_i + \frac{\Delta x^2}{2} \left(\frac{\partial^2 u}{\partial x^2} \right)_i + \frac{\Delta x^3}{3!} \left(\frac{\partial^3 u}{\partial x^3} \right)_i + \frac{\Delta x^4}{4!} \left(\frac{\partial^4 u}{\partial x^4} \right)_i + \cdots$$

$$u_{i-1} = u_i - \Delta x \left(\frac{\partial u}{\partial x} \right)_i + \frac{\Delta x^2}{2} \left(\frac{\partial^2 u}{\partial x^2} \right)_i - \frac{\Delta x^3}{3!} \left(\frac{\partial^3 u}{\partial x^3} \right)_i + \frac{\Delta x^4}{4!} \left(\frac{\partial^4 u}{\partial x^4} \right)_i + \cdots$$

Re-writing the above two equations, we get:

Forward difference : $\left(\frac{\partial u}{\partial x} \right)_i = \frac{u_{i+1} - u_i}{\Delta x} + O(\Delta x)$

Backward Difference : $\left(\frac{\partial u}{\partial x} \right)_i = \frac{u_i - u_{i-1}}{\Delta x} + O(\Delta x)$

Central Difference : $\left(\frac{\partial u}{\partial x} \right)_i = \frac{u_{i+1} - u_{i-1}}{2\Delta x} + O(\Delta x^2)$

Note that central difference takes second derivative of delta x so the value of error is lesser than compared to forward or backward difference. For example if forward or backward difference gives an error of 0.001 then second derivative will be 0.001 x 0.001 which gives 0.000001. This error value is a lot lesser.

You can learn more about these methods in books dedicated to Finite Difference Method and solution techniques.

INTRODUCTION TO CFDOF

In this chapter, you will learn about the basic use of CfdOF workbench. This workbench is not the part of native FreeCAD but you can use Addon Manager to use this workbench. The CfdOF workbench acts as front-end of OpenFOAM CFD toolkit. OpenFoam is the open source free CFD software developed by OpenCFD Ltd. Various features supported by CfdOF workbench in FreeCAD are given next.

Flow physics
- Incompressible, laminar flow (simpleFoam, pimpleFoam)
- Support for various RANS, LES and DES turbulent flow models
- Incompressible free-surface flow (interFoam, multiphaseInterFoam)
- Compressible buoyant flow (buoyantSimpleFoam, buoyantPimpleFoam)
- High-speed compressible flow (HiSA)
- Porous regions and porous baffles
- Basic material database
- Flow initialization with a potential solver
- Solution of arbitrary passive scalar transport functions

Meshing

- Cut-cell Cartesian meshing with boundary layers (cfMesh)
- Cut-cell Cartesian meshing with baffles (snappyHexMesh)
- Tetrahedral meshing using Gmsh, including conversion to polyhedral dual mesh
- Post-meshing check mesh
- Support for dynamic mesh adaptation for supported solvers

Post-processing and monitoring

- Postprocessing using Paraview
- Basic support for force-based function objects (Forces, Force Coefficients)
- Basic support for probes

Other features

- Runs on Windows 7-11 and Linux
- Unit/regression testing
- Case builder using an extensible template structure
- Macro scripting

ACTIVATING CFDOF WORKBENCH

The option to activate CfdOF workbench is available in the Addon manager. The procedure to activate this workbench is given next.

- Click on the **Addon manager** tool from the **Tools** menu. The **Addon Manager** window will be displayed; refer to Figure-4.

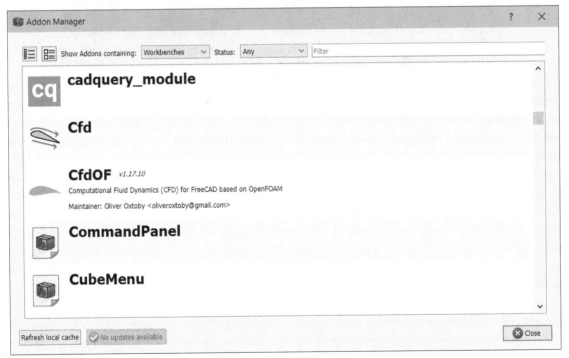

Figure-4. Addon Manager options

- Select the **CfdOF** option from the **Addon Manager** and click on the **Install** button. A message box will be displayed after installation is complete; refer to Figure-5.

Figure-5. Installation message

• Click on the **Close** button to exit the message box. Close the **Addon Manager** and restart the software.

INSTALLING CFDOF RELATED SOFTWARE

For running CfdOF workbench, you need to install some necessary software to efficiently perform the analysis. After installing the CfdOF addon, you will be asked to restart the software. The procedure to install related software is given next.

• Restart the FreeCAD software in administrator mode. To do so, right-click on FreeCAD from Start menu and select **Run as administrator** option from right-click shortcut menu; refer to Figure-6.

Figure-6. Right-click shortcut menu

• Select the **Preferences** option from the **Edit** menu. The **Preferences** dialog box will be displayed.
• Select the **CfdOF** option from the left area. The options will be displayed as shown in Figure-7.

Figure-7. CfdOF options

- Click on the **Browse** button for **OpenFOAM install directory** field and select/create a folder location for installing the OpenFOAM software; refer to Figure-8.
- After selecting desired folder, click on the **Select Folder** button from the dialog box. The path will be added in the **OpenFOAM install directory** field.

Figure-8. CFD software install folder

- Click on the **Install OpenFOAM** button from the **OpenFOAM** area at bottom in the **Preferences** dialog box. The dialog box for installing OpenFOAM will be displayed; refer to Figure-9.

Figure-9. OpenFOAM installation dialog box

- Click on the **Next** button and follow next instructions to install the software. Make sure to set same location for installation in the dialog box; refer to Figure-10. After setting location, click on the **Install** button and install the software.

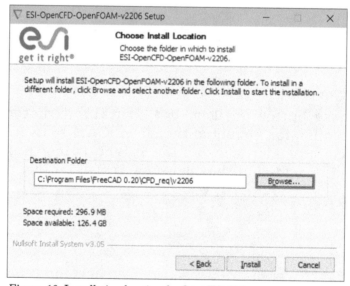

Figure-10. Installation location for OpenFOAM

- Similarly, install ParaView, cfMesh, and HISA software.
- After installing all requisite software, set path for executable of gmsh and ParaView.
- Click on the **OK** button from the **Preferences** dialog box to apply settings.

PREPARING MODEL FOR CFD

In terms of CAD, there are two ways to prepare model for CFD. The model in which internal CFD is to be performed is generally a hollow tube with lid faces at both the ends (outlet/inlet) making it airtight. For external CFD, you can use any solid model to check aerodynamics and other CFD aspects. In this step, you need to close any openings in the model and make sure there is continuity in the model for fluid flow.

An example on performing CFD is given next.

PERFORMING CFD ON A MODEL

The common steps involved in performing CFD analysis on a model are given next.

- Create/open the model on which you want to perform CFD analysis; refer to Figure-11.

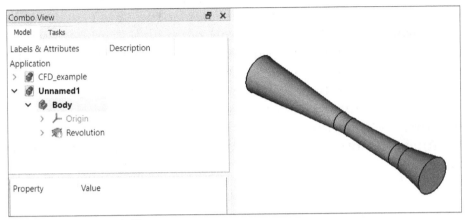

Figure-11. Model for CFD

- Select the **CfdOF** option from the **Workbench** drop-down in the **Toolbar**. The tools to perform CFD analysis will be displayed.
- Click on the **Analysis container** tool from the **CfdOF** toolbar. The options related to CFD will be added in the **Model Tree**.

Applying Boundary Conditions

- Select the face of model as shown in Figure-12 and click on the **Fluid boundary** tool from the **CfdOF** toolbar. The **CFD boundary condition** dialog will be displayed in the **Tasks** pane.

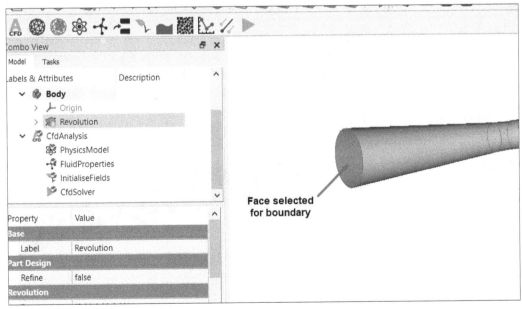

Figure-12. Face selected for inlet

- Select the **Inlet** option from the **Boundary** drop-down to define fluid inlet condition. Select desired option from the **Sub-type** drop-down to define how fluid flow will be defined. Select the **Uniform velocity** option to define velocity of fluid flowing uniformly through inlet. Select the **Volumetric flow rate** option to define flow of fluid by volume of fluid passing through inlet per second. Generally, unit for volumetric flow rate is m^3/s. Select the **Mass flow rate** option from the drop-down to define fluid flow rate in kg/s or lbs/s. Select the **Total pressure** option from the **Sub-type** drop-down to define pressure at which fluid will enter the domain. Select the **Static pressure** option from the drop-down to specify pressure of the fluid when it is not moving.

- Set desired parameters in the edit boxes of **CFD boundary condition** dialog based on sub-type selected. After setting desired parameters, click on the **OK** button to apply the boundary condition.

- Similarly, apply the outlet CFD boundary condition for other face; refer to Figure-13.

Figure-13. Outlet cfd boundary condition

- Select the faces to be used as wall for analysis and specify parameters as shown in Figure-14.

Figure-14. Faces selected for wall

- After setting parameters, click on the **OK** button to apply wall boundary conditions.

Creating CFD Mesh

- Select the body from the **Model Tree** in **Combo View** and click on the **CFD Mesh** tool from the **Toolbar**; refer to Figure-15. The **CFD Mesh** dialog will be displayed in the **Tasks** panel of **Combo View**; refer to Figure-16.

Figure-15. CFD mesh tool

Figure-16. CFD Mesh dialog

- Set desired value of mesh element size in the **Base element size** edit box and click on the **Write mesh case** button. The status of mesh case writing will be displayed in the **Status** section of **CFD Mesh** dialog.
- Once you get the Mesh case written successfully message in the **Status** section of dialog, click on the **Run mesher** tool from **Meshing** section of dialog. Note that it may take some time for mesher to generate the cfd mesh. After the meshing is complete, the status **"Meshing completed"** will be displayed in the **Status** section of dialog.

- Click on the **Load surface mesh** button from the **Visualisation** section in **CFD Mesh** dialog to check preview of CFD mesh. You can also use the **Paraview** tool to check mesh.

Mesh refinement

The **Mesh refinement** tool is used to control mesh element size/type in specific region. The procedure to use this tool is discussed next.

- Select meshed object from the **Model Tree** and click on the **Mesh refinement** tool from the **Toolbar**; refer to Figure-17. The **Mesh refinement** dialog will be displayed; refer to Figure-18.

Figure-17. Mesh refinement tool

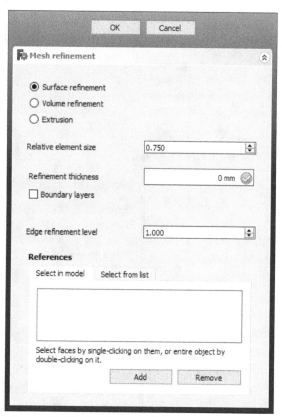

Figure-18. Mesh refinement dialog

- Specify desired parameters in the dialog to refine the meshed object and click on the **OK** button. The meshed object will be refined; refer to Figure-19.

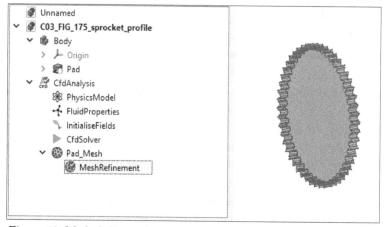

Figure-19. Meshed object refinement

Setting Physics Model

- Select the CfdAnalysis container from the **Model Tree** in **Combo View** and click on the **Select models** tool from the **Toolbar**. The **Select physics model** dialog will be displayed in the **Tasks** pane; refer to Figure-20.

Figure-20. Select physics model dialog

- Set desired parameters in the dialog to define type of the study. Here, we will be using steady, single phase, incompressible, and viscous fluid analysis with RANS turbulence. After setting parameters, click on the **OK** button.

Adding Fluid Properties

- Click on the **Add fluid properties** tool from the **Toolbar**. The **Fluid properties** dialog will be displayed; refer to Figure-21.

Figure-21. Fluid properties dialog

- Select desired option from the **Predefined fluid library** drop-down to define the fluid to be used in analysis. In our case, we are using Water as fluid. After setting parameters, click on the **OK** button.

Setting Initial Internal Flow Conditions

- Click on the **Initialise** tool from the **Toolbar**. The **Initialise flow field** dialog will be displayed; refer to Figure-22.

Figure-22. Initialise flow field dialog

- Set desired parameters in the dialog and click on the **OK** button to apply flow conditions.

Solving the Analysis

- Click on the **Solver job control** tool from the **Toolbar**. The **Analysis control** dialog will be displayed in the **Tasks** pane; refer to Figure-23.

Figure-23. Analysis control dialog

- Click on the **Write** tool from the **Case setup** section of the dialog. The case file will be written and status will be displayed in the **Status** section of the dialog.
- Click on the **Run** tool from the **Solver** section of the dialog to perform solution of analysis. If your mesh is not updated before analysis then a message box will be displayed asking you to remesh the case. Select **Yes** button from the message box.
- Once the analysis is solved, the result will be displayed; refer to Figure-24.

Figure-24. CFD analysis result

- Click on the **Paraview** button from the **Results** section of **Analysis control** dialog. Results of the analysis will be displayed; refer to Figure-25.

Figure-25. Result of analysis in ParaView

FOR STUDENT NOTES

Index